AF294057

Environmental Science

Series editors: R. Allan · U. Förstner · W. Salomons

Ricardo Casale · Claudio Margottini (Eds.)

Floods and Landslides: Integrated Risk Assessment

With 150 Figures and 30 Tables

Springer

Editors

Dr. Riccardo Casale
European Commission
Directorate General XII
Science, Research and Development
Rue de la Loi 200
1049 Brussels
Belgium

Dr. Claudio Margottini
ENEA CR Casaccia
Via Anguillarese, 301
00060 S. Maria di Galeria (Rome)
Italy

ISSN 1431-6250
ISBN 978-3-642-63664-6

Library of Congress Cataloging-in-Publication Data
Floods and landslides : integrated risk assessment / Riccardo Casale, Claudio Margottini, (eds.).
p. cm. -- (Environmental science, ISSN 1431-6250)
Includes bibliographical references and index.
ISBN 978-3-642-63664-6   ISBN 978-3-642-58609-5 (eBook)
DOI 10.1007/978-3-642-58609-5
1. Floods--Congresses. 2. Floods--Europe--Case studies--Congresses. 3. Landslides--Congresses. 4. Land-
slides--Europe--Case studies--Congresses. 5. Risk assessment--Congresses.
I. Casale, Riccardo. 1963- . II. Margottini, Claudio. III. Series: Environmental science (Berlin, Germany)
GB1399.F53 1999
363.34'9372--DC21   98-32410
                                                                                CIP

Cover Design: Struve & Partner, Heidelberg
Dataconversion: Büro Stasch, Bayreuth

SPIN: 10664319   32/3020 – 5 4 3 2 1 0 – Printed on acid-free paper

Preface

Scientific reserach on natural disasters is now implementing new methodologies and aproaches as consequences of:

1. mutating impact of extreme natural events in response to societal change like land use, lifelines, communications, transportation, etc.;
2. development which claim for new urban and industrial territories, often in hazard prone areas;
3. impact of modern society on natural climate variability (climate change) and consequently, on spatial and temporal frequency of extreme events related to hydrological cycle.

As consequence of the above mentioned items it is necessary to develop a global approach to territory in order to understand reciprocous influence between climate dynamics with their extreme consequences like floods and landslides and socio-economic development. In such a way, and in response to societal change, scientific research on floods and landslides is beginning to loose the classical monodisciplinary approach and it is starting to be a science of the hydrological processes.

The course on "Floods and Landslides: integrated risk assessment" held in Orvieto (Italy) 19-26 may 1996, has been organised by the European Commission specifically with the aim of transferring to young European scientists these new views, in order to contribute to a future scientific community, capable to face with the future environmental problems.

Riccardo Casale and Claudio Margottini

Remarks on the European Commission School on Floods and Landslides (Orvieto, May 1996)

Floods and landslides are natural phenomena that, in close relationship with other processes such as erosion, earthquakes and volcanic eruptions, shape the landscape and control the evolution of the Earth surface. However, these phenomena are the origin of increasing damages and casualties, because our society is characterised by high vulnerability to natural hazards, due to its intrinsic complexity and false sense of sureness. Owing to these reasons, strategies and actions are required to prevent and to mitigate natural hazards, especially by means of non-structural measures which can reduce the risk factors. The main goals of this approach are to allow the citizens to safeguard their own lives, and to protect goods and properties including the ones of high cultural and historical value. Strong scientific support is needed for these actions to be designed and planned, in order to gain knowledge of the fundamental mechanisms dominating processes such as floods and landslides.

The rising of floods and flash floods and the occurrence of widespread shallow slope failures are often simultaneous events due to the same meteorological triggering process. Thus, floods and landslides turn out to be two different aspects of a single issue: the coping with hydrogeological hazards. It is also worthy to note that generally the investigation of these processes is undertaken by following a common approach:

- Analysis of the meteorological triggering factor;
- Examination of the geomorphological characteristics;
- Assessment of the anthropogenic impact on the natural systems where the process occurs.

Finally, fundamental understanding and accurate prediction of these processes and of their impact on our society requires close co-operation between researchers and research groups in different disciplines. An important aspect in developing good practice and developing a better understanding between different research and professional groups is contributed by courses such as that organised in Orvieto by DGXII in May 1996. In fact, education of the different intervenors in emergency plans is of most importance for motivation and better understanding of the mission to be pursued.

The course was organised based on a logical presentation of natural hazards and risk. The analysis focused on:

- Identification of the triggering processes;
- Understanding of the natural phenomena evolving as natural hazards;
- Good practice and actions required to mitigate the hazards;
- Analysis of case studies.

The first section of the course provided basis for the fundamental understanding of the processes and for the use of available techniques and tools available for quantitative representation and description of these processes. Modelling and monitoring techniques were introduced with emphasis on remote sensing technology (i.e. weather radar and satellite) for precipitation measurement, mathematical modelling and computer systems which exploit currently available technology, geographical information systems. Concepts of redundancy and system robustness were explicitly identified in relation to the development of warning systems. The multi-disciplines nature of modern sophisticated warning/dissemination systems was noted, e.g. hydrologists, earth scientists, geologists, geomorphologists, computer scientists, mathematicians, geographers and system engineers etc. and social scientists. Risk and vulnerability concepts were addressed in the second section. Consequences of floods and landslides were highlighted, in terms of both economic and social impacts. Mitigation measures were identified and described.

Finally examples were given on how to incorporate planning activities in a wider multi-hazards approach. Several case studies were organised to describe real-life situations of hazard identification, risk assessment, remedies and mitigation measures evaluation. The format of the Course was successful enough to attract attention from a wider audience than expected. It was clearly well fitted to integrate and improve training and education programmes at European level. Operating in this way, theoretical knowledge and field experience can be made directly available to technicians, administrators, researchers and post-graduate students. Efforts should be undertaken to extend these training experiences not only in Europe but also in developing countries.

Sergio Fattorelli and Mario Panizza

Contents

Contributors

Angeli, M.-G.
CNR-IRPI, Via Madonna Alta, 126, 06100 Perugia, Italy

Bandis, S.C.
Department of Civil Engineering, Aristotle University, Thessaloniki, Greece

Bauer, B.
Department of Geography, University of Vienna, Universitätstr. 7, 1010 Vienna, Austria

Bellos, C.
Democritus University of Thrace, 67100 Xanthi, Greece

Brandolini, P.
Institute of Geography, University of Genoa, Corso Europa, 26, 16123 Genoa, Italy

Closson, D.
University of Liège, Department of Physical Geography and Quaternary, Hydrology and Fluvial Geomorphology, Institut de Géographie, Bat B. 11, 2, Allée du 6 Août, Sart Tilman, 4000 Liège, Belgium

Correia, F.N.
Instituto Superior Técnico (IST), DECivil, Av. Rovisco Pais, 1096 Lisboa Codex, Portugal

Crespi, M.
Experimental Centre of Hydrology and Meteorology, Via Marconi, 55, Teolo (Padova), Italy

Dalla Fontana, G.
Università degli Studi di Padova, Dipartimento Territorio e Sistemi Agro-Forestali, Sezione Risorse Idriche e Difesa del Suolo, Agripolis, Via Romea, 35020 Legnaro (Padova), Italy

Da Ros, D.
Università degli Studi di Padova, Dipartimento Territorio e Sistemi Agro-Forestali, Sezione Risorse Idriche e Difesa del Suolo, Agripolis, Via Romea, 35020 Legnaro (Padova), Italy

Delmonaco, G.
Geological Dynamics and Territory Section, ENEA CR Casaccia, Via Anguillarese, 301, 00060 S. Maria di Galeria, Rome, Italy

Deroanne, P.
University of Liège, Department of Physical Geography and Quaternary, Hydrology and Fluvial Geomorphology, Institut de Géographie, Bat B. 11, 2, Allée du 6 Août, Sart Tilman, 4000 Liège, Belgium

Dikau, R.
Department of Geography, University of Bonn, Meckenheimer Allee 166, 53115 Bonn, Germany

Dutto, F.
CNR-IRPI, Strada delle Cacce, 73, 10135 Turin, Italy

Fantucci, R.
Via del Lago, 51, 01027 Montefiascone (Viterbo), Italy

Fattorelli, S.
Università degli Studi di Padova, Dipartimento Territorio e Sistemi Agro-Forestali, Sezione Risorse Idriche e Difesa del Suolo, Agripolis, Via Romea, 35020 Legnaro (Padova), Italy

Flageollet, J.-C.
European Centre on Geomorphological Hazards, 28 rue Goethe, 67000 Strasbourg, France

Gilard, O.
Cemagref, Division Hydrologie-Hydraulique, 3 bis quai Chauveau, C.P. 220, 69336 Lyon cedex 09, France

Givone, P.
Cemagref, Division Hydrologie-Hydraulique, 3 bis quai Chauveau, C.P. 220, 69336 Lyon cedex 09, France

Graça Saraiva, M. da
Instituto Superior de Agronomia (ISA), DECivil, Av. Rovisco Pais, 1096 Lisboa Codex, Portugal

Margottini, C.
Geological Dynamics and Territory Section, ENEA CR Casaccia, Via Anguillarese, 301, 00060 S. Maria di Galeria, Rome, Italy

Mason, P.J.
Department of Geology, Royal School of Mines, Prince Consort Road, London SW7 2BP, United Kingdom

Moore, J. MCM.
Department of Geology, Royal School of Mines, Prince Consort Road,
London SW7 2BP, United Kingdom

Moore, R.J.
Institute of Hydrology, Maclean Building, Crowmarsh Gifford, Wallingford,
Oxfordshire OX10 8BB, United Kingdom

Mortara, G.
CNR-IRPI, Strada delle Cacce, 73, 10135 Turin, Italy

Nosengo, S.
Earth Science Department, University of Genoa, Corso Europa, 26, 16123 Genoa, Italy

Oberlin, G.
Cemagref, Division Hydrologie-Hydraulique, 3 bis quai Chauveau, C.P. 220,
69336 Lyon cedex 09, France

Palladino, A.F.
Department of Geology, Royal School of Mines, Prince Consort Road,
London SW7 2BP, United Kingdom

Panizza, M.
Università degli Studi di Modena, Dipartimento di Scienze della Terra,
Largo S. Eufemia, 19, 41100 Modena, Italy

Pasuto, A.
CNR-IRPI, Corso Stati Uniti, 4, 35127 Padova, Italy

Petit, F.
University of Liège, Department of Physical Geography and Quaternary, Hydrology
and Fluvial Geomorphology, Institut de Géographie, Bat B. 11, 2, Allée du 6 Août, Sart
Tilman, 4000 Liège, Belgium

Pissart, A.
University of Liège, Department of Physical Geography and Quaternary, Hydrology
and Fluvial Geomorphology, Institut de Géographie, Bat B. 11, 2, Allée du 6 Août, Sart
Tilman, 4000 Liège, Belgium

Pittaluga, F.
Earth Science Department, University of Genoa, Corso Europa, 26, 16123 Genoa, Italy

Ramella, A.
Civil Protection Service, Municipality of Genoa, Piazza Ortiz, 8, Genoa, Italy

Razzore, S.
Civil Protection Service, Municipality of Genoa, Piazza Ortiz, 8, Genoa, Italy

Serafini, S.
Geological Dynamics and Territory Section, ENEA CR Casaccia, Via Anguillarese, 301, 00060 S. Maria di Galeria, Rome, Italy

Silva, F.N. da
Instituto Superior Técnico (IST), DECivil, Av. Rovisco Pais, 1096 Lisboa Codex, Portugal

Silvano, S.
CNR-IRPI, Corso Stati Uniti, 4, 35127 Padova, Italy

Soldati, M.
Università degli Studi di Modena, Dipartimento di Scienze della Terra, Largo S. Eufemia, 19, 41100 Modena, Italy

Solway, L.
Cedar View House, Netherby Park, Weybridge, Surrey KT13 0AG, United Kingdom

Tiedemann, E.
Stefan-Rotthaler-Str. 5, 85368 Moosburg, Germany

Todini, E.
Department of Earth and Geological-Environmental Sciences, Bologna University, Via Irnerio, 57, 40126 Bologna, Italy

Trocciola, A.
ENEA CR Casaccia, Via Anguillarese, 301, 00060 S. Maria di Galeria, Rome, Italy

Part I

Introduction –
Overview, Technical Papers

Chapter 1

Landslide Hazard –
A Conceptual Approach in Risk Viewpoint

J.-C. Flageollet

1.1
Introduction

Even non specialists have a general idea of what constitutes a landslide. Cruden recently reminded us of the precise definition: "a movement of mass of rock, earth or debris down a slope" (Cruden 1991). It is a collection of phenomena differentiated by the type of movement process involved: a fall, either free or with rebounds, a roll, a slide, a flow, which may be either fluid or viscous. And we know that one or more of these processes can happen in different places or successively in a landslide (Dikau et al. 1995).

Landslide *risks* are governed by the concept which is applicable to other natural hazards and we use this concept to assess the risk by confronting the *vulnerability* with the *hazard* (Flageollet 1993; Ayala-Carcedo 1993). If we want to make a full assessment

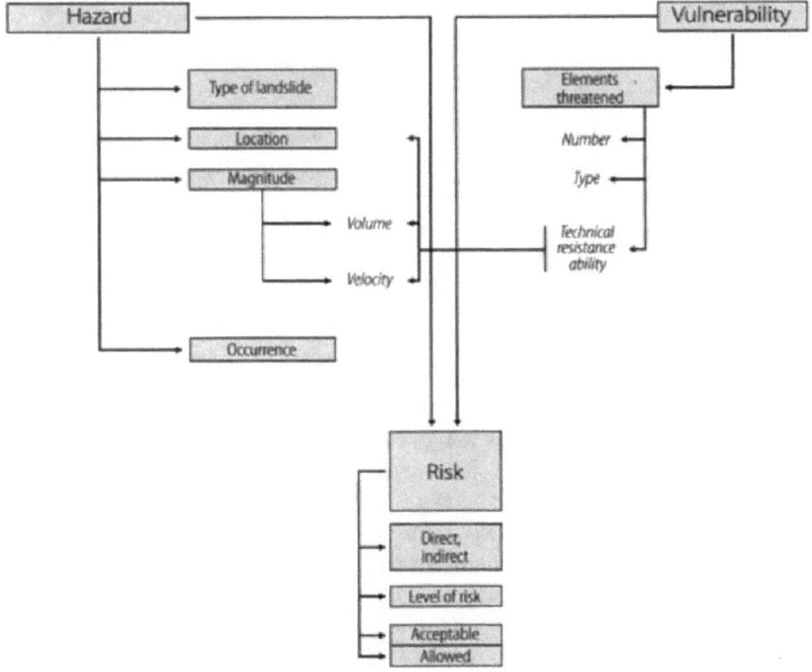

Fig. 1.1. Landslide hazard and risk conceptual approach

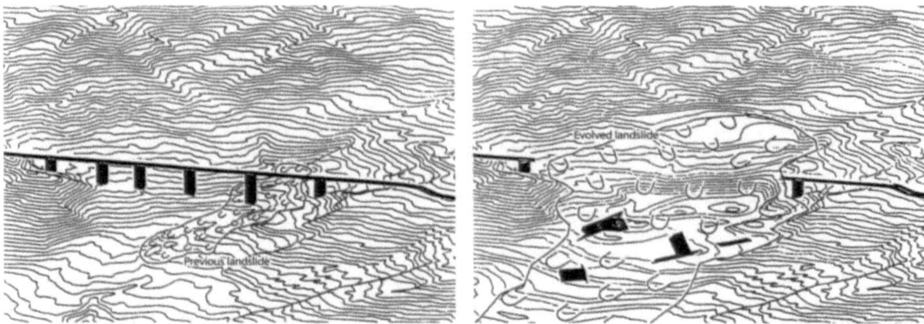

a RISK: damage to the project. A road is constructed on an unstable slope and is destroyed by the evolving geomorphological process (slide)

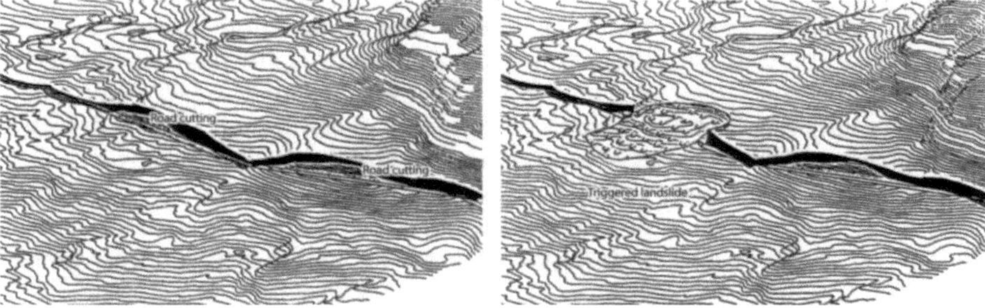

b DIRECT RISK: on the project itself. The construction of the road by cutting a stable slope triggers a landslide which involves the road itself

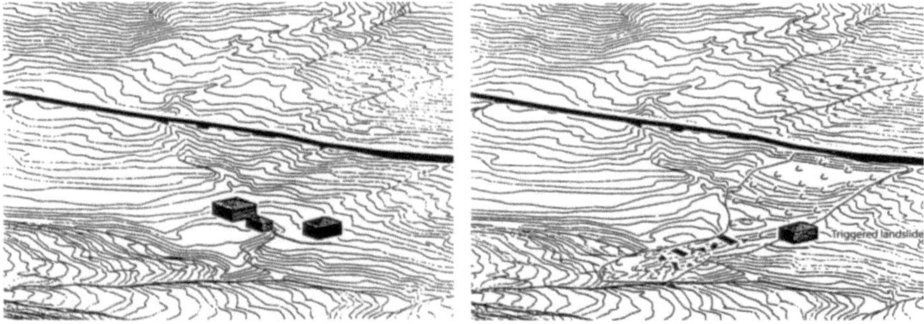

c INDIRECT RISK: on the surrounding settlements. The construction of the road on a stable slope triggers a landslide which damages some preexisting houses

Fig. 1.2. Concept of landslide risk in the assessment of the impact of human activities on the environment (From Cavallin et al. 1994)

of the *hazard* we must know the displacement mode, which means the type of movement, its location, its magnitude in terms of speed and volume and its occurrence, that's to say the probability that it will happen (Fig. 1.1) (Hansen 1984).

Vulnerability is often assessed in terms of the probable cost of the damage the hazard has done or may do to various types of installation. We have to consider how resistant each element may be to this particular type of movement, so that we can make an accurate assessment of its vulnerability (Leone 1996).

The risk may be to an existing element or to one that is planned in the future (Fig. 1.2) (Cavallin et al. 1994). This being said, we shall look at what is specific to landslides in terms of planning research. First we will look at the hazard, secondly the vulnerability and thirdly the risk assessment (Sitar and Rogers 1993).

1.2
The Hazard's Characteristics

1.2.1
Location

This may be difficult to determine when we have to predict the starting point, the trajectory and the stopping point. This is particularly difficult for block falls and rock falls (Bozzolo et al. 1988).

Slope stability calculation models have two objectives: on a stable slope, define the most plausible rupture surface and estimate the rupture safety margin; assess the modifications induced by works (Fig. 1.3) (Bourdeau et al. 1988). On an unstable slope, calculation in equilibrium limit is useful for estimating the respective role of every parameter and testing the performance of the planned reinforcements.

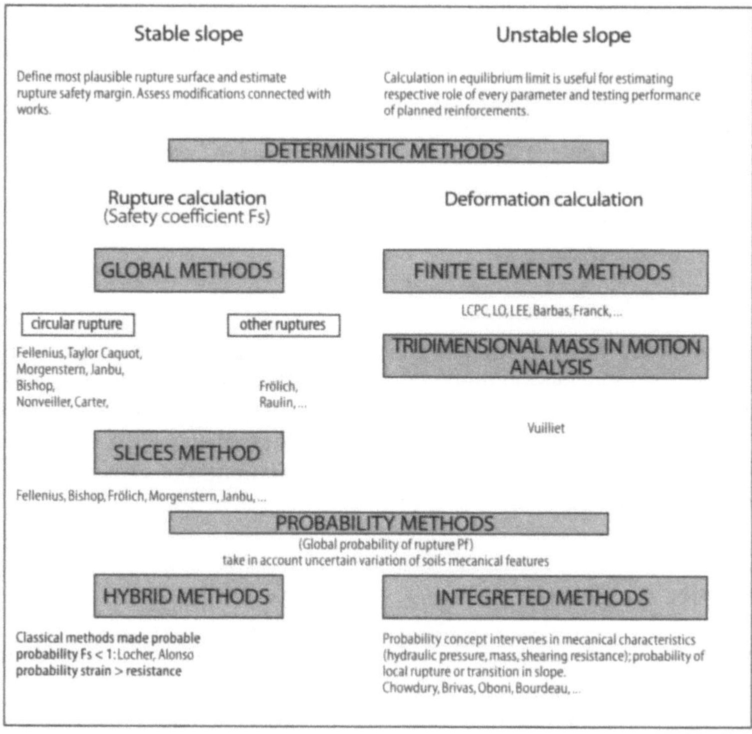

Fig. 1.3. Slope stability calculation models

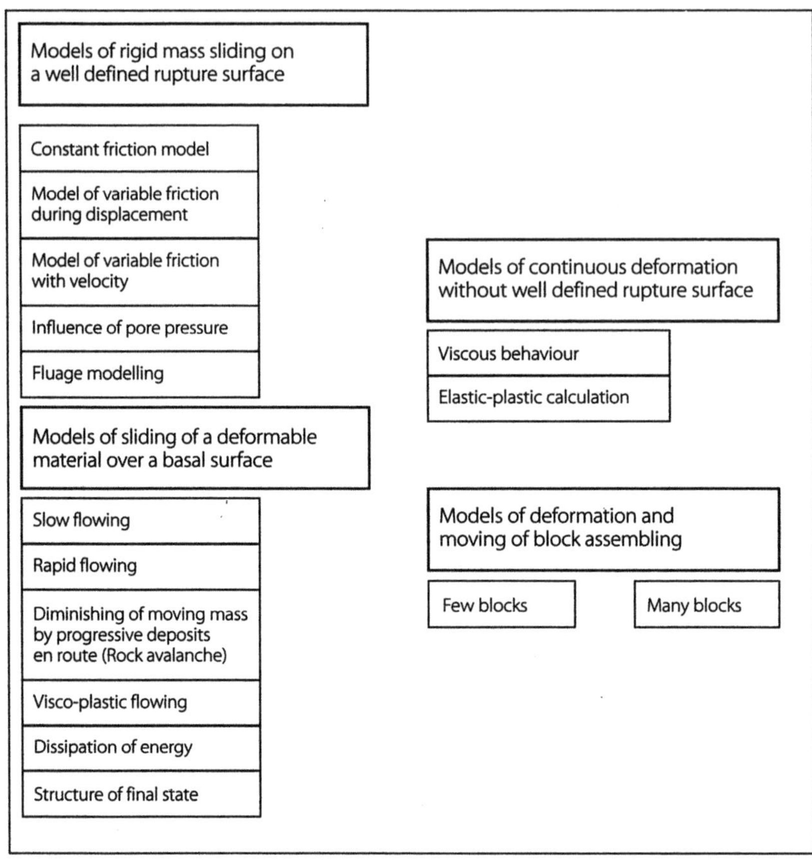

Fig. 1.4. Rock behaviour modelling

Behaviour models must start from hypothesis in regards to the rigidity of the moving mass and its deformation mode (Fig. 1.4). In block trajectory models we try to calculate the angle of the block at the point of impact when it rebounds, and the elasticity of the slope at the point of impact, among other parameters (Fig. 1.5) (JPA Consultant 1994). The design of these models uses observations and experiments which are still relatively rare (Fig. 1.6).

1.2.2
Magnitude

It is hardly possible to estimate the volume of potential movements, particularly sliding.

We can sometimes predict the limits of a rock mass which threatens to fall from the reference marks and structural accidents (Durville 1991; Maquaire 1990). This was the case at Randa in Switzerland before the second rock fall, on 19th May 1991.

Fig. 1.5. Trajectory models; **a** Single frame pictures of the motion of rocks as they approach and leave the contact points (From Bozzolo 1988); **b** Comparison between the calculated trajectories (SASS) and those observed through examination of video film on the slope of the Bedrina area (From Bozzolo 1988)

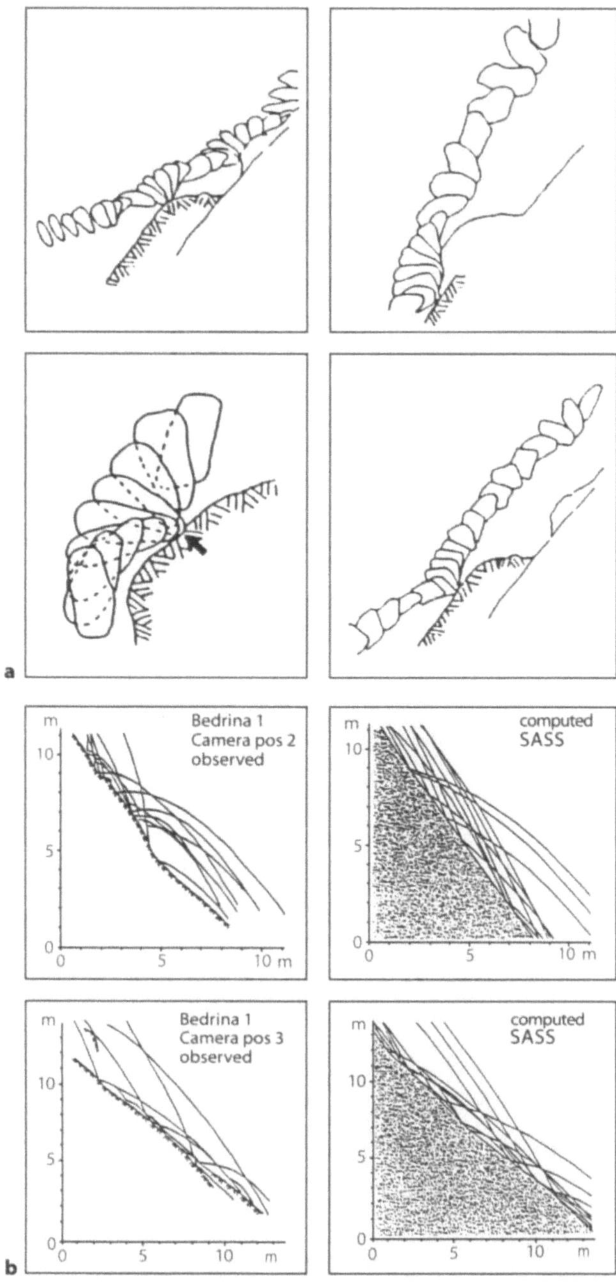

The difficulty in calculating the volume of declared landslides arises mostly from the fact that we know very little about the surface of the rupture, its shape or depth. We can make deductions from the surface morphology and this will give us partial an-

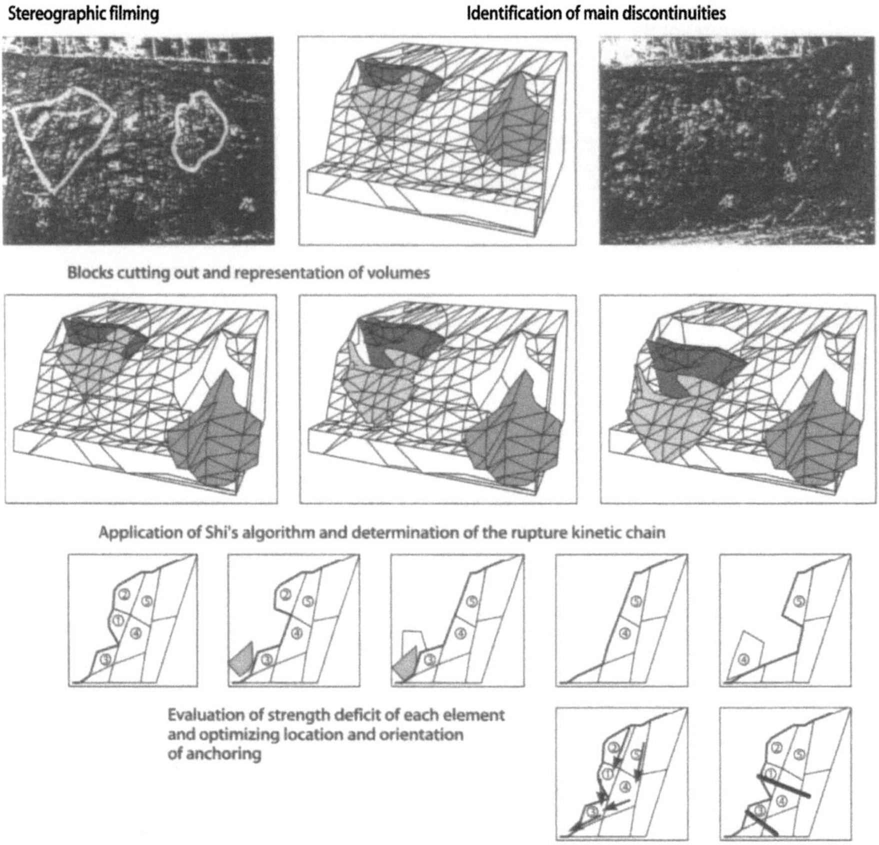

Fig. 1.6. Blocks cutting out of a rocky slope (From JPA Consultant 1994)

swers, but above all we must do a great deal of direct investigation using geotechnical or geophysical methods (Fig. 1.7).

1.2.3
Measuring Speeds

The faster the movement, the more dangerous it is, because it's not possible to raise the alarm or evacuate the area. Generally we try to measure the landslide's speed of movement in order to understand the significance of an acceleration or a slowing down, particularly in relation to rainfall and fluctuations in the piezometric level. We also try to give a prognosis on the development of the movement or suggest an alert model. There are various tools and methods for measuring the speed of movements. The pro-

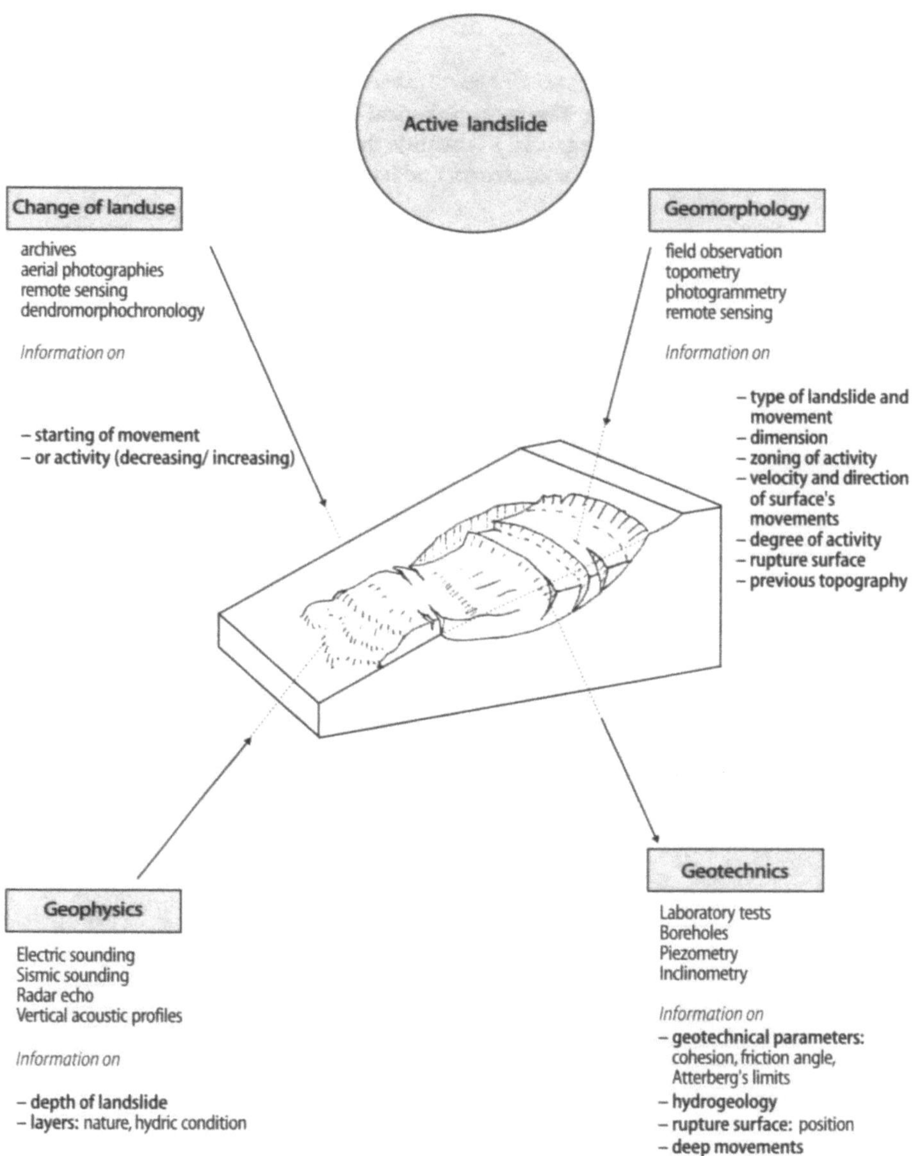

Fig. 1.7. Objectives and technics of investigation

cess is less expensive if we confine ourselves to surface measurements, but these are not as useful as depth measurements.

Changes in the dynamic mode and variations in magnitude lead us to specify the mechanical origin of stresses on buildings and infrastructures in an area affected by an earth movement. We make use of the prompting concept.

1.2.4
The Prompting Concept

The prompting mode (Fig. 1.8): The morphological effects of an earth movement are not the same throughout its length. In a landslide the movement occurs in accordance with a dominant, which is vertical upstream and lateral thereafter. At its foot the move-

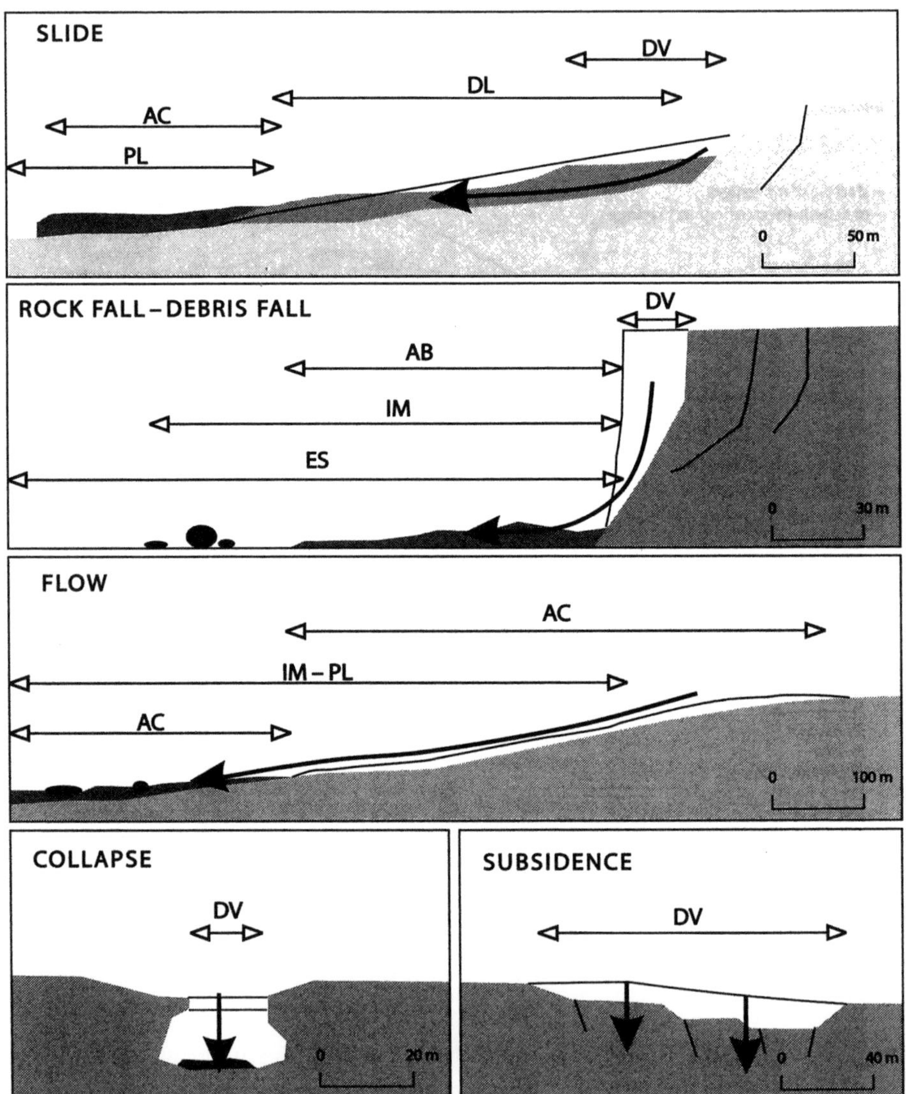

Fig. 1.8. Types of prompting. *DV* displacement mainly vertical; *DL* displacement mainly lateral; *IM* impact; *PL* lateral pushing; *AB* ablation;   → direction of movement;   ↔ effect zone of prompting (After Leone 1995)

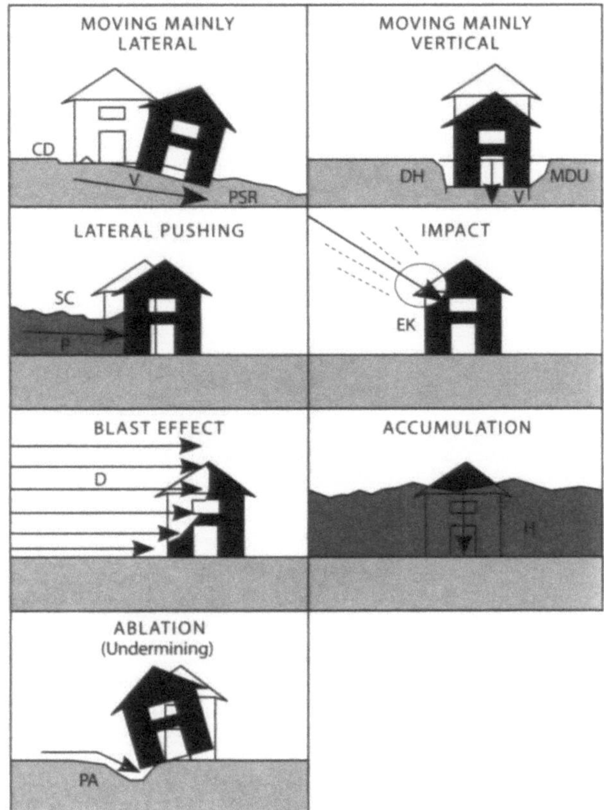

Fig. 1.9. Illustration of main types of landslide sollicitations and criteria for measurement of their intensity; *V* velocity; *CD* field of deformation; *PSR* depth of rupture surface; *DH* pressure; *SC* geometry of contact surface; *EK* kinetic energy; *D* horizontal distance; *H* hight; *PA* depth of ablation (From Leone 1995)

ment is replaced by a pushing movement and an accumulation. In a block fall or debris fall there is accumulation, impact (shock) and a blast effect. In a flow, ablation is followed by push, impact and accumulation. The strength of the destructive effect of these prompting modes depends on their individual magnitude (Leone 1996).

Prompting magnitude (Fig. 1.9): This is determined by the deformation field which is measured using parameters of speed, depth, distance, kinetic energy etc.

1.2.5
Occurrence

The estimation or calculation of the return time for the triggering or reactivation of a landslide over a wide area is particularly interesting, when we are able to establish connections with natural or anthropogenic causes (Corominas 1993). This calls for inventories and statistics over a long period and a critical in-depth analysis of the reliability and meaning of the data prior to using them for correlation purposes.

Here we give in Fig. 1.10 an example of the methodological steps leading to the discovery of relationships which may establish a connection between rainfall and landslides (Maquaire 1990; Gostelow 1991).

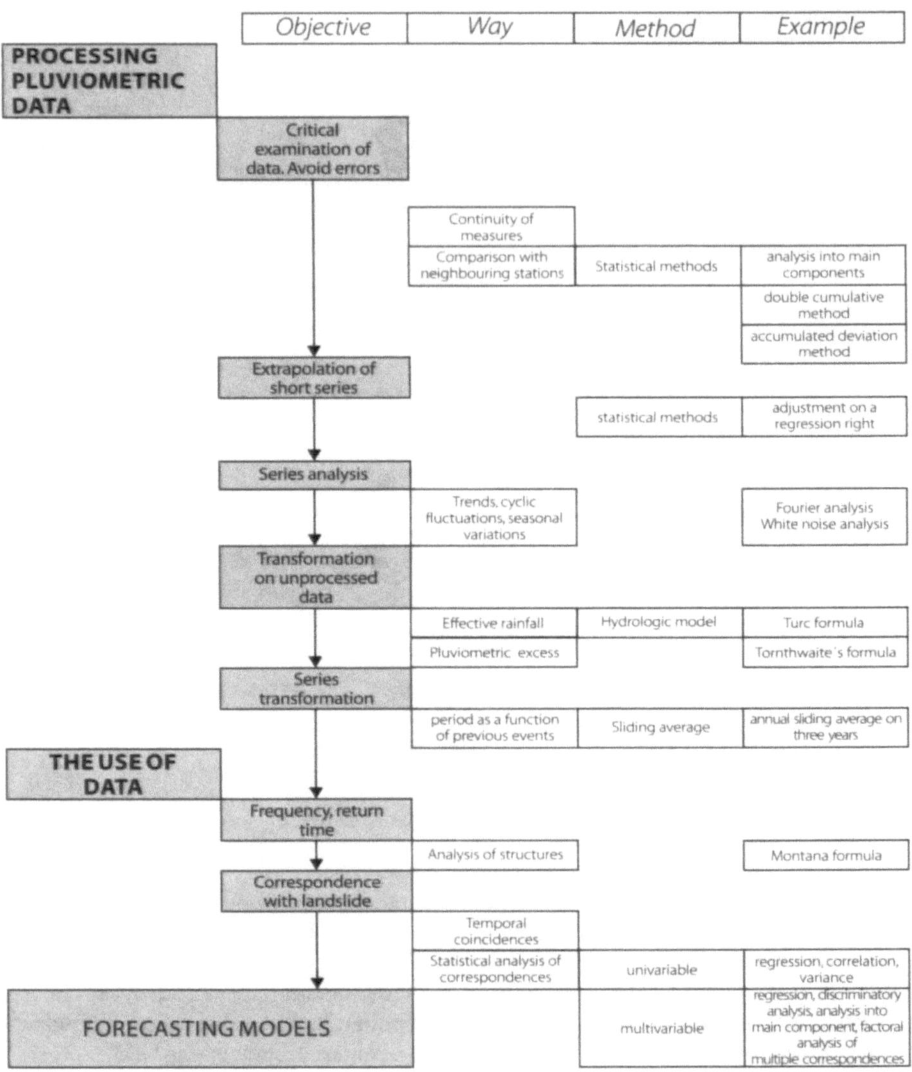

Fig. 1.10. Rainfall and landslides

1.3
Vulnerability to Landslides

We have to emphasize the need to adapt two essential conceptual aspects of vulnerability:

- The relationship between prompting and structural damage;
- The type of damage and the costs arising from it.

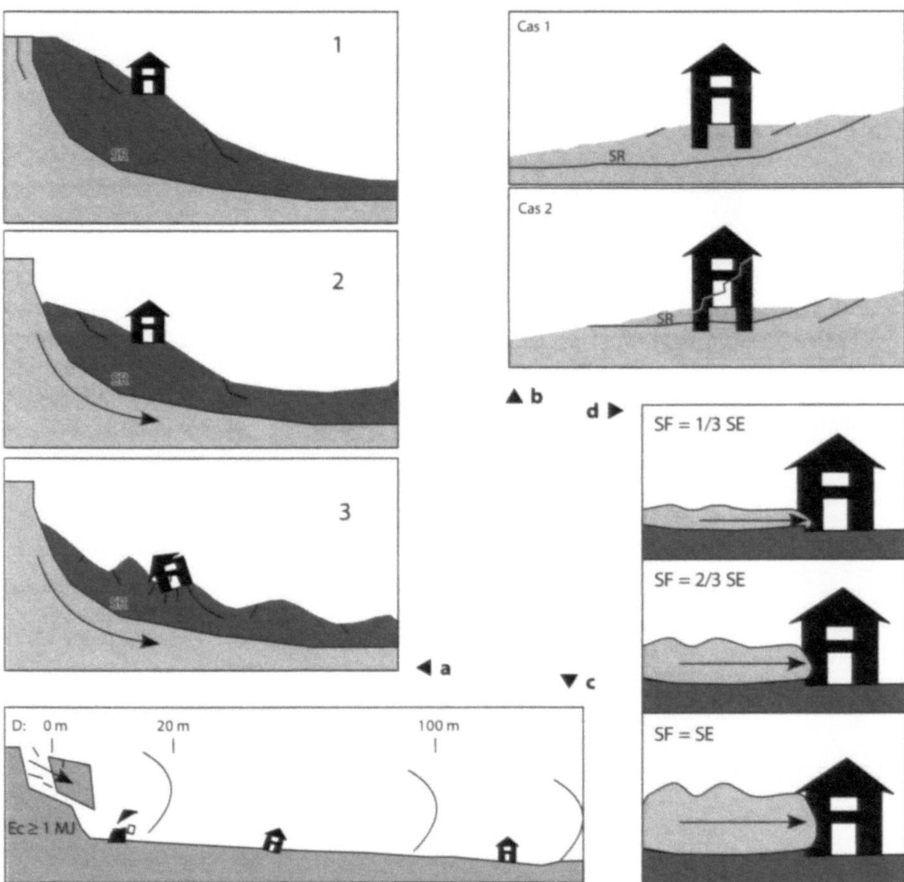

Fig. 1.11. a Importance of field deformation resulting of movement. *1.* initial condition; *2.* weak field, conservation of the initial topography; *3.* strong field, disrupted topography. **b** Relations of rupture surface with foundations of a construction. *1.* construction spared; *2.* construction damaged. **c** Blast effect depending on the distance (*D*) from the source. **d** Shape of contact in case of surface movement. *1. SF* surface movement; *2. SE* exposed surface (After Leone 1995)

Prompting and structural damage. Here (Fig. 1.11) we present a simple example of the effects on a building, but it can be applied to a road, a gas pipeline etc. The extent of the damage depends on the magnitude of the surface deformation, the depth of the rupture surface, the contact between the movement or flow with the building and the blast effect. However, the technical resistance of the building is also a factor; we know that wooden chalets can better withstand deformation arising from a slide than buildings of brick or prefabricated cement panels.

Types of damage and their cost (Fig. 1.12). In addition to the structural damage we must take account of interruptions in economic activity and the immediate loss of housing and we must try to assess the direct or indirect cost if we are to calculate the risk in all of its aspects in real terms.

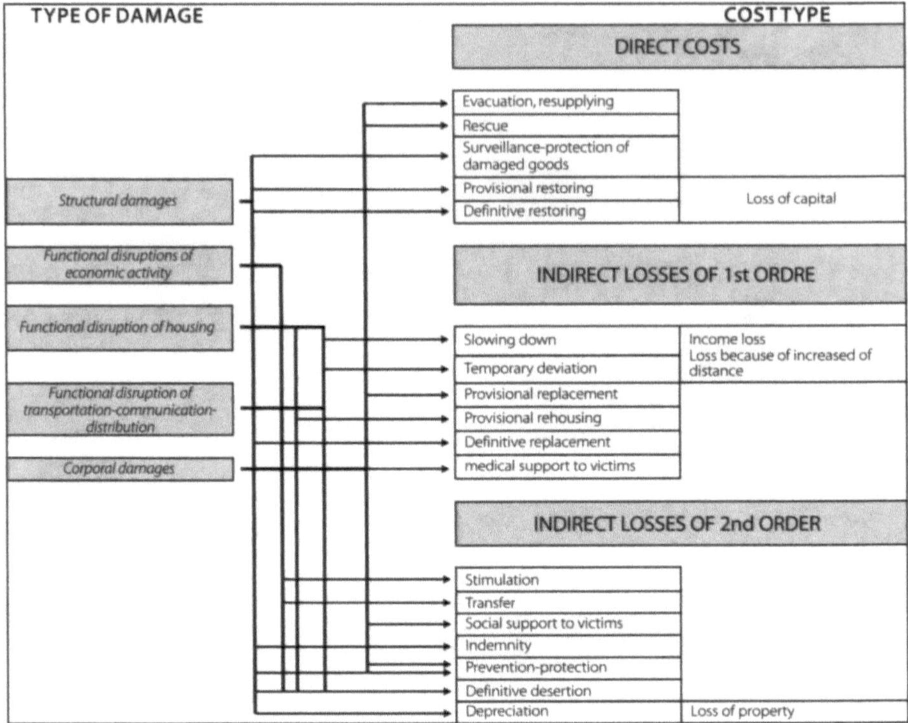

Fig. 1.12. Types of economic losses with regard to main damages (After Leone 1996)

1.4
The Risk

1.4.1
Component Defining Landslide Risk

The two following schemes (Fig. 1.13) indicate all components which finally define landslide risk and lead to specific risk and to global risk.

1.4.2
Mapping over Extended Areas

In general, maps on a scale at 1 : 100 000 or more locate and classify mass movements in order to determine the extent of the area and its distribution. The mapping of zones or sites as yet unaffected by landslides but likely to be so by extrapolation beyond mapped landslide boundaries is more complicated, but better adapted to prevention.

Mapping methods are direct and indirect (Hansen 1984).

Direct mapping can be either inventorial or geomorphological, including landslide hazard zoning. This type of mapping is the work of an experienced and knowledge-

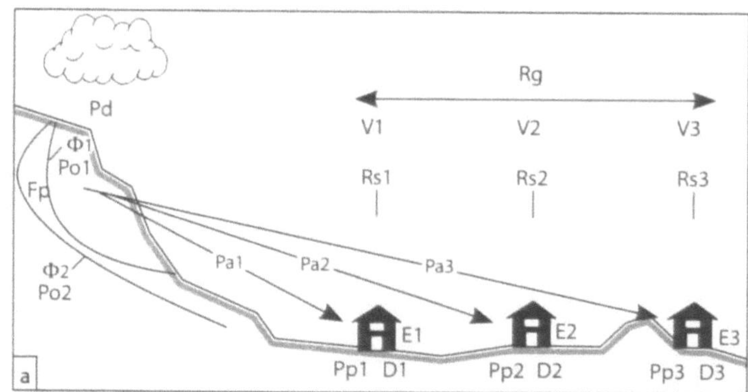

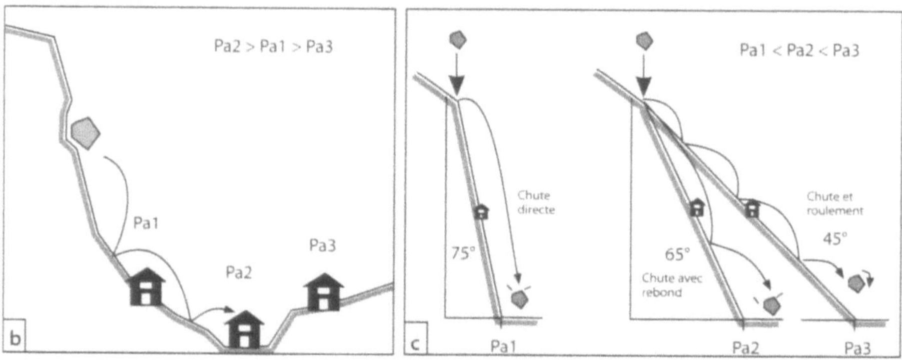

Fig. 1.13. a Illustration of the components defining landslide risk. *E* element exposed to …; *Φ* magnitude of hazard; *Rs* specific risk; *Rg* global risk; *Pd* probability of occurrence of a triggering factor; *Fp* permanent factors; *Po* occurrence probability of the hazard; *Pa* probability the hazard reaches the exposed element; *Pp* presence probability of the exposed element; *D* level of potential damages; *V* value of exposed element. **b, c** Probability for a rock to reach an exposed element. **b** according to the position of the element; **c** according to the trajectory (After Leone 1995)

able geologist or geomorphologist; as we are obliged to depend on the mapmaker's personal judgement, it is difficult to compare maps and areas as the quality of the results is sometimes questionable.

Indirect mapping extrapolates slope stability assessments beyond the landslide boundary using selected parameters and multivariate analysis, either simple statistical analysis with few variables, combined with factor weighting (or not), or complex statistical techniques with many variables (Carrara 1994). The selected parameters include the slope and the type of terrain, i.e. a minimum of predisposition factors for the simplest of maps, for example. "Calculated" maps such as "landslide susceptibility maps", or "computer-drawn landslide susceptibility maps", hazard and risk are mapped, but vulnerability is assessed in a very simple manner.

In spite of the multiplicity of parameters and controlling factors involved in mapping zones susceptible to landslides, and regardless of the objectivity of statistical methods, these maps rarely determine the exact location of failure and reactivation,

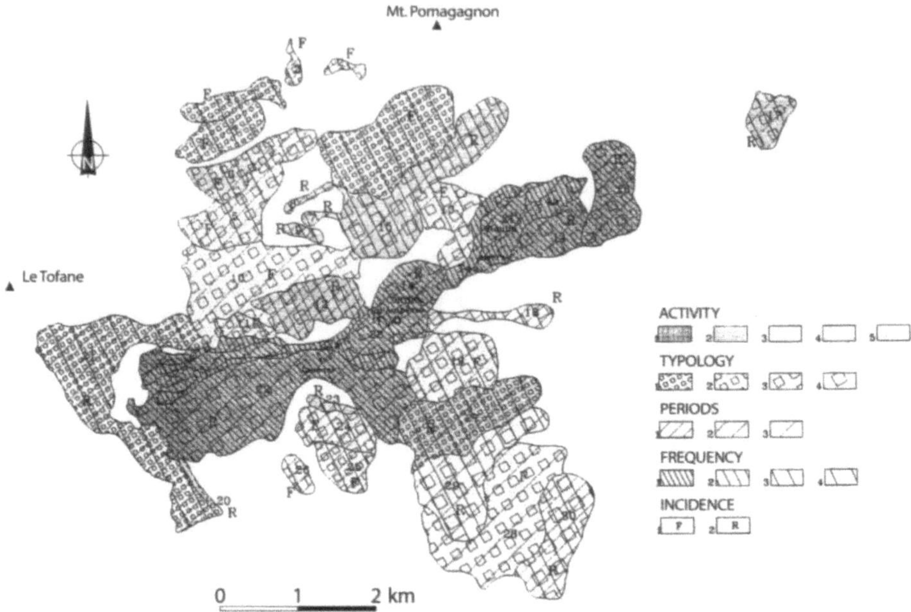

Fig. 1.14. Landslides map of the Cortina d'Ampezzo area. *Activity: 1* active-continuous; *2* active-intermittent; *3* dormant-episodic; *4* dormant -singular; *5* inactive. *Periods: 1* Hol.-historic; *2* Hol.-recent; *3* Hol.-ancient. *Frequency: 1* very high; *2* high; *3* medium; *4* long term. *Typology: 1* complex; *2* fall; *3* slide; *4* flow. *Incidence: 1* first time failure; *2* removement failure (After Panizza et al. 1994)

as local variations and weaknesses in the lithological or geological structure cannot be detected and localised on this scale (nor even on a larger scale). It is perhaps for this reason that less research in being done in this field as opposed to in the 1980s, and this limits their role in prevention, a role more informative than operational.

There has been progress in inventory mapping recently (Flageollet 1994). It consists of associating the classification of landslides with their location and occurrence. Natural and historical archives have been used in this research. Old landslides in a given area have been inventoried, dated and mapped, including occurrence beside the location; the work has renewed the content of prevention inventory maps and has revived interest in them (Fig. 1.14).

1.4.3
Mapping on a Regional Basis

As compared to the tract basis there is no fundamental difference in objectives and methods, but the scale (1 : 25 000 to 1 : 50 000) allows us, for example, to go deeper into the knowledge of occurrence, and to tackle the concept of vulnerability and risk, in particular with regard to the form in which it is mapped.

Vulnerability and risk are generally mapped at middle scale (1 : 25 000 to 1 : 50 000), one reason being that at this scale a landslide can be mapped sufficiently and accurately, with regard, for example, to boundaries, zoning of displacements or morphological changes. A further reason, is that the risk can be assessed for a community or one or more communes, and not at the individual level.

1.5
Conclusion

Fragmentary and superficial knowledge is not enough if hazard and vulnerability are looked at in view of the extended risk concept. This concept demands an in-depth study, both scientific and economic, which involves exchanges between specialists, and the highest degree of multidisciplinary cooperation. However, the cartographic representation is generally simplified compared to the lengthy work of estimation and calculation which we have undertaken. This representation varies considerably according to the scale of the territory in consideration. Maps or risk exposure plans in France, for example, bring the hazard, the vulnerability and finally the risk down to three or four levels, and this may be in conflict with a very close and accurate evaluation of the risk. This problem is not solely concerned with landslides.

References

Ayala-Carcedo FJ (1993) Socioeconomic impacts and vulnerability in slope movements. Report of the proceedings of the US Spain workshop on natural hazards, Barcelona, pp 235–254
Bourdeau et al.(1988) Stability of natural slopes: recent evolutions of calculation methods. In: Bonnard C et al. (1988) Landslides, AA Balkema, Rotterdam, pp 541–548
Bozzolo D et al. (1988) Rockfall analysis. A mathematical model and its test with field data. In: Bonnard C et al. (1988) Landslides, AA Balkema, Rotterdam, pp 555–560
Carrara A (1994) Landslide hazard mapping: aims and methods. Documents du BRGM, 83, Orléans, pp 142–151
Cavallin A et al. (1994) The role of geomorphology in environmental impact assessment. Geomorphology 9, pp 143–153
Corominas J (1993) Landslide occurrence. A review of the spanish experience. Report of the proceedings of the US Spain workshop on natural hazards, Barcelona, pp 175–194
Cruden DM (1991) A simple definition of a landslide. Bulletin of the International Association of Engineering geology 43:27–29
Dikau R et al. (1995) Landslide recognition. J. Wiley
Durville J L (1991) Mécanismes et modèles de comportement des grands mouvements de versants. In: Prévention des mouvements de grande ampleur, Bulletin of the International Association of Engineering geology, N° 45, Paris, pp 25–42
Flageollet J-C (1993) Knowledge of landsliding for prevention and rescue. International Conference on natural risk and civil protection, Proceedings preprint. EC DG XII, pp 149–151
Flageollet J-C (1994) The time dimension in the mapping of earth movements. In: Casale R et al. (1994) Temporal occurrence and forecasting of landslides in the European Community. Final Report, vol 1, EUR 15805 EN, pp 9–20
Gostelow TP (1991) Rainfall and landslides. In: Almeida – Texeira et al. (1991) Prevention and control of landslides and other mass movements, E.C. report EUR 12918 EN, pp 139–161
Hansen A (1984) Landslide hazard analysis. Slope instability, Wiley & Sons, pp 523–601
JPA Consultant (1994) Etude d'avant projet sommaire de mise en sécurité des carrières de Saint Nabor (67) à partir de l'état existant en Mai 1994. Rapport principal, Direction Régionale de l'Environnement, Strasbourg, 75 pp
Leone F (1995) Concept de vulnérabilité appliqué à l'évaluation des risques générés par les phénomènes de mouvements de terrain. Thèse doctorat université Joseph Forier, Grenoble 1, spécialité Géographie, 274 pp
Leone F (1996) Concept de vulnérabilité appliqué à l'évaluation des risques générés par les phénomènes de mouvements de terrain. Thèse de doctorat Université Joseph Fourier de Grenoble, Laboratoire de la Montagne Alpine, 274 pp
Maquaire O (1990) Les mouvements de terrain de la côte du Calvados, Recherche et Prévention. Document du BRGM N° 197, BRGM, Orléans France, 431 pp

Panizza M et al. (1994) Research in the Cortina d'Ampezzo area. In: Temporal occurrence and forecasting of landslides in the European Community, Final Report, EUR 15805 EN, vol. II, pp 741–768
Sitar N, Rogers C (1993) On scale depending of data for analysis of landslide hazard using GIS and expert systems. Report of the proceedings of the US Spain workshop on natural hazards, Barcelona, pp 195–208

Chapter 2

Flood Hazard Assessment and Mitigation

S. Fattorelli · G. Dalla Fontana · D. Da Ros

2.1
Introduction

A flood is an overflowing of water from rivers onto land not usually submerged. Floods also occur when water levels of lakes, ponds, reservoirs, aquifers and estuaries exceed some critical value and inundate the adjacent land, or when the sea surges on coastal lands much above the average sea level. Nevertheless, floods are a natural phenomenon important to the life cycle of many biota, not the least of which is mankind. Floods became a problem as humans began establishing farms and cities in the bottom-lands of streams and rivers. In doing so, they not only exposed their lives and properties to the ravages of floods, but also exacerbated floods by paving the soil and constructing the stream channels. Over time, continued urbanisation of natural floodplains has caused great annual losses of both wealth and human life. In this way, in many countries and regions of the World, floods are the most costly hazards in terms of both loss of human lives and material damage.

The spatial scale at which this kind of phenomenon arises can be highly variable. Flooding can affect large continental areas, such as in the case of the flooding frequently occurring in Asia (in 1991 as a consequence of flooding the final death toll surpassed 139 000 in Bangladesh alone), but it can also occur as very localised events. Although this kind of events may not hit the headlines of newspapers like major earthquakes or big disasters do, their cumulative effect is much greater: on the global scale, storms and floods are the most destructive of natural disasters and cause the greatest number of deaths. This agrees with the results of a survey conducted by the Centre for Research on the Epidemiology of Disasters of the Université Catholique de Louvain (Belgium) relative to the major natural disaster occurring at a global scale during the period 1966–1990 (Fig. 2.1). The figure also shows that floods are the natural hazard affecting the largest number of people. Figure 2.2 displays the distribution of the fatalities consequent to weather events (including floods, storms, drought, high and low temperature, landslides and avalanches, etc.) for the period 1989–1993 (Fig. 2.2), on a global scale, based on statistics yearly reported by the World Meteorological Organization (WMO).

Floods and inundation in Europe are mainly caused by storms, high tides and surges along the North European coasts, or by severe thunderstorms that affect the South of Europe and produce intense rainfall, sudden rises in river levels and flooding in urban areas, with consequent damages and disruption. Table 2.1 reports some of the major weather related disasters in the years 1960–1994. Regarding the most recent years, it is necessary to remember the big disasters that involved Northern Italy (November 1994) and North and Central Europe (January 1995) that caused substantial damages to private and public property, enormous social disruption and several losses of life.

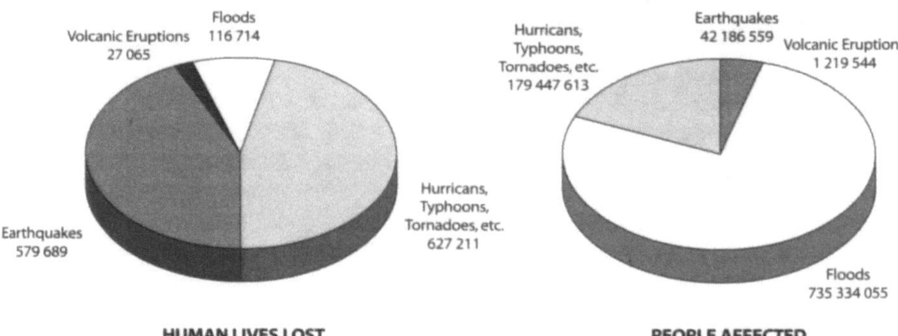

Fig. 2.1. Major disasters in the World during the period 1966–1990 (Source: Centre for Research on the Epidemiology of Disasters, Université Catholique de Louvain, Belgium)

Fig. 2.2. Yearly average number of the reported fatalities as a result of weather events during the period 1989–1993 per WMO regions (Limbert 1995)

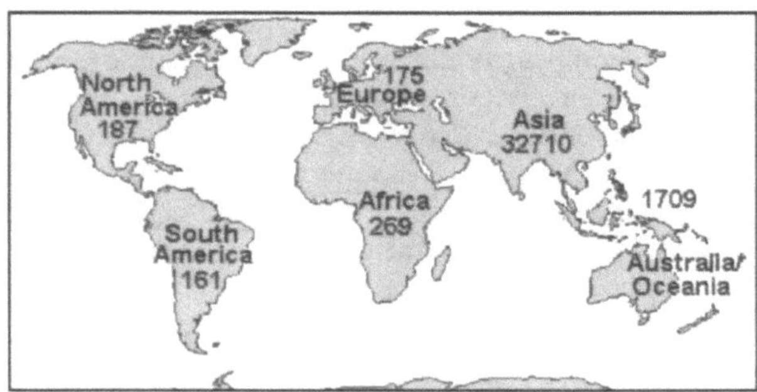

The structure of the paper is as follows: at first general characteristics of the flood phenomenon are recalled and a tentative typology of the flood hazards in Europe is presented. The following section is devoted to the methodologies developed for flood risk assessment. In the last part of the paper, an overview of the principal measures adopted to cope with floods is provided, with a special emphasis on flood forecasting and warning systems that represent the most innovative solution to the problem of the reduction of flood impact.

2.2
General Characteristics of Floods

Spatial and temporal scales of flood events are generally linked to the corresponding scales of the flood generating rainfall. According to the classic model formulated by Austin and Houze (1972), so-called synoptic areas with an extension greater than 10 000 km² usually contain large mesoscale areas (LMSA), also called 'bands' ranging in extension from 1 000–10 000 km². LMSAs contain, in turn, smaller mesoscale areas (SMSA), or cluster of convective cells with an extension of 50–1 000 km². Size of rain cells within SMSA varies from about 1–10 km². All structures within the synoptic area possess their own motion, development, build up and decay. The life span of these struc-

Table 2.1. Some of the major weather related disasters in Europe since 1960 (Source: Munich Reinsurance Company, Munich, Germany)

Date	Event	Area	N° of dead	Overall losses (million US ^{a})	Insured losses (million US ^{a})
Feb. 1962	Storm surge	Germany	347	600	10
Nov. 1966	Floods	Italy	113	1 300	
Feb. 1967	Winter gale	Germany	40	300	50
Nov. 1972	Winter gale	Central Europe	54	420	200
Jan. 1976	Winter gale	Europe	82	1 300	500
Nov. 1981	Winter gale	Denmark	9	250	95
Nov. 1982	Winter gale	France	14	350	224
Aug. 1983	Floods	Spain	42	1 250	433
Jul. 1984	Hailstorm	Germany		1 000	500
Jan. 1985	Winter damage	Central Europe		350	200
Aug. 1986	Hailstorm	Switzerland		105	85
Jan. 1987	Winter damage	Great Britain	34	700	385
Feb. 1987	Floods	USSR	110	550	
May 1987	Floods	Poland		500	
Jul. 1987	Floods, landslides	Italy	44	500	
Jul. 1987	Heat wave	Greece	1 000		
Jul./Aug. 1987	Floods	Switzerland		700	175
Oct. 1987	Winter gale	Great Britain/ France	17	3 700	3 100
Nov. 1989	Floods	Spain	12	375	
Jan./Feb. 1990	Winterstorms	Europe	230	15 000	10 000
Sept. 1993	Floods	Italy, France	15	1 000	
Jan. 1994	Floods	Great Britain		160	
Nov. 1994	Floods	Italy	70	3 000	

^{a} Referred to value in 1992.

tures decreases with their extent, rain intensity within smaller structure is always higher than in the region outside the structure. Thus, the time distribution of rainfall observed in a point gauge is a result of several characteristics of a rainfall field and several processes going on within these fields, such as spatial variation or raincells, their development and decay, the distance between them, and the velocity and direction of their movement.

As to the spatial and temporal scales of the phenomenon, the smaller the basin area, the larger is the rainfall of given duration and probability of exceedence. As a result, the specific peak discharge increases with decreasing the basin area. This appearance is a direct consequence of time-space distribution of precipitation, that is the more intense the average precipitation above a given value, the lesser is its duration and smaller is its areal coverage. The smaller the catchment area, the more rapid is its flood flow concentration, and the faster it passes a location. Similar impact is produced by the average river basin slope and the river channel slope, namely the greater these slopes the faster are the flood flow concentration and its travel time. Therefore, the average duration of the flood of given exceedence probability increases with an increase of the size catchment area, with all the other basin condition being equal. These general characteristics of floods play an important role in flood forecasting, warning, impact and prevention (Yevjevich 1994).

2.2.1
A Tentative Typology of European Flood Hazards

Penning-Roswell and Peerbolte (1994), as a result of the tentative of assessing the nature of the problem and the seriousness of the flood hazard situation in the different countries of the European Union, proposed a classification of the main types of flooding and flood risk in Europe. Their work was used as a guidance for the classification proposed in the following, which tries to cover the types of floods that occur in Europe.

Winter rainfall floods. Westerly depressions with well-developed warm fronts bring winter precipitation, mainly in Central and Northern Europe. When these precipitations are heavy, continuous and prolonged, they can lead to soil saturation and consequent high volumes of runoff. As a consequence, rivers may flow out of banks, causing flooding. Floods which occurred in England in 1994 were an example of this type of flooding, followed by a December during which four times the monthly December average rainfall fell. A major example of this type of flood events occurred during the Winter of 1995 and interested wide areas in the Netherlands, in Germany (the Rhine valley) and in France.

Summer convectional storm induced floods. Heavy convectional thunderstorms can sometimes generate intensive storms and floods. They are most likely to occur in the summer. In particular, in Southern European areas where prolonged hot periods of the summer months can end with sudden storms. Furthermore, the seriousness of this type of event is compounded by the fact that even if the storm event can be localised they can lead to severe flash floods affecting highly developed urban areas. The floods in France during 1992, in Vaison-la Romaine, and the floods that occurred in the coastal streams adjacent to Lisbon are examples of this type of problem. In general, this type of phenomena affects many European Mediterranean areas from the Iberian peninsula to Southern France, Italy and Greece. Meteorological conditions (intensive precipitation) associated with the combination of local topography (steep orography) along with geological and land use conditions (presence of wide impermeable areas) may lead to very localised flash-flooding with consequential huge damage of properties and loss of human lives.

Convective frontal storm induced floods. Frequent meteorological conditions over Western and Southern Europe are characterised by extended low pressure, associated to cold fronts which travel from the west Mediterranean sea towards the continent. In these situations mesoscale convective systems can develop, resulting in extreme rainfall, lasting more than 24 h. The air mass can also be subjected to orographic enhancement upon reaching the slopes of the mountainous chains. The floods that affected Genoa (Italy) in the autumn of 1993 and 1994, or the flooding that occurred in the Piemonte Region (Italy) in the autumn of 1994 are examples of this type of phenomenon.

Snowmelt floods. Rapid snowmelt can sometimes cause flooding, especially in the spring when warm southern airstreams can influence Alpine or upland areas and create sudden snowmelt accompanied frequently by heavy rainfall. This phenomenon is usually very localised and in very steep watersheds can produce flash-floods, since flood water velocity can be high. This problem affects urban areas occupying bottom valleys. An example is represented by the flood that occurred in Perth (Scotland) in 1993.

Urban sewer flooding. Inadequate sewerage systems can lead to serious flooding problems in urban areas, since even normal intensive rainfall events can create abnormal flooding. This problem is obviously minor when viewed at a continental scale, but can be locally serious and regards not only many older cities which have insufficient storm sewerage systems but even some new urban areas, because of unplanned developments.

Sea surge and tidal flood threat. One of the major problems of flooding affecting many European coastal areas is related to the sea surge and tidal effects. Moreover, associated with this problem is the phenomenon of coastal erosion which may consequently lead to flooding. Especially in Northern Europe many coastal areas are low lying and the combination of high tides, low atmospheric pressure and strong onshore winds producing tidal surges consist in a major threat. Coastal flooding can also be caused by many other factors including, seepage of sea water through natural or artificial embankments, breaches of these embankments, or exceptionally through ocean swell phenomena causing waves to overtop the embankments. The problem has worsened because of the land subsidence phenomenon. The floods in the Netherlands and the United Kingdom in 1953 are an example of this type of phenomena.

Dam-break flood risk. Flood problems can also arise from the breaking of dams and dikes. These events are generally characterised by low probability, but the consequences of flood risk associated are very high, and clearly any floods that have this cause are likely to be very serious.

2.3
Current Approach for Estimation of Flood Risk

An essential element of an effective program for reducing riverine flood hazard is accurate risk assessment. Here risk assessment is loosely defined to include the estimation for a particular river reach of the probabilities associated with various flood levels, as well as, of various functions of flood level, such as expected flood damages. With existing methods of flood risk assessment it is possible to estimate the current risk of flooding, as well as, the risk which would exist under various hypothetical future scenarios of floodplain development and mitigation activities.

Availability of historic flood data allows us to estimate the peak flows for given probabilities (or return periods). Several probability distribution functions have been suggested and widely used to fit peak-flow data, such as two- and three-parameter Log-normal, Pearson, Log-Pearson, Extreme Value Type 1, Generalized Extreme Value, Wakeby, etc. Flood frequency relationships can be developed by different approaches. Three cases are generally identified:

1. Gauged site with a peak-discharge record,
2. ungauged site on a gauged stream and
3. ungauged site on a ungauged stream.

In the first case, one of the several theoretical probability distributions can be fitted directly to the available flood discharge record, while in the other cases a more or less complex technique should be employed to transfer flood information from gauged sites to ungauged sites of interest, mainly by regional analysis.

2.3.1
Regional Floods Frequency Analysis

Nowadays, there is increasing agreement over the convenience and necessity of adopting a regional flood frequency analysis approach, not only for evaluating flood characteristics at ungauged sites, but also to reduce the uncertainty in parameter estimation at gauged sites with small samples, thus improving the reliability of quantiles of large return periods. Hence, to estimate the T-year flood at sites with little or no recorded flood data, it is possible to use a regional regression model to predict the relevant quantile from catchment characteristics or to use the index flood approach in which, a dimensionless regional frequency curve is rescaled at the site of interest by a local estimate of the scaling factor. Dalrymple (1960) introduced the index flood method as a way to derive a regional flood frequency curve, and the method was extended in the UK Flood Studies Report (Natural Environment Research Council, NERC 1975) which also presented regression models for the estimation of the index flood, the mean annual flood, from catchment characteristics. More recently, attention has been focused on the use of the index-flood approach to improve the estimation of the T-year flood at sites with some recorded flood data (Wallis 1980; Greis and Wood 1981; Wallis and Wood 1985). In these applications the scaling factor is computed from the sample of flood peaks at the site. In recent years, a regional method for flood estimation based on the TCEV distribution with a three-level regionalization scheme was applied in Italy (Rossi and Villani 1994) in the framework of the VAPI research program conducted by the National Group for the Prevention of Hydrological Disasters (CNR-GNDCI).

2.3.2
Flood Risk Estimation by Using Rainfall-Runoff Models

Flood risk assessment for complex settings generally requires the use of a hydrologic simulation model, which simulates river flow based on an assumed temporal and spatial distribution of precipitation. Hydrologic simulation models enable the analyst to simulate streamflows, for actual conditions in the watershed, or for proposed or future conditions. Hence it is possible to use these models to evaluate the effectiveness of various flood mitigation strategies.

In several countries the design storm method is the dominant method for estimating flood probabilities based on hydrologic simulation modelling. The rationale of this method is as follows. Statistical analysis is performed on historical rainfall data to produce intensity-duration-frequency relationships, which are estimates of the probability distribution of rainfall intensity for various durations. These relationships are used to specify the intensities of design storms, which are in turn used as input to a hydrologic simulation model. It is assumed that the flood peak resulting from a given design storm has the same exceedance probability as the storm used to simulate it.

The assumption that the storm and resulting flood have the same probability is critical to the design storm method. It implies that average rainfall intensity is the dominant source of randomness in the generation of flood flows. In many simple applications this is a reasonable assumption. But in most complex settings which are typical of flood mitigation applications, other factors contribute significantly to randomness. For example, the severity of flooding resulting from a given storm typically depends on the initial soil moisture conditions. In the case of simulations involving reservoirs, their flood mitigation effectiveness in a given event will depend on the initial reservoir levels. In many watersheds, flood magnitudes are highly dependent on the spatial and temporal distribution of rainfall, particularly as the distribution controls the relative timing of the peak flows in large confluent tributaries. Clearly factors other than average rainfall intensity can affect the probability distribution of flood peaks, and hence the critical assumption of the design storm method is often non valid.

The weaknesses of the design storm method are especially critical when it is used to evaluate complex strategies for flood mitigation. In most applications, design storms are very different from actual storms, with respect to both spatial and temporal structure. Furthermore, the same artificial storm structure is used in simulating successively more extreme events, with only the intensity changing. This provides a severely limited evaluation of any given strategy for flood mitigation. Consider for example, a complex reservoir system with the capacity for real-time control. Simulations with design storms would hardly test such a system. All simulations would be similar in terms of the spatial and temporal development of the flood event, when in fact tremendous diversity is possible in actual flood events.

An alternative approach which has sometimes been used to estimate flood risk is based on continuous simulation. While event-type models are developed with the purpose of simulating flood hydrographs for the individual storm event, continuous-time models can simulate the watershed behaviour even during dry periods. In this approach one or more long historical rainfall records are used as input to a continuous-time hydrologic simulation model to produce a long streamflow record. The latter record is then analyzed as if it were an historical streamflow record obtained at a gauging station, providing an estimate of the probability distribution of peak flood discharges. Continuous simulation enables the modeller to explicitly account for many of the important factors which are ignored in the design storm method. Continuous-time hydrologic simulation models were specifically devised to model the temporal variation of soil moisture, both during and between storm events. These models allow for explicit representation of reservoir operations and can account for spatial and temporal variability in rainfall, depending on the resolution of the available rainfall data in space and time. Finally, the use of historical rainfall results in a large diversity of hydrologic responses, and hence provides a good test of alternative strategies for flood mitigation. However, a serious drawback of continuous simulation lies in the limitation of digitized available hourly rainfall records. These generally do not provide sufficient data to accurately define the upper tail of the probability distribution of peak flood discharges, based on conventional method. Furthermore, probability distributions for alternatives incorporating complex flood mitigation strategies often violate the distributional assumptions of conventional methods. In some cases – see, for instance, Bradley and Potter (1991) – the uncertainty can be very large. Note that this latter problem is much less likely with the design storm method, since the use of the same storms in all of the hydrologic simulations forces consistency across simulations. Approaches based on applying continuous simulation and using information from extreme storms which have occurred in the meteorologically homogeneous region containing the watershed of interest have been presented recently to overcome the above problems (Bradley and Potter 1992).

2.3.2.1
Rainfall-Runoff Modelling

In the previous section some issues on using hydrological models for flood risk assessment have been illustrated. Mathematical models of hydrologic systems are used to represent hydrologic system operation, and to predict system output. A mathematical model provides an approximation of the actual system by a set of relations linking system inputs and outputs through a system state vector. Reviews of mathematical models used in rainfall-runoff models have been reported elsewhere (Nemec 1986; Todini 1989; Fattorelli 1982; among others). Here, attention is focused on aspects of conceptual and physically-based hydrological models.

The origins of hydrological models are closely related with the development of civil engineering in the nineteenth century. In the design of roads, canals, railways and city sewers a requirement common to all was a reliable method of estimating storm runoff. Obviously models could only be developed as understanding of the hydrological cycle evolved and could only be applied when measurements of rainfall, streamflow and other variables became available. The rational formula (Mulvaney 1850) was one of the first event-type models relating storm runoff to rainfall. Later the development of the unit hydrograph concept (Sherman 1932), and its subsequent evolution to the instantaneous unit hydrograph provided the basis for the storm response models of Nash (1957) and Dooge (1959). From these beginnings current methods of estimating catchment response to individual storm events have developed and grown in complexity.

Many rainfall-runoff models are based on the unit hydrograph method; however the application of the method requires the determination of effective rainfall, namely the portion of rainfall that is not lost via evapotranspiration and infiltration. In an attempt to achieve a better physical interpretation of this phenomena, hydrologists tried to represent the different components of the hydrologic cycle by using conceptual schemes or studying the physical behaviour of the basin.

Among the conceptual, continuous-time models, the most common used and known are the Stanford Watershed Model (Crawford and Linsley 1966; Fattorelli 1982; Cazorzi et al. 1984), the National Weather Service Model (derived from the Sacramento model, Peck 1976), the SSARR (Rockwood and Nelson 1966), etc., which represent in different ways the response mechanisms of various phenomena and the interconnections between the various subsystems that constitute the watershed. Figure 2.3 shows a block diagram of the Stanford Watershed Model, representative of this group of models.

Rainfall-runoff models presented above are referred as conceptual models, because they invoke simple conceptualisation of watershed dynamics. Despite their simplicity, many conceptual models have proven quite successful in representing an already measured hydrograph.

A severe drawback of these modelling systems, however, is that their parameters are not directly related to the physical conditions of the catchment. Accordingly, it may be expected that their applicability is limited to areas where runoff have been measured for some years and where no significant change in catchment conditions have occurred.

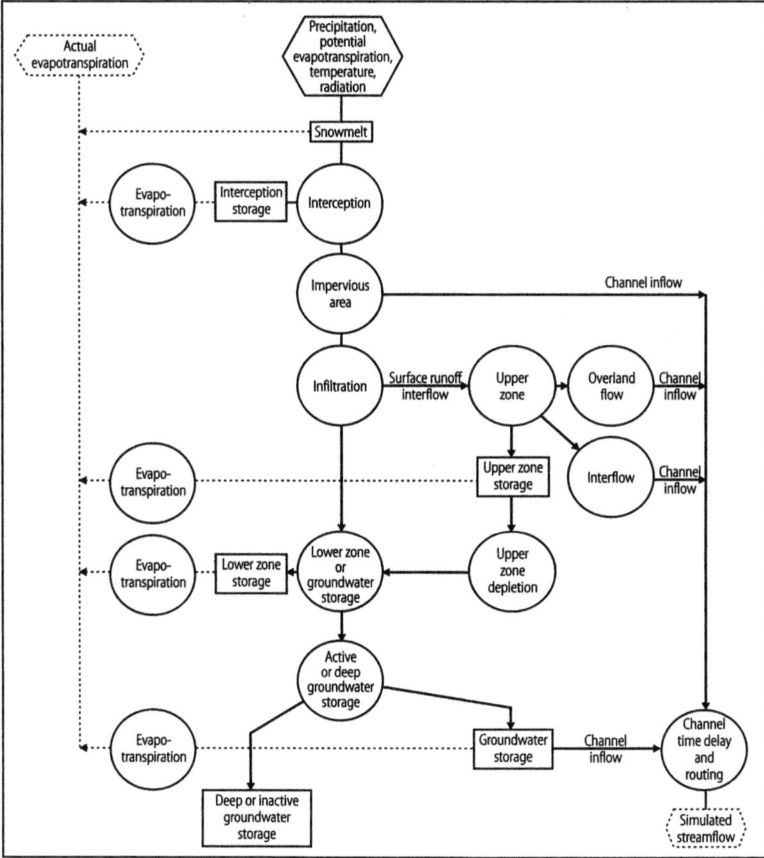

Fig. 2.3. Schematic representation of the various components of the Stanford IV model (After Crawford and Linsley 1966)

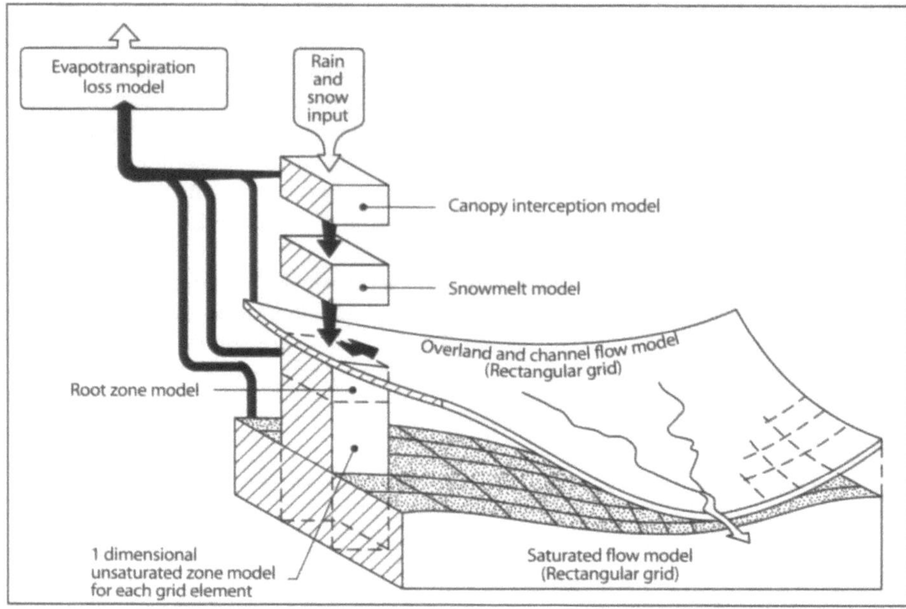

Fig. 2.4. Schematic representation of the SHE model (After Beven et al. 1987)

To provide a more appropriate tool for the type of studies where discharge data are not directly available (for instance, impact on flood risk due to land use/climate change), considerable efforts within hydrological research have been directed toward development of distributed physically based catchment models. These models use parameters which are related directly to the physical characteristics of the catchment and operate within a distributed framework to account for the spatial variability of both physical characteristics and meteorological conditions. These models aim at describing the hydrological processes and their interaction as well as where they occur in the catchment and therefore, offer the prospect of remedying the shortcomings of the traditional rainfall-runoff models.

The most representative of these models is the SHE, Système Hydrologique Européen (Abbott et al. 1986) that represents a collaborative research project of the Danish Hydraulic Institute (DK), the Institute of Hydrology of Wallingford (UK) and SOGREAH (F) (Fig. 2.4). For these developments the advent of digital computers has been very significant. Furthermore, parallel to the development of computer hardware has been the development of computer programming languages, data-base systems, graphic and image processing tools, remote sensing techniques which has made the use of large amounts of data easier and has enhanced the analysis and interpretation of the results.

The advent of Digital Terrain Models (DTM) and Geographic Information Systems (GIS) has had a enormous impact on the capability in the description of hydrologic processes at basin scale. GISs are highly specialised database management systems for spatially distributed data, that represent a powerful tool in dealing with the spatial nature of most hydrological variables and parameters. The capabilities offered by these systems in the analysis of spatial data of hydrological interest can be exploited for better understanding of the hydrological processes and of the correlation and interaction among different hydrological variables.

The benefits in the use of these systems include: improved data access, the ability to perform complex spatial queries, the ability to access multiple databases simultaneously, more consistent and reproducible results (as compared to manual methods), etc. For example, spatial and/or attribute searches are readily accommodated by a GIS that allows the user to extract information from established areal units (e.g. a region, a country), as well as, from interactively defined areas. Figure 2.5 gives an example of simulation of a flood hydrograph obtained by the application of a distributed hydrological model (TOPMODEL, Beven and Kirby 1979) in which the topographic analysis of the watershed has been performed based on the DTM of the basin.

2.4
Defence Against Floods

The defence against floods is defined as any set of human activities which decrease loss of lives, property and production due to floods in comparison with the consequences of floods if those activities were not undertaken. Many measures can be used by societies to cope with floods. They are usually classified into two major groups: structural measures and non-structural measures. Furthermore, these measures can be combined together in order to maximise the effects of the alleviation of the flood risks.

Structural measures of flood management can be defined as the measures that alter the physical characteristics of the floods. They usually involve engineering works as dams, reservoirs and retarding basins, channel and catchment modifications, leveebanks, flood proofing, etc.

On the other hand non-structural measures alter the exposure of lives and properties to flooding (flood forecasting and warning, flood insurance, planning controls, public information and education, etc.).

In the present paper the various action that can be taken to cope with flood risk are classified as follows (Askew 1991): flood control, flood proofing, planning control, emergency plans and flood forecasting. The first two belong to the structural measures, while the latter are considered non-structural measures. Each of these topics worthy of a separate paper, and there is extensive literature on the whole subject. In the following just a few comments will be made on each, while the topic of real-time flood forecasting and warning systems will be examined in more detail.

2.4.1
Flood Control Measures

They comprise major engineering works that can temporarily store or divert the flow of water and thus lowering the flood peak, such as dams, reservoirs and retarding basins, levee banks, catchment modifications, etc. If they are used wisely, these measures can greatly reduce the level of flooding, even to a level where the river remains within its banks. However, it is important to underline that structural measures cannot eliminate flooding, they can only reduce its impact, and, above all, they can be very costly and disruptive to the environment.

a

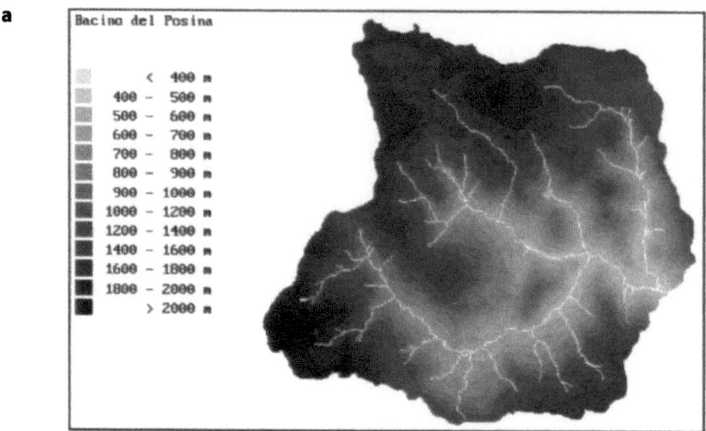

b

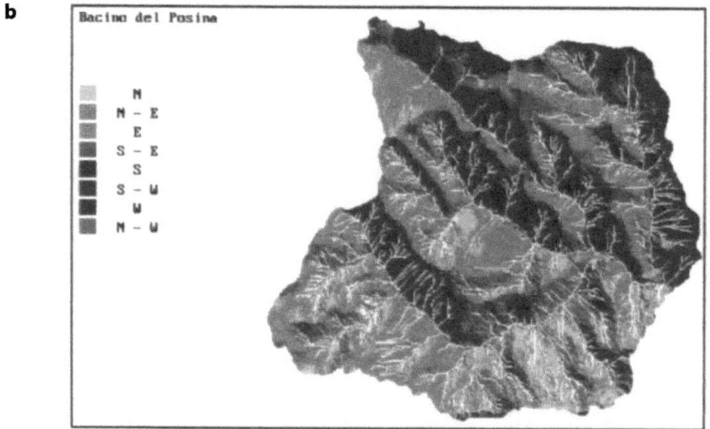

c

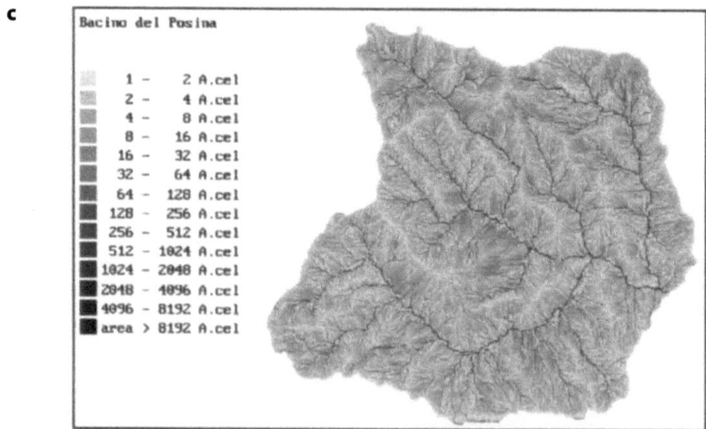

Fig. 2.5 a–c. Use of DEM for the application of the distributed model TOPMODEL for the simulation of flood hydrograph. **a** DEM of the Posina River basin (116 km²); **b** drainage directions; **c** drainage area (expressed in number of cells)

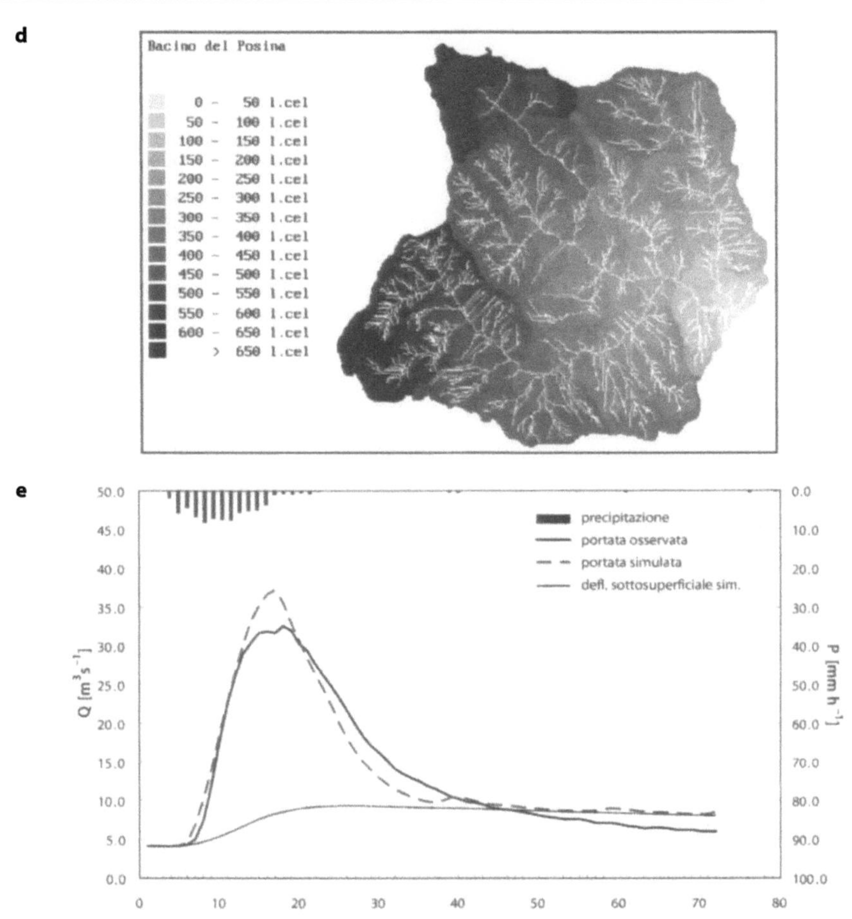

Fig. 2.5 d,e. Distance from the outlet (**d**)(expressed in number of cells); **e** example of hydrological simulation (precipitazione: rainfall; portata osservata: observed flow; portata simulata: simulated flow: defl. sottosuperficiale sim.: observed baseflow) (After Da Ros 1996)

2.4.2
Flood Proofing

Flood proofing consists in the modification of buildings and structures and their immediate surrounding to reduce damage in flooding. This can be achieved by appropriate building codes which control the building standards and characteristics as part of a program to minimise flood losses. Building codes may be voluntary or developed by public authorities and they can establish the minimum elevations for footings and floors, minimum specification for foundations or walls, etc. Flood proofing provides individual property owners with a means of reducing their risk of damage, although its effectiveness may be limited in the case of very serious floods.

2.4.3
Planning Control

Hydrologists can estimate the likelihood of a flood inundating an area to a given depth and flood risk maps can therefore be prepared showing high-risk areas. Moreover, the total designation of flood prone areas into zones of different exposure to flooding may be used to enlighten potential users of their exposure to damage, to identify zones for insurance purposes, as well as, to underlie compulsory statuary limits on land use in the exposed areas. On the basis of flood risk maps, communities can adopt legal tools with which to control the extent and type of future development which will be permitted in the river valleys.

2.4.4
Emergency Plans

Public authorities may arrange plans to be implemented upon receipt of a warning. Planning includes the identification of responsible authorities, flood warning levels, targets and dissemination channels, evacuation, relief and rescue forces, repair and maintenance equipment and materials, emergency flow control mechanisms and training requirements. Emergency plans should be developed and reviewed from time to time in consultation with all concerned and should be clearly announced to the public. Regarding the latest aspect, it is worthy to remember that floods, after all, are periodic events, so that in between these events people forget their impact. For this reason it is important to increase awareness to the public of this problem.

2.4.5
Flood Forecasting and Warning Systems

One definition of hydrologic forecasting is the following: "that branch of science and engineering which deals with the assimilation and analysis of hydro-meteorological data and information, and the input of such information into hydrologic modelling and prediction procedures to arrive at forecast the present and future states of the various components of the hydrologic cycle, especially the streamflow conditions in streams and rivers". Thus, hydrological forecasting involves the application of hydrological and meteorological principles in an engineering and, most often, a systems framework.

The importance of real-time flood forecasting is related to the evidence that the effectiveness of emergency response is a function of the advanced warning given and of the reliability of the prediction. The earlier and more precise the forecast, the greater the chance of saving lives and properties.

Real-time hydrological forecast systems generally consist of several components which include computer hardware and software to perform data handling functions and enable users to calibrate models and use the models to generate forecasts. These systems are indicated as integrated when meteorological and hydrological predictions are coupled together and a decision support system is incorporated within, thus helping decision makers to obtain and use data and models to solve problems of on-line flood hazards forecasting.

The components of an Integrated Hydrological Forecasting System (IHFS) can be categorised as follows:

- Real-time data acquisition and preprocessing;
- Meteorological and hydrological modelling and forecasting;
- Forecast analysis;
- Warning dissemination.

The first three components of the system are analysed in detail below.

2.4.5.1
Real-Time Data Acquisition and Pre-Processing

Data may be in the form of inputs which drive the system, such as precipitation or temperature, or observations which can be used to check, correct or update model outputs, such as river stages or discharges and reservoir status. Many sources of data may be available for a particular hydrometeorological variable, on the basis of availability of on site and remote sensors; the capability to integrate and capitalise such information qualifies the system (Fattorelli and Capovilla 1989; Borga et al. 1991). Traditional data acquisition systems are based upon conventional telemetering equipment; the connections between the peripheral stations and the dedicated units may be achieved through dedicated or commutated telephone lines, radio links and geostationary satellites. Modern IHFSs also include procedures capable to integrate remote and on site sensors to accomplish data preprocessing.

The case of the precipitation field estimation is particularly interesting as it regards the integration of several data sources: these estimates may originate from raingauge networks, weather radar (Collier 1989) and satellite (Engman and Gourney 1991). The promise of radar rainfall estimation for real-time hydrological applications lies mainly on two key features of the radar sensing process. These are the detailed spatial coverage (e.g. 1×1 km within a 200 km radius) and the capability for remote sensing. Spatial coverage of rainfall rates implies improved site specific flood predictions and warnings. The remote sensing capability promises longer forecast lead times of reliable flow predictions in hydrological basins under the radar umbrella for approaching storms. It also allows for more timely warnings since the rainfall data are available for a large area in a way which in many cases is more efficient than that allowed by the communication network associated with operational on site raingauge sensors. There are, however, problems as to the correct measurement of the reflectivity and in the calibration procedures which are required to transform measurements of the electromagnetic radar echo into rainfall rate. The combination of these two sources of error gives a rather complex problem, which involves so many conditions that one figure could hardly cover all operational situations (Fattorelli et al. 1995). In order to prevent these kinds of errors, the adjustment of radar maps with raingauges data remains the most widely used method to correct radar data, although this method suffers from limitations related to the fundamental differences in the sampling characteristics of the two sensors. Figure 2.6 represents an example of the application of this technique to an area covered by the weather radar of the Veneto region, in Italy (Borga et al. 1995).

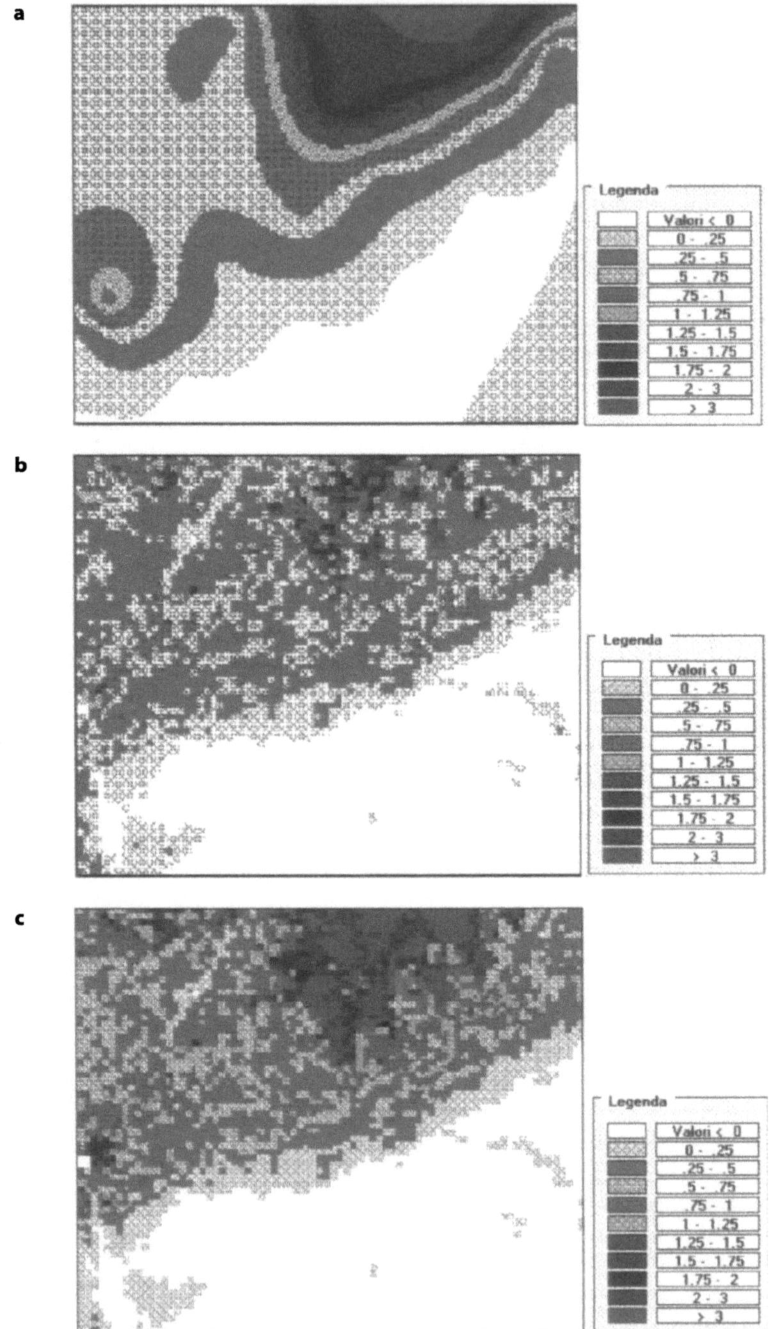

Fig. 2.6. Hourly rainfall field (threshold table units in mm h^{-1}) estimated based on: **a** raingauge data (*Spline* technique); **b** weather radar; **c** radar data combined with raingauge data

2.4.5.2
Meteorological and Hydrological Modelling and Forecasting

An integrated meteorological-hydrological system for real-time use is characterised by its capability to incorporate the precipitation forecast into the hydrological modelling. The strength of such an approach lies in cases where the time scale of the direct runoff and flood propagation processes is comparable to the response time of the precipitation process. Particularly interesting cases are those for which the maximum response time of the drainage basin is about 6 h or less (these cases are commonly referred to as flash floods).

Experience shows that the ability to short term forecast rain rate fields depends on the synoptic situation. Research in the last two decades pointed out the feasibility of the operational implementation of advection models for rainfall forecasting. These are based on simple linear extrapolation procedures used to project the current radar rainfall forwards, according to the advection velocity, to form forecast fields at successive lead times. Recent researches in the area of real-time precipitation forecasting also include the lumped one dimensional model of Georgakakos and Bras (1984), its two-dimensional formulation by Lee and Georgakakos (1990), improved by using advection by French et al. (1993). The scale of interest of such types of models are lead times up to a few hours and areal extents in the order of 10 000 km^2.

As to the rainfall-runoff modelling, a WMO survey of hydrological forecasting systems in Europe indicated that in 17 countries a total of 74 forecasting procedures are used, of which 35 are correlations, 16 are routing procedures and 23 are of another nature, mainly falling into conceptual moisture accounting or system approach categories (Nemec 1986). This variability is partially explained by the different characteristics of the watersheds to be controlled. In very large rivers the minimum required 6–8 h forecasting lead time is generally met by monitoring upstream levels and/or discharges; in this case correlation procedures are well suited. In smaller catchments one has to take advantage of concentration time to increase the forecasting lead time, and therefore the rainfall-runoff process must be modelled by using soil moisture accounting procedures. In other watersheds, (e.g. in Southern Europe, particularly on mountain areas) the size of the catchments and their steepness are such that one has to include also rainfall forecasts in order to gain as much lead time as possible. On the basis of this conceptualisation, Table 2.2 identifies general types of flood forecast systems and their characteristics (Feldman 1994). Of course, the more backwards one has to move into this chain, the more uncertain the forecast will be, given the increased number of uncertainties introduced by the different system components. It is, therefore, essential to go backwards no more than needed and to add to the system those stochastic components required for the minimisation of the residual error variance.

Errors on the model input and inadequacies of the model structure lead to the constraint that a 'true' real-time procedure should be able to operate within an adaptive mode, namely the model has to be conveniently updated on the basis of current hydrometeorological conditions. A feedback control structure would therefore characterize the model in order to achieve such a requirement.

Georgakakos (1986) brings up extensive tests of a hydrometeorological model complemented by a (non-linear) state updating processor. Sensitivity analysis using data from significant past flash-floods has indicated that short-term model predictions

Table 2.2. Types of flood forecast systems (After Feldman 1994)

System	Forecast lead time	Sophistication of forecast procedures	Expense	Potential uncertainty
Real-time measurement of streamflow and routing/ correlation to key locations	Shortest	Shortest	Least	Least
Same as above but including precipitation in correlation to key locations	Short	Simple	Little	Little
Real-time measurements of precipitation, streamflow temperature, etc. plus a watershed runoff model	Long	More complex to very complex	More to very	More
Same as above but including meteorological model	Longer	Very complex	Very	More
Same as above, but including weather (precipitation, temperature, etc.) forecasts	Longest	Most complex	Most	Most

were not significantly sensitive to the runoff generating model component when real-time updating was performed. In addition, the sensitivity analysis has suggested that the model performance is better when all available meteorological and hydrological data are utilised by the state estimator. These results indicate the need for strong integration of meteorological and hydrological predictions, in this way allowing to reduce the uncertainty generated by the lack of knowledge about soil moisture conditions.

2.4.5.3
Forecast Analysis

After a forecasted crest stage is available, a local flood warning system is of little value if no one knows what action has to be taken to save lives and properties. Therefore, the IHFS should include provisions for local response plans. A major ingredient of the decision making process is the assessment of the uncertainty which affects the forecast; within IHFS, it is possible to include explicitly such knowledge by analysing several different 'scenarios'. These are simulations which will eventually evolve from the initial observation of possible disastrous incoming storms; different ground effects, conditional on the observation, arise from the uncertainty upon the general advection velocity of the storm, storm area coverage, and soil moisture patterns. Each 'scenario' will produce a different pattern of flooding events in the drainage system.

Particularly in real-time forecasting and warning system framework, multicriteria screening and stochastic optimisation packages should be used in order to analyse the different alternatives resulting as a consequence of the different future 'scenarios' generated by the chain rainfall, rainfall-runoff models. The results should be presented to the decision maker in a graphical form using, as much as possible, a Digital Elevation Model which should provide a realistic view of the damages resulting from the alternative decisions (for instance, through flood and vulnerability maps) (Walsh 1993; Todini 1992).

Moreover, given that not only real-time hydro-meteorological acquired data concur into the decision making process but also a large amount of geo-related information, it is essential for the whole system to be based upon a Geographical Information System (GIS). A GIS holds two essential types of data: location data, which serve the purpose of visual representation of geographical features; and descriptive data, consisting of supplementary information about the displayable geographic features that are necessary to carry out spatial analysis on them. In this way, a GIS can integrate a number of socio-economic variables together with the more conventional physics or geographic characteristics of a region allowing to elaborate vulnerability maps based on various criteria (social, economic, etc.) as the result from the combination of land use and the sensitivity of the various land uses to flooding.

Multiple components Decision Support Systems are now becoming available to aid decision makers in real-time situations. A DSS has to provide the flow of information for defining the decision model, performing the multicriterion decision process and presenting in a synthetic and graphical form the alternative choices and the evaluation of the expected damages or benefits arising from their decisions (Tecle et al. 1988). Incorporation of a GIS within the DSS appears particularly useful, by increasing the utility of both GISs and DSSs to assist decision makers in a real-time forecasting and warning system.

References

Abbott MB, Bathurst JC, Cunge JA, O'Connell PE (1986) An introduction to the European Hydrological System – Système Hydrologique Européen, SHE 1: History and philosophy of a physically based distributed modelling system, J Hyd 87:45–59

Askew A (1991) Learning to live with floods, Nat Resour 27(1):4–9

Austin P M, Houze RA (1972) Analysis of the structure of precipitation patterns in New England, J Appl Meteo 11:926–934

Beven KJ, Kirkby MJ (1979) A physically-based variable contributing area model of basin hydrology, Hyd Sci Bull 24(1):43–69

Beven KJ, Calver A, Morris EM (1987) The Institute of Hydrology Distributed Model, Inst. of Hydrology, Rep. N° 98, Wallingford UK

Borga M, Capovilla A, Cazorzi F, Fattorelli S (1991) Development and application of a real-time flood forecasting system in the Veneto Region of Italy, Wat Res Manag 5:209–216

Borga M, Da Ros D, Fattorelli S, Vizzaccaro A (1995) Influence of various weather radar correction procedures on mean areal estimation and rainfall-runoff simulation, Proceedings of III International Symposium on Hydrological Applications of Weather Radar, August, 20–23, 1995, S Paolo Brazil, pp 146–157

Bradley AA, Potter KW (1991) Flood frequency analysis for evaluating watershed conditions with rainfall-runoff models, Wat Res Bull 27(1):83–91

Bradley AA, Potter KW (1992) Flood frequency analysis of simulated flows. Wat Res Research, 28(9):2375–2385

Cazorzi F, Dalla Fontana G, Fattorelli S (1984) Simulazione idrologica del bacino del Cordevole, Quaderni di Ricerca, Regione Veneto

Collier CG (1989) Applications of weather radar systems. Ellis Horwood Limited

Crawford NH, Linsley RK (1966) Digital simulation in hydrology: Stanford Watershed Model IV, Tech Rep 39, Dept. Of Civil Eng., Stanford Univ., Stanford California

Da Ros D (1996) Utilizzo di stime radar di precipitazione nella simulaione dei deflussi di piena, PhD Dissertation, University of Padova

Dalrymple T (1960) Flood frequency analysis. US Geol Surv Wat Supply Pap, 1543–A

Dooge J CI (1959) A general theory of the unit hydrograph, J Geophysic Research 64(2):241–256

Engman ET, Gourney RJ (1991) Remote sensing in hydrology, Chapman and Hall, London, 225 pp

Fattorelli S (1982) Modelli deterministici di bacino, Valutazione delle Piene, CNR, Progetto Finalizzato Conservazione del Suolo, pp 198–212

Fattorelli S, Capovilla A (1989) Real-time systems for water resources management, Proc. of International Conference on Global Natural Resources monitoring and assessments: preparing for the 21st Century, September 24–30, Fondazione Cini, Venezia, pp 354–362

Fattorelli S, Casale R, Borga M, Da Ros D (1995) Integrating radar and remote sensig techniques of rainfall estimation in hydrological applications for flood hazard mitigation, European Commission, DG XII, Report EUR N° 16494

Feldman AD (1994) Assessment of forecast technology for flood control. In: Rossi G, Harmancioglu N, Yevjevich V (eds) Coping with floods, NATO ASI Series, Series E: Applied Sciences, vol 257, Kluver Academic Publisher, pp 445–458

French MN, Krajewski WF, Andrieu H (1993) Rainfall forecasting using remote sensing and a physically-based stochastic model. Proc. of Fourth International Conference on Precipitation, April 26–28, 1993, Iowa City Iowa

Georgakakos KP (1986) A generalized stochastic hydrometeorological model for flood and flash-flood forecasting, part I – formulation. Wat Res Research 22(13):2083–2095

Georgakakos KP, Bras RL (1984) A hydrologically useful station precipitation model, 1, formulation. Wat Res Research 20(11):1585–1596

Greis NP, Wood EF (1981) Regional Flood frequency estimation and network design, Wat Res Research 17:1167–1177

Lee TH, Georgakakos KP (1990) A two-dimensional stochastic-dynamical quantitative precipitation forecasting model. J Geophysic Research 95(D3):2113–2126

Limbert DWS (1995) Human and economic consequences of weather events during 1994. WMO Bulletin, pp 364–375

Nash JE (1957) The form of the instantaneous unit hydrograph, Int Assoc Sc Hydrol, Publ. vol 45, N° 3, pp 114–121

Nemec J (1986) Hydrological forecasting. Design and operation of hydrological forecasting systems. D Reidel, Dordrecht

NERC (1975) Flood studies report. Vol I–V, Nat Env Research Council, London

Peck EL (1976) Catchment modeling and initial parameter estimation for the National Weather Service River Forecast System, NOAA Technical Memo NWS HYDRO-31, National Weather Service, US Dep Commerce, Silver Spring Maryland

Penning-Roswell E, Peerbolte B (1994) Concepts, policies and research. In: Penning-Roswell E, Fordham M (eds) Floods across Europe. Flood hazard assessment, modelling and management. Middlesex University Press, pp 1–17

Rockwood DM, Nelson ML (1966) Computer application to streamflow synthesis and reservoir regulation, IV Int. Conference on Irrigation and Drainage

Rossi F, Villani P (1994) Regional flood estimation methods. In: Rossi G, Harmancioglu N, Yevjevich V (eds) Coping with floods. NATO ASI Series, Series E: Applied Sciences, vol 257, Kluver Academic Publisher, pp 193–217

Sherman LK (1932) Streamflow from rainfall by unit-graph method. Engineering News Record 108, pp 501–505

Tecle A, Fogel M, Duckstein L (1988) Multicriterion analysis of forest watershed management alternatives. Wat Res Bull 24(6):1169–1178

Todini E (1989) Flood forecasting models. Excerpta 4:117–162

Todini E (1992) From real-time flood forecasting to comprehensive flood risk management decision support systems. Proc. of 3rd International Conference on Floods and Flood Management, November 24–26, Florence Italy, pp 313–326

Wallis JR (1980) Risk and uncertainties in the evaluation of flood events for the design of hydrologic structure. In: Guggino A, Rossi G, Todini E (eds) Piene e Siccita. CLUP, Milano

Wallis JR, Wood EF (1985) Relative accuracy of log-Pearson III procedures. J Hydraulic Eng 111:1043–1065

Walsh MR (1993) Toward spatial decision support system in water resources. J Wat Res Planning and Managment, ASCE

Yevjevich V (1994) Floods and society. In: Rossi G, Harmancioglu N, Yevjevich V (eds) Coping with floods. NATO ASI Series, Series E: Applied Sciences, vol 257, Kluver Academic Publisher, pp 3–9

Chapter 3

The Recognition of Landslides

R. Dikau

3.1
Introduction

The first aim and purpose of the European community research project TESLEC, "The Temporal Stability and Activity of Landslides in Europe with Respect to Climatic Change" (1994–96) (Dikau et al. 1996b; Schrott and Pasuto 1997) was to prepare the technical manual "Landslide Recognition" which presents the main characteristics of different landslide types (Dikau et al. 1996a). The manual is based on a classification of a previous European project, EPOCH, "The Temporal Occurrence and Forecasting of Landslides in the European Community" (1991–1993) (Casale et al. 1994; Soldati 1996).

Given the possibility of an increase of landslide frequency with changing climate and human development, it is essential that these phenomena are recognised by a broader spectrum of people and not left only to the highly trained individual. The purpose of the manual, therefore, is to assist in the education of landslide recognition with the aim of helping the reader to distinguish different types within the field. The emphasis of the manual is on simple description. Clear methods of representation have been introduced in the form of brief descriptions and diagrams, as well as, an aid to the interpretation of maps and photographs. A short account of the planning and engineering implications are also included and where possible representative statistics and behavioural data are incorporated. The manual has been prepared to bring together a range of materials which accurately depict and distinguish the various landslide types recognised by the International Community. It is also the result of cooperation between the European teams of the TESLEC project with contributions from 31 scientists. In the following chapter a short summary and introduction of Chapter 1 of the technical manual "Landslide Recognition" will be given. For more detailed information the original publication (Dikau et al. 1996a) is recommended.

3.2
Classification of Landslides Proposed by the TESLEC Project

In an advanced stage of planning, where terrain locations of construction and their alternatives must be chosen, or remedial measures developed to bring slopes to a safer condition, the first step is to recognise the existence of a landslide and to distinguish its type, activity and causes. It is a premise of this study that the detection of mass movements should be based upon thorough geomorphological and geological investigations and that the diversity of landslides is emphasized. Therefore, it is necessary to discuss the variability of form, behaviour, volumes involved, speed, movement whether single, cyclic or pulsed and material which may fail as one body or disaggre-

Table 3.1. Classification of slope movements used in the TESLEC project based on Casale et al. (1994) (From Dikau et al. 1996a)

Type	Rock	Debris	Soil
Fall	Rock fall	Debris fall	Soil fall
Topple	Rock topple	Debris topple	Soil topple
Slide (rotational)	Single (slump) Multiple Successive	Single Multiple Successive	Single Multiple Successive
Slide (translational) Non-rotational Planar	Block slide Rock slide	Block slide Debris slide	Slab slide Mudslide
Lateral spreading	Rock spread	Debris spread (no case reported in Europe)	Soil spread
Flow	Sackung (rock flow/sagging)	Debris flow	Soilflow
Complex (with runout or change of behaviour downslope, note that nearly all forms develop complex behaviour)	E.g. rock avalanche	E.g. flow slide	E.g. slump-earthflow

N.B. Compound: a landslide which consists of more than one type e.g. rotational-translational slide. This should be distinguished from a complex slide where one form of failure develops into a second form of movement i.e. a change of behaviour downslope by the same material.

gate with movement. As the probability of landsliding changes, due to changing climate or increasing human activity it becomes more important to recognise the potential event. There are, however, morphological indicators which a trained geomorphologist or engineer experienced in geomorphology and geology, would recognise, these can be catalogued, classified and mapped. A primary task, therefore, is to develop a manual of such indicators and mapping techniques, providing a basic understanding to landslide recognition.

There are numerous classifications within the landslide literature, each dependent on a different landslide factor or author's objective. Most definitions give a guide to the processes, as well as, the type of material involved in the displacement (e.g. Varnes 1978; Cruden and Varnes 1996). Hutchinson (1988), who has prepared the most comprehensive scheme, classified slope movements into eight categories containing several subdivisions (see Tables 3.1, 3.2, 3.3 and 3.4). These are based on morphology, mechanism, type of material and rate of movement. In the manual, the European classification developed by the EPOCH project "The Temporal Occurrence and Forecasting of Landslides in the European Community" is used, because it is simple and suitable for European conditions upon which the manual is based. From this classification the following types were used: fall, topple, slide (rotational), slide (translational), lateral spreading, flow and complex which result from the important element in a classification, to recognise the mechanism of failure.

A fall usually denotes the free-fall movement of material from a steep slope or cliff, whereas a topple which is very similar to a fall in many respects, normally involves a

Table 3.2. A classification of landslide mechanisms, compatible with Hutchinson (1988) and Casale et al. 1994). Each category should be read continuously from left to right (After Brunsden 1985; from Dikau et al. 1996a)

Type	Form of initial failure surface		Subsequent deformation
Fall			
Detachment from a	a planar b wedge c stepped d vertical	failure surface	Free fall, may break, up, roll bounce, slide, flow down slopes below. May involve fluidisation, liquefration, cohesionless grain flow, heat generation, chemical, rate effects or other secondary mechanisms.
Topple			
Detachment from	a single b multiple	a pre-existing discontinuities b tension failure surfaces	As above.
Slide			
Rotational movement (sliding) on a	a single b successive c multiple	circular failure surface	Toe area may deform in a complex way. May bulge, override, flow, creep. May be retrogressive.
Non-rotational compound movement (sliding) on a	a single b progressive c multi-storied	non-circular i. listric ii. bi-planar	Often develops a graben at the head. May have a toe failure of different type.
Translational movement (sliding) on a	a planar b stepped c wedge d non-rotational	failure surface	May develop complex runout after disintegrating. As for falls and flows.
Spread			
Lateral spreading of ductile or soft material which deforms in	a a layer beneath hard rock b a weak inter stratified layer c a collapsible structure		Can develop sudden spreading failure in quick clays. Slope opens up in blocks and gulls or fissures. Liquefaction can occur and the whole slope spreads either as a totally collapsed flow or with 'floating' blocks and grabens.
Flow			
Debris movement by flow on	a a natural, complex i. unconfined ii. channelised	topographic surface	Flow will involve complex runout from source. May be catastrophic. May move in sheets or lobes and involve viscous or rheological mechanisms.
Movement by creep on		any hillslope	Creep may be gravity, seasonal, pre-failure or progressive.
Rock flow (sagging, Sackung) movement on	a single sided b double sided	mountain slope a of rotational b compound form listric bi-planar	May be slow gravity creep or early stages of landsliding, but not displaying toe deformations other than bulging.
	c stepped	discontinuity	May involve toppling.
Complex			
Movements involving two or more of the categories above.			

Table 3.3. Classification of the processes that 'cause' landslides (After Brunsden 1993; from Dikau et al. 1996a)

	External process group	Causal processes	Description	Specific effect on stability state of the slope system (examples only); changes in:
1.	Weathering physical, chemical and biological	Physical properties	Changes in particle size	Density, unit weight strength, strength, permeability, vertical and spatial strength variation, total stress, critical depth, friction, cleft water pressure
		Chemical properties	Cation exchange, cementation	
		Horizonation	Internal layers and basal surface, development of weak discontinuities	
		Regolith thickness	Weathering > erosion, accumulation in hollows and foot of slope, soil ripening	
2.	Erosion fluvial, glacial and coastal, material removal from the face or base of slope	Geometrical change	Relief, height, length, angle, aspect	Total stresses, permeability, strength
		Unloading	Removal of lateral support, expansion, swelling, fissuring, strain softening, stress concentration	
3.	Ground subsidence	Undermining	Mechanical eluviation of fines, solution, leaching, removal of cement, seepage erosion, backsapping, piping	Strength, physical support, consolidation, water concentration, pore water pressure
4.	Deposition of fluvial, glacial and mass movement. New material added to face or top of slope	Loading	Solifluction, mudsliding, rock fall, deltaic addition, talus accumulation	Water content, weight, strength, stress
		Undrained loading		Underconsolidation, pore pressure
5.	Shocks and vibrations, seismic activity	Vertical and horizontal movements	Shocks of varying frequency, magnitude, intensity, duration, disturbance to intergranular bonds and cements, water table change	Horizontal stress, strength, excess pressures
6.	Air fall of loess, tephra	Mantling with fine regolith. Addition of fine components to soil	New slope created with strong discontinuity beneath	Strength, water content, water pressures
7.	Water regime change; geomorphological or meteorological	Perched water tables, surface saturation, water table and pressure change	Piping, flooding, lake bursts; 'Wet' rainfall years, intense precipitation, snow and ice melt, drawdown	Water pressures Excess pressures, water table, pore pressure, weight
8.	Complex 'follow-on' or run out processes after initial failure, bank collapse, seismic slope failure	Liquefaction, remoulding, fluidisation, air layer lubrication, cohesionless grain flow		Strength, water distribution, consolidation, friction

Table 3.4. Some features used for the recognition of landslides (After Brunsden 1985; from Dikau et al. 1996a)

Deposits

1. Transverse ridges at the head, longitudinal ridges in track, concentric ridges at toe running transverse to morphology;
2. Transverse and radial fractures at toe. Reidel shears at margins. Crevasse-style cracking over breaks of slope;
3. Spreading at toe. Valleys partially or wholly blocked. Rivers diverted;
4. Materials locally derived but displaced below outcrop. May include blocks of intact stratigraphy, stratigraphy repeated downslope.

Active movement

1. Scarps and fractures possess sharp edges and open mode. No secondary filling;
2. Main units show secondary fracture and pressure ridges;
3. Surfaces show polishing and striations, fresh appearance;
4. Drainage deranged, ponds;
5. No soil development and only fast growing vegetation;
6. Considerable distinction between form, roughness, texture, vegetation, between slide and non-slide areas;
7. Tilted vegetation.

Inactive movement

1. Scarps and fractures weathered and indistinct, cracks infilled;
2. No secondary failures of pressure ridges (often subdued);
3. Surfaces weathered, vegetated;
4. Integrated drainage but may have irregular pattern and sudden infilled depressions;
5. Soil cover and well vegetated or cultivated;
6. Different to distinguish margins and textures exception on air photographs;
7. New growth on trees and vertical growth of post-slide trees.

---

pivoting action rather than a complete separation at the base of the failure. Movements occurring on a distinct slide or shear surface are termed slides. These may be subdivided into rotational and translational according to the form of the failure surface. Rotational slides involve a semi-circular shear surface, whilst translational failures usually occur on planar slip surfaces. Lateral spreading is characterised by the low-angled slopes involved and the unusual form and rates of movement. Flows normally behave as a fluidised mass in which water or air are significantly involved. The complex failures are principally a combination of two or more of the previously described movements. In reality, nearly all landslides involve more than one type of movement either acting concurrently in different parts of the failure (compound landslides) or evolving downslope over time into a different process (i.e. initial failure to subsequent deformation (complex landslides)). Thus, many rotational slides develop into a flow form at the toe, described by some authors as a slump-earthflow. A rockslide or a rock fall may also advance into a flow form, known as a rock avalanche or sturzstrom. This type of failure is a very destructive, high velocity run-out of rock debris due to one of the following processes; fluidization, cohesionless grain flow, heat or steam generation, frictionite production or strength changes caused by the rate of shear. Another complicated form of landslide is the debris flow. This begins and is fed by debris slides, rotational slides, bank collapse, bed erosion and falls. The result is a run-out which varies in water and sediment concentration from mudflow to a mixture similar to 'wet concrete'. It may also be on a free rectilinear slope confined in a valley or so catastrophic that it overwhelms the topography. Many other examples of complex failures can be given and so the advice to the reader is to recognise that, although the single processes are described here, the reality will be combinations of movement types.

The landslide classification has also recognised that the type of material should be defined and where appropriate the chapters of the manual have been divided into rock, debris and soil. Debris and soil are distinguished by the particle size of the material involved in the movement. Debris is material coarser than 2 mm but usually describes an assortment of material including clasts incorporated into a matrix. Soil is finer than 2 mm and rock is a coherent, consolidated mass normally of significant proportions and extent.

The terminology involved in describing a landslide is based upon the suggested classifications of the International Geotechnical Societies' UNESCO Working Party on World Landslide Inventory (WP/WLI 1993; UNESCO 1993). The author feels that all European scientists should adopt the International Terms for landslide features; dimensions; activity; distributions and types.

References

Brunsden D (1985) Landslide types, mechanisms, recognition, identification. In: Morgan CS (ed) Landslides in the South Wales Coalfield, Proceedings Symposium, 1–3 April, 1985. The Poly. of Wales, pp 19–28

Brunsden D (1993) Mass movement; the research frontier and beyond: A geomorphological approach. Geomorph 7:85–128

Casale R, Fantechi R, Flageollet JC (eds) (1994) The temporal occurrence and forecasting of landslides in the European Community. Comm Europ Comm Prog EPOCH, Contract 90 0025, Final Report

Cruden DM, Varnes DJ (1996) Landslide types and processes. In: Turner AK, Schuster RL (eds) Landslides: Investigation and mitigation. Transp Res Board, Spec Rep 247:36–75

Dikau R, Brunsden D, Schrott L, Ibsen M (eds) (1996a) Landslide Recognition. Wiley & Sons

Dikau R, Schrott L, Dehn M, Hennrich K, Rasemann S (eds) (1996b) The temporal stability and activity of landslides in Europe with respect to climatic change (TESLEC) Final Report Part I and II, European Community CEC Environment Program Contract N° EV5V-CT940454

Hutchinson JN (1988) Morphological and geotechnical parameters of landslides in relation to geology and hydrology. General Report. In: Bonnard C (ed) Landslides. Proc. 5th. Int. Symp. on Landslides, vol 1, pp 3–35

Schrott L, Pasuto A (eds) (1997) The temporal stability and activity of landslides in Europe with respect to climatic change (TESLEC). Geomorphology

Soldati M (ed) (1996) Landslides in the European Union. Geomorphology, p 15

UNESCO (1993) Multilingual landslide glossary. The International Geotechnical Societies' UNESCO Working Party for World Landslide Inventory. BiTech Pubs, Canada

Varnes DJ (1978) Slope movements: type and processes. In: Schuster RL, Krizek RJ (eds) Landslides analysis and control. Transp Res Board, Spec Rep 176:11–33

WP/WLI (International Geotechnical Societies' UNESCO Working Party for World Landslide Inventory) (1993) A suggested method for describing the activity of a landslide, Bulletin International Association of Engineering Geology 47:53–57

Climate Change Impact on Frequency and Distribution of Natural Extreme Events: an Overview

G. Delmonaco · C. Margottini · S. Serafini

4.1
Introduction

Natural disasters repeatedly occur with a certain frequency in time and space, nevertheless their progressive intensification during the last decades with effects of amplification and accumulation, as well as, the predominance of geographic localisation may be considered as a consequence of anthropogenic dynamics and development policies of socio-economic systems. The analysis of correlation between disasters and climate change becomes fundamental to better understand causes and effects, in order to calibrate decisions and human interventions rather than assuming disasters as an unpredictable occurrence (Margottini 1994).

During the last decades a large number of disasters have occurred with an increase of economic losses in the face of a short-term global meteoclimatic scenario which can be considered as constant, as a strict consequence of a noticeable increase of vulnerability in environmental systems.

Hydrometeorological events and geomorphological evolution are a consequence to well detected mechanisms linked to the atmospheric circulation and earth's geodynamics (climate variability); the analysis of the different typologies of disasters and, generally, the evolution of the natural environment, should take into account natural causes (climatic, geophysical, etc.); during the last century some human effects are likely to overlap the natural aspects of climate variability, enhancing or sometimes attenuating the effects of natural fluctuations of climate (climate change).

The reliability of climate change analysis and the assessment of their impact on the territory depend strongly on the knowledge of climate evolution in the past and on natural and anthropogenic effects manifested during climate fluctuations.

Climate is usually defined as average weather conditions over a period of time and possibly over a certain geographical region, in terms of the mean of meteorological parameters. Climate system is influenced by internal components as the atmosphere, the oceans and the land features as hydrology, snow and ice cover, vegetation and so on. The external forcing components include the sun, brightness, sunspots, the Earth's rotation; human activities as the well known greenhouse effect may result as an external component. The correlation among the above mentioned components is very difficult and sometimes not easy to assess; for example it is still extremely complex to define a strict relationship between solar activity and consequent impacts on climate even if extreme cold events were recorded in the past when sun anomalies occurred (i.e. 'Maunder Minimum' and 'Little Ice Age'). Nevertheless, some climatic reconstructions are not sufficiently reliable to demonstrate a correlation between cold/warm events and global climate conditions and many controversies exist with actual 'global

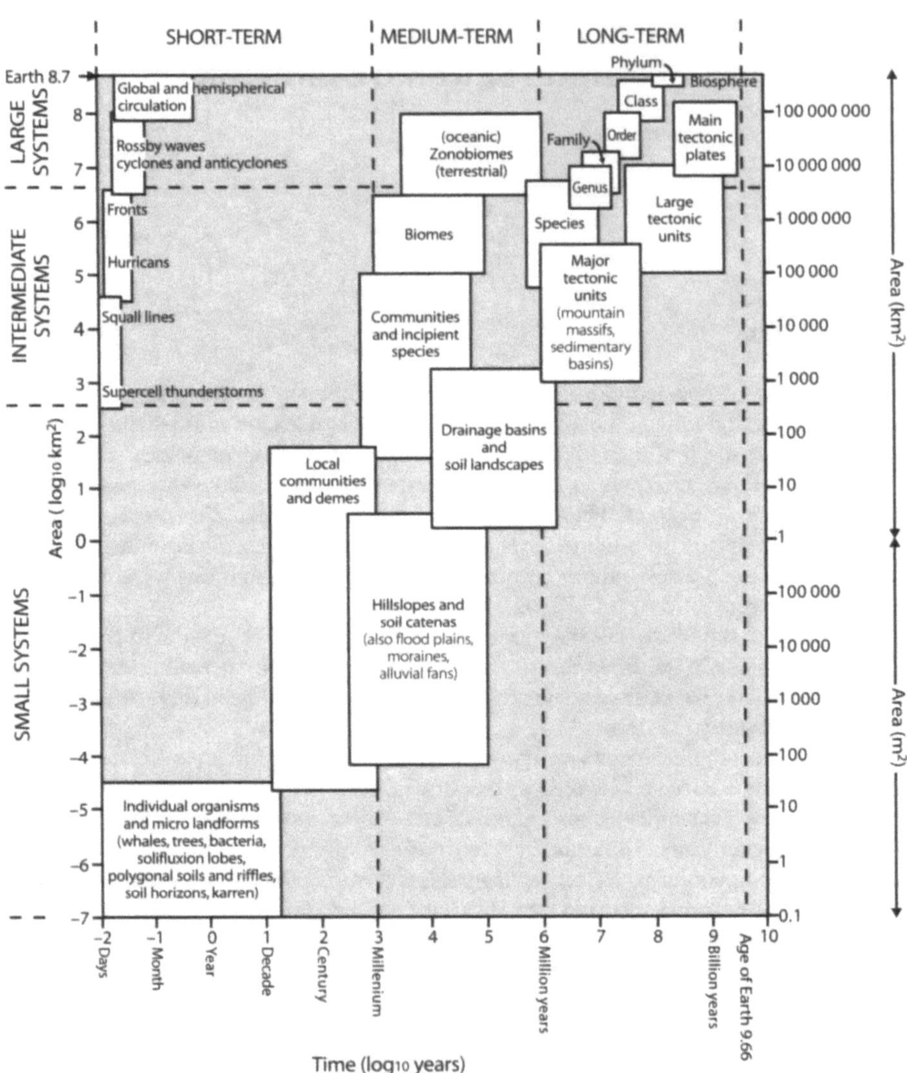

Fig. 4.1. Terrestrial systems: size and 'life-span' (Hugget 1991)

warming' models (i.e. during the 'Little Ice Age' the sea level in Venice rose while a lowering had been expected (Camuffo and Enzi 1994).

Moreover, climate changes can occur over time scales ranging from less than an hour, as in short-term meteorological events, to million years, as in prolonged phases of global warming and cooling, following the above mentioned causes as well as over spatial scales from local to continental context (Fig. 4.1).

Natural disasters often belong to large-scale geographical distribution and short-time meteorological behaviour; understanding their relationship with Global

Circulation Models or general trends can be rather problematic; the only solution is likely to investigate triggering mechanisms of events, to reconstruct long-term series of data, in order to understand the relationship with past meteoclimatic scenarios.

Paleoclimatology may help in extending, as much as possible in the past, the space and time-scale of climate variability/climate change to determine whether present day situation reflects natural climate variability rather than human induced, mainly through the analysis of short-term series of data (100–200 years). Useful information is coming from geological investigation on recent stratigraphic sequences, dendrochronology, palinology, paleontological and vegetational analysis in deep-sea cores, isotope analysis from corals, speleothems, ice cores, historical and documentary researches on climatic evidences and extreme events occurrences. Historical researches in some geographical regions are able to reconstruct the climate variability for the last 1 000 years, which can be considered the most useful for determining the width of natural fluctuations; instrumental records, usually restricted for less than 150 years, are not sufficient enough to give a complete picture of climate fluctuations and, moreover, could be affected by anthropogenic influences.

4.1.1
External Forcings

The cycles within the Solar System influence the complex processes leading to variations in solar output which occur at different time scales. The peak at 3–7 d is associated with the synoptic disturbances mainly at middle latitudes, high-frequency fluctuations of solar radiation of about 25–28 d period, are linked to sunspots alignment with the Earth (Smith et al. 1983), while a full 11-year solar or sunspot cycle has been recognized, due to sunspot activity (Hugget 1991). The Hale cycle, also known as the double sunspot or heliomagnetic cycle, with a period of 22 years involves a reversal of the Sun's magnetic field and sunspot variations influencing the cycles of terrestrial magnetic activity. Low frequency oscillations are the Gleissberg cycle, with a period of 78–80 years, due to auroral evidences, and the 178.73-year period of the Sun's orbital motion (Hugget 1991).

All the above cycles seem to be connected with climatic fluctuations. The 11-year solar cycle have been detected in the analysis of meteorological series in parameters as temperature and precipitation, the 22-year cycle seems to influence the occurrence of droughts, rainfalls and tropical cyclones; the Gleissberg cycle influence seems to affect long-term temperature trends while the 178.73-year cycle can be detected in many historical European meteorological series as a long-term pattern of precipitation and temperature (Hugget 1991).

The Earth's climate is influenced also by other extraterrestrial motions inducing short-term gravitational forcing such as: the planetary orbital motions, the Sun's baricentric orbit and lunar tidal cycles. The lunar nodal precession with a period of 18.6 years is particularly interesting on climate influence; some relations with it have been found in the frequency of thunderstorms, in the fluctuation of air and sea temperature, rainfall in United States, in sea-level changes and in the frequency of droughts and floods in some regions of the Southern Hemisphere. The same 18.6-years period represents, just as well, the main rotation of the Earth's axis associated with the lunar

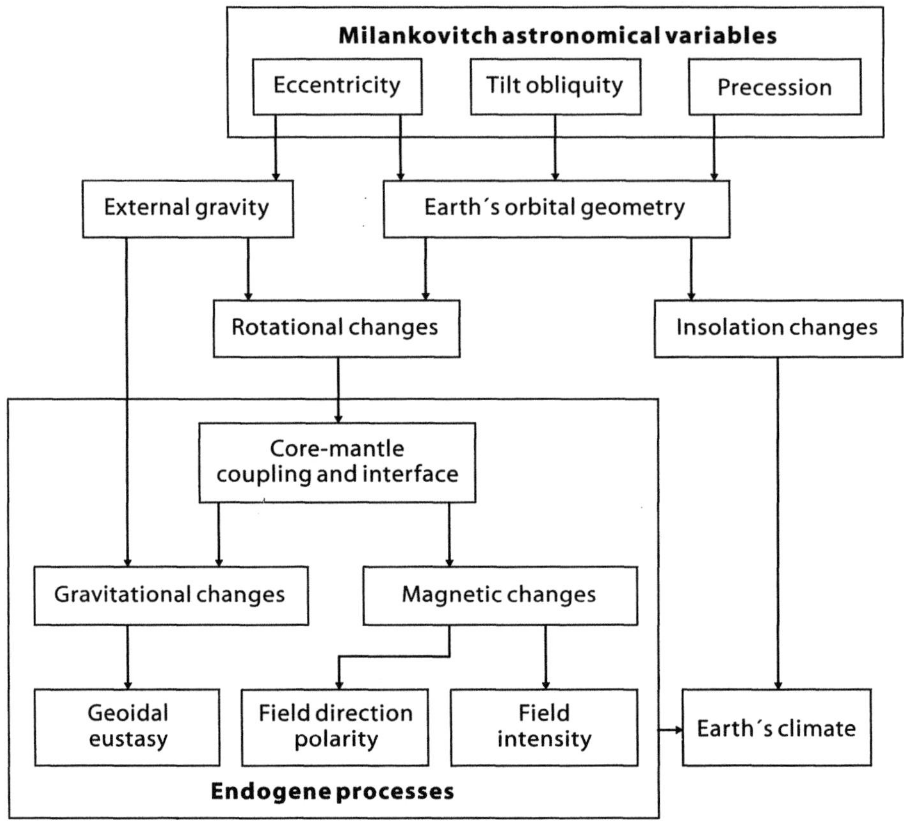

Fig. 4.2. Croll-Milankovitch forcing of terrestrial processes (Hugget 1991)

nodal cycle influencing the amount of radiation received by the Earth due to gravitational interactions within the Solar System (Hugget 1991).

Long-term climate change is influenced by medium-term gravitational forcing as ellipticity, precession and obliquity which occur with periods ranging from 10 000–500 000 years. The eccentricity has a short component of 100 000 years and a long cycle of about 400 000 years; the tilt cycle has a period of 41 000 years and the precessional cycle has two components with periods of 23 000 and 19 000 years. The influence on climate has been detected in the collection of proxy data, i.e. paleotemperature records of deep-sea sediment cores. All the above periodicities are likely to be connected with Quaternary glaciations, following the Croll-Milankovitch Theory, which connects cycles of climatic change with all the frequencies corresponding to orbital forcings. Nevertheless, the world climate system is linked to many variables which respond in a very complex way to orbital forcing, as demonstrated by empirical and theoretical models developed in the last decades; in addition the whole response of climate is influenced as well, by endogene processes which may assume a role most likely greater than the external driving force in climatic fluctuations (Hugget 1991) (Fig. 4.2).

4.1.2
Internal Forcings

Apart from astronomical reasons, Earth's climate is conditioned by other factors internal to the system like carbon dioxide variation, methane and dust, volcanic activity, vapour plume and meteoritic dust, relief, arrangement of land and sea, True Polar Wander, Earth's rotation rate and large changes of obliquity (Hugget 1991).

Variations in carbon dioxide concentration in the atmosphere have been detected from air bubbles entrapped in Arctic and Antarctic ice cores (Barnola et al. 1987; Genthon et al. 1987). The purest record available covering the last glacial cycle has been recovered from the Vostok ice cover in which carbon dioxide levels vary from between 190 and 200 ppm by volume during a glacial stage and between 260 and 280 ppm by volume during an interglacial stage (Barnola et al. 1987). This high correlation between temperature and carbon dioxide concentration would be expected if carbon dioxide had played an important role in climatic forcing. Further insight into the forcing of the climate system by carbon dioxide came out of a multivariate statistical analysis of temperature and atmospheric carbon dioxide content from the Vostok ice core and several other variables. The analysis have highlighted that carbon dioxide changes influence the world climate system, possibly orbital induced; this would also explain the synchronism of major glaciations in the Northern and Southern Hemispheres, a point left unexplained by the Croll-Milankovitch theory (Hugget 1991).

It seems that methane, another greenhouse gas but, unlike carbon dioxide, emanating chiefly from the land, may have contributed to glacial-interglacial temperature changes (Chappellaz et al. 1990), as well as, the dust content of the atmosphere may have affected glacial and interglacial climates (Hugget 1991).

Volcanic activity has played in the past and is still playing an important role on short-term climatic variation. In fact, volcanic eruption, although considered among geophysical disasters, may have great impacts on climatic trends by means of release of large amounts of dust and gases into the atmosphere that determines a decrease in temperatures. Examples of this process occurred in the past, as drawn from $\Delta^{18}O$ negative isotopical rates some 75 000 years ago of investigations performed in the Arabic Sea, Baffin Bay, Devon Island and Camp Century in Greenland (Tiedeman 1992); such a $\Delta^{18}O$ decreasing to which a temperature decreasing corresponds, has been associated to insulation changes caused by the explosion of Toba in Sumatra. More recently, in the last two centuries, the well known "year without a summer" in 1816 when, as consequence of the eruption of Tambora of 1815, in all of the Northern Hemisphere unexpected low temperatures were recorded; values of 1816 constitute the lowest temperatures in Ginevra since 1753–1960, and at New Haven (Yale College) where the records start from 1779; 1816 is the second coldest year in Philadelphia and, finally, one of the coldest in the historical records of Vienna, Rome and Edinburgh (Margottini, in printing).

Volcanic eruptions with large emission of dust and ashes as Tambora (1815), Krakatoa (1883), Santa Maria (1902), Katmai (1912), Cerro Azul (1932), Agung (1963) and El Chicon (1982) seem to have caused temperature decreasing of 0.2–0.5° C on Earth's surface. Nevertheless, the effects of volcanic eruptions are supposed to be responsible of perturbations of natural ecosystems, maybe without modifying general trends; some experimental data, on the other hand, as the double content of sulphurs with respect to the normal amount, recorded in ice samples coming from the Antartic representing the period 1450–1850, known as 'Little Ice Age', seems to promote great importance on the effects of volcanic dust on climatic mechanisms of the atmosphere (Margottini, in printing).

4.1.3
Climate of the Last 1 000 Years

The reconstruction of the climate of the last 1 000 years constitutes one of the most important and actual arguments, in order to better understand whether the present tendencies of climate are due to human effects or to natural and cyclical trends, or even if anthropogenic influences are superimposed over natural climatic fluctuations. This issue is largely debated since the meteorological parameters are being recorded only for a short period, limited to a temporal span from few decades to about two centuries, and varying from one country to another.

Understanding high frequencies of fluctuations, on the decade to century time scale, is necessary to assess changes before possible anthropogenic impacts on the environment and need reliable records, from natural and human archives, at high resolution, from season to year. Such archives include the analysis of records in tree rings, corals, ice core, varved sediments, historical documents.

For example, a significant number of annually resolved, precisely dated temperature histories from tree rings are available. Nevertheless, these records are still too sparse to provide a reliable and complete analysis on global change, but only for a regional context. In addition, tree rings records reflect changes in warm-season (growing season) temperature and represent interannual and decadal time-scale climate variability with a good reliability as compared with instrumental records. However, the extent to which multi-decadal, century and longer time-scale variability is expressed can vary, depending on the length of individual ring-width or ring density series that make up the chronologies, and the way in which these series have been processed to remove non-climatic trends. Coral records are available from regions not represented by tree rings and usually have annual resolution with an extension in time of few hundred years. The interpretation of ice core records from polar ice sheets and tropical glaciers may be in some cases limited by the noise inherent to snow depositional processes and provides records of accumulation change on annual basis. Varved sediments may be widely available even though their analysis is limited to few regions. Historical documents constitute a largely used tool for understanding climatic variations in absence or inadequate instrumental records, especially in Europe, but, on the other hand, a large diffusion of printed documents occurs only from the end of the 15th century and the lack of written records involve large regions of the world.

Finally, each natural archive of past climate generally only provides information about a particular season of the year when the influence of climate on the archive material is strongest; thus paleoclimatic reconstruction of climate over the last millennium requires careful retrieval of the archive, calibration of the climatic signal it incorporates, and reconstruction of the past climatic record which can only be provided by comparison with other independent sources of information that may support or verify the reconstruction. In this way, it may be possible to assess whether the observed climatic change were of more than local significance, and to identify large-scale changes which may be related to causes (forcing factors) or global (or at least hemispheric) significance (Bradley 1994).

Many records from various regions of the world indicate that the period A.D. 1 000–1 300 was relatively warm, defined by Lamb (1965) as "Early Medieval Warm Epoch". In Europe, this period was characterised by mild winters and very warm summers with estimated temperature of 0.5–1.0° C higher than present time. Precipitation was more abundant in winter while summers were characterised by dry conditions. For other regions of the world, the scarce information gathered indicates warm conditions for some areas of the Northern Hemisphere where nevertheless some cold sub-periods were recorded, while not enough information is available for the southern hemisphere to establish climate conditions (Schuurmans 1981).

The period between 17–19th centuries is known as "Little Ice Age" and the available information on climate condition is much more abundant than the previous Medieval Warm Epoch. Both winter and summer temperature on the average were lower than present, respectively of 1.0 and 0.5° C, while precipitation amount can be estimated around 90% with respect to the present century. In Northern Europe winters with frozen rivers and channels occurred 30–40 times per century, with respect to a 2–3 times frequency during the last century and a world-wide extension of abnormal conditions seems to be indicated by reported advances of glaciers at different parts of the world (Schuurmans 1981). Anyway, this period shows a large variability of climate conditions, from year-to-year up to a century time-scale; the 18th century was characterised by

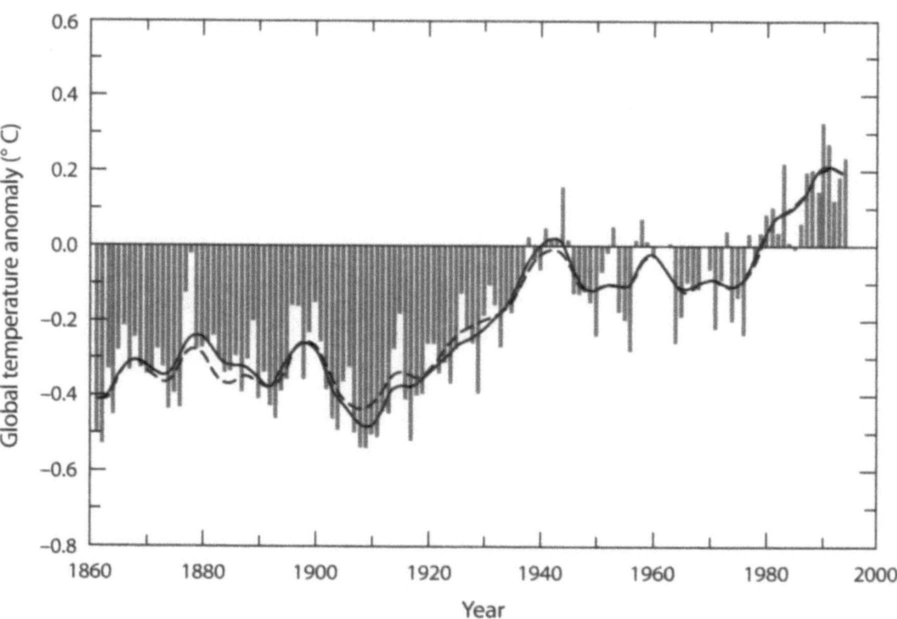

Fig. 4.3. Combined land-surface air and sea surface temperatures (° C) 1861–1994, relative to 1961–1990. The solid curve represents smoothing of the annual values shown by the bars to suppress sub-decadal time-scale variations. The dashed smoothed curve is the corresponding result from IPCC (1992) (IPCC 1995)

relatively warm conditions so that, it can be concluded that the term "Little Ice Age" is not appropriate since it should imply uniformly cold conditions.

According to the back analysis of the last 500 years a general tendency of cooler conditions are shown, while a further back in time displays warm episodes with a magnitude similar to present it should be difficult to associate univocally the increasing temperature of the last decades as a consequence of anthropogenic impacts on climate and not to a natural oscillation of it in a wider view of its cycles; as a matter of fact instrumental data on meteorological parameters are referred, at the beginning of observations, to one of the coldest periods of the last 1 000 years and the temperature increase of the 20th century can be confined in two short intervals, in the 1915–1935 period and in the 1970s and 1980 (Fig. 4.3).

4.2
Impacts of Climate Change

Extreme events, as above reported, may be considered as a direct consequence of the environmental evolution since they are linked to general mechanisms due to the atmospheric circulation and to Earth's geodynamics.

Mountain environments are potentially vulnerable to the impacts of a possible global warming because the combination of enhanced sensitivity to climatic change with implications for a wide range of natural systems as hydrology and slope systems.

Few assessments of the impacts of climate change have been conducted in mountain regions, in contrast to other ecosystems such as coastal zones, tropical rainforests, arid areas, due to the fact that the dominant feature of mountains (i.e. topography) is so poorly resolved in most general circulation climate models (GCMs) that it is difficult to use GCM-based scenarios to investigate the potential impacts of climate change. There also is a significant lack of comprehensive multidisciplinary data for impact studies, which is one of the prerequisites for case studies of impacts on natural or socio-economic systems (Kates et al. 1985; Riebsame 1989; Parry et al. 1992).

Mitigation and adaptation strategies to counteract possible consequences of abrupt climate change in mountain regions require climatological information at high and temporal resolution. Unfortunately, present-day simulation techniques for predicting climate change on a regional scale are by no means satisfactory. This is because GCMs generally operate with a rather low spatial resolution over the globe ($\approx$300 km) in order to simulate climate trends over a statistically significant period of time. As a result, entire regions of sub-continental scale have been overlooked in terms of their climatological specificities, making it difficult to predict the consequences on mountain hydrology, glaciers or ecosystems in a specific mountain region (Giorgi and Mearns 1991; Beniston 1994). The situation is currently improving with the advent of high resolution climate simulation in which the spatial scale of GCMs is on the order of 100 km (Beniston et al. 1995; Marinucci et al. 1995). However, even this resolution is generally insufficient for most impact studies.

A number of solution exist to help improve the quality of climate data used in impact assessments and economic decisionmaking.

Options include statistical techniques of downscaling from synoptic to local atmospheric scales (Gyalistras et al. 1994); coupling of mesoscale, or limited area, models (LAMs) to GCMs (Giorgi et al. 1990, 1992; Marinucci and Giorgi 1992; Marinucci et al. 1995); and use of paleoclimatic and geographical analogies (e.g., La Marche 1973; Webb et al. 1985; Davis 1986; Schweingruber 1988; Graumlich 1993; Luckman 1994).

Any meaningful climate projection for mountain regions and, indeed, for any area of less than continental scale, needs to consider processes acting on a range of scales, from the very local to the global. The necessity of coupling scales makes projections of mountain climates difficult.

4.2.1
Impacts on Water Cycle

In spite of limitations in the quality of historical data set and inconsistencies in projections between GCMs, particularly for precipitation (Houghton et al. 1990), assessments of the potential impacts of climate change on water resources, including snowfall and storage, have been conducted at a variety of spatial scales for most mountain regions (Oerlemans 1989; Rupke and Boer 1989; Lins et al. 1990; Slaymaker 1990; Street and Melnikov 1990; Nash and Gleick 1991; Aguado et al. 1992; Bultot et al. 1992; Martin 1992; Leavesley 1994). For example, high-resolution calculations for the Alps (Beniston et al. 1995) using a nested modelling approach (GCM at 1° latitude/longitude coupled to a LAM at 20-km resolution) indicate that in a doubled-CO_2 atmosphere wintertime precipitation will increase by about 15% in the Western Alps; this is accompanied by temperature increases of up to 4° C. Summertime precipitation generally decreases over the alpine domain, with July temperatures on average 6° C warmer than under current climatic conditions. Such numerical experiments, while fraught with uncertainty, nevertheless provide estimates of possible future regional climatic conditions in mountains and, thereby, allow more detailed impact assessments.

Climate change may be characterised by changes in seasonal or annual precipitation, proportions of solid to liquid precipitation, or frequencies of extreme events. Whatever the directions and magnitude of change, mountain communities and those downstream need to be prepared to implement flexible water management strategies that do not assume that recent patterns will continue. Events in recent history may provide useful guidelines for developing such strategies (Glantz 1988).

Climate-driven hydrology in mountain regions is determined to a large extent by orography itself; mountain belts produce regional-scale concentration of precipitation on upwind slopes and rain-shadows effects in the lee of mountains and in deep intermontane valley systems, often giving rise to high mountain deserts. Many of the more elevated mountain chains of the world intercept large atmospheric moisture fluxes and produce belts of intense precipitation. Along the southern slopes of the Himalayas, enhancement of monsoonal conditions results in some of the highest annual average precipitation in the world, such as at Cherrapunji, India. The spectrum of variability of hydrological regimes ranges from the predominantly rainforested slopes of Papua-Nuova Guinea to the ice-fields of the Patagonian Andes. Climate change will affect the relative importance of these two extreme regimes, as well as, the total moisture flux and how it is delivered temporally. Mountains such as the Andes and the Himalayas are source regions for some of the world's largest rivers, such as the Amazon, the Ganges, Irrawady, and Yangtse, and the discharge characteristics of these rivers and their shifts under changed climatic conditions will be largely modified precipitation regimes (IPCC 1995).

4.2.2
Impacts on Extreme Events

It is uncertain whether a warmer global climate will be accompanied by more numerous and severe episodes of extreme events because current GCM capability to simulate extremes and their altered frequency of occurrence in a changed climate is extremely limited. Anyway, physical arguments indicate that global warming will cause an increase of evaporation from the ocean, moreover a warmer atmosphere can carry more moisture; this leads to a larger amount of precipitation. Global warming will further induce higher temperature differences between the land and the sea surfaces causing an increased transport of precipitable water to the continents and an increase of conventional rainfall (Vellinga et al. 1995). Enhanced occurrences of intense storms, accompanied by high precipitation and/or winds, would inevitably have significant repercussion on a number of sensitive environmental and socio-economic systems. If the severity of storms or the intense precipitation were to increase in frequency in a changed climate, populations living in mountains would be faced with significant social and economic hardships.

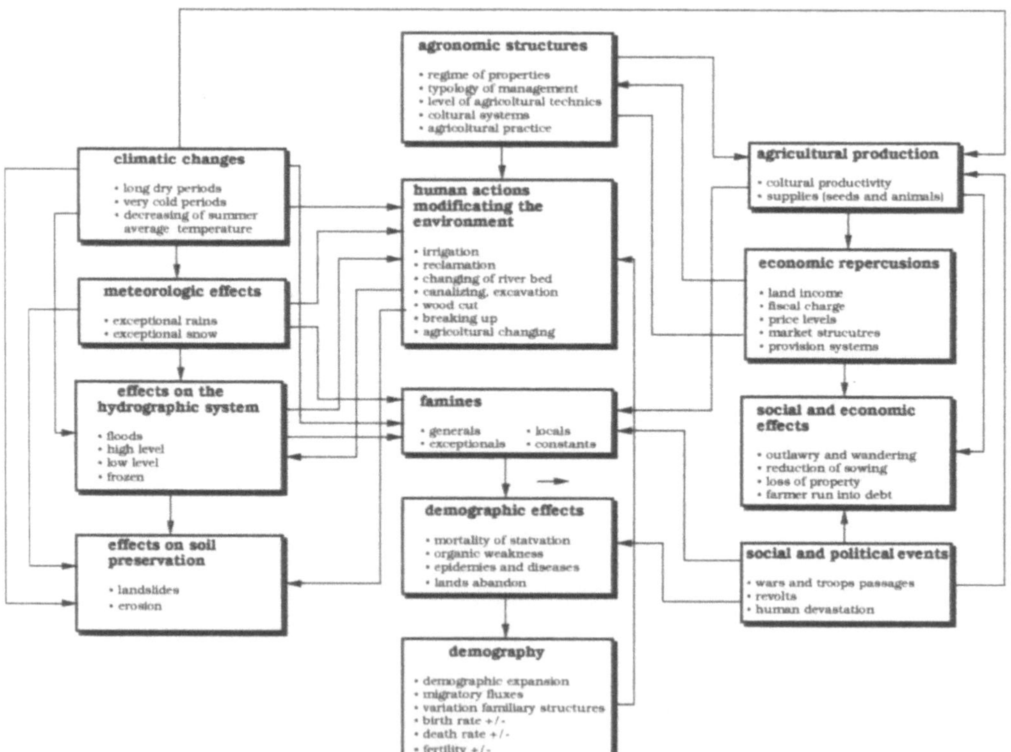

Fig. 4.4. Interactive model between climatic inputs and anthropogenic and environmental reactions in the Po plain area, North Italy, 1550–1650 (ENEA-SGA 1987)

Other impacts associated with extreme events include fire. Forest fires are likely to increase in places where summers become warmer and drier, as has been projected for the Alps (Beniston et al. 1995). Prolonged periods of summer drought would transform areas already sensitive to fire into regions of sustained fire hazard with consequential major socio-economic impacts in sensitive regions close to major population centres.

For the past, several European researches have been carried out dealing with the influence of climatic modifications on the natural environment and on impacts of extreme events on human activities in the last 1 000 years (ENEA and SGA 1987; Bradley and Jones 1992; Frenzel 1992). In fact, Europe can be considered the area where witnesses of climatic fluctuations are more clear and irrefutable; natural evidences are enriched by historical sources following an interactive model (Fig. 4.4) between climatic fluctuations and repercussions on natural and human environments (ENEA and SGA 1987). This model should take into account as much as possible the flexibility of the socio-economic systems as a direct consequence of climatic fluctuations that occurred in the past, especially in the lack of instrumental data of meteorological and climatical parameters.

In the pre-industrial age, both long-term and medium short-term environmental effects triggered by climate, are to be detected in order to recognize a cause-effect genetic link as predisposing factor to disasters. The former represents the protection of the society from negative economic situations due to a climatic variation, while the latter can be detected in the endurance of such negative situations and in the consequent adaptation of social systems to new conditions, as well as, in the long-term effects that climate changing have induced on the environment.

The effects on the environment can be recognized with direct information on climate (i.e. prolonged droughts, cold wave, decrease of average temperature), on meteorological effects (i.e. heavy rainfall, snowfall), on hydrological effects (i.e. floods, inundations, rivers and lakes level variations) and on effects on slopes and soils (i.e. landslides, erosion).

The analysis of demographic elements, agricultural eco-systems and food consume (social, economic and demographic effects) can clarify (ENEA and SGA 1987) the impact that climatic imputs have generated on socio-economic systems of a certain territory.

Demographic pressure, new cultivations and deforestations, crops yield, food consume of urban populations, crisis of subsistence are the main elements that, as validated by past and ongoing researches, may contribute to the reconstruction of climatic trends.

Moreover, the investigation of human reactions to climatic variations can be inferred from other significant elements as: the identification of agricultural species in use in different epochs (i.e. vine cultivation and connected times of grape harvests, introduction of maize cultivation in Italy during the cooling of 1600–1700, etc.), the changing of hydraulic systems due to flooding or low water, land reclamations.

On the above mentioned investigation model of impacts of climatic change the drought in Italy during the period 1158–1160, has been carried out on qualitative and quantitative data (ENEA, SGA 1987).

Different and independent sources report a long drought which affected North and Central Italy with dangerous effects on economy (Fig. 4.4). The drought began in the spring of 1158 and finished in April 1160. In Northern Italy no rainfall was recorded from the beginning of May 1158 to the last week of March 1159 while in the following period precipitation was scarce. In the area of Brescia several water-courses dried up and the water level of rivers in Lombardia were so low that the passage of Frederick Barbarossa's army was possible. Winter and autumn were so dry that harvest was compromised; this caused a general increase of prices and a lack of food. In Pisa in 1158 a great fire caused victims and destroyed many towers. During the winter of 1159 in Genoa, wells dried up.

The absence of information for the rest of Central and South Italy is due to a general lack of contemporary literary sources containing environmental data.

4.2.3
Impacts on Slopes

According to recent and authoritative studies (IPCC 1995), slope stability can be heavily influenced in the light of possible future climate change: the latitude and altitude of different mountain systems determine the relative amount of snow and ice at high elevations and intense rainfall at lower elevations. Because of the amount of precipitation and relief, and the fact that many of these mountains are located in seismic active regions, the added effect of intense rainfall in low- to middle-altitude regions is to produce some of the highest global rates of slope erosion. Climate change could alter the magnitude and/or frequency of a wide range of geomorphologic processes (Eybergen and Imeson 1989). The following examples provide an indication of the nature of the changes that might occur with specific changes in climate.

Large rockfalls in high mountainous areas often are caused by groundwater seeping through joints in the rocks. If average and extreme precipitation were to increase, groundwater pressure would rise, providing conditions favourable to increased triggering of rockfalls and landslides. Large landslides are propagated by increasing long-term rainfall, whereas small slides are triggered by high intensity rainfall (Govi 1990; Bandis et al. 1996). In a future climate in which both the mean and the extremes of precipitation may increase in certain areas, the number of small and large slides would correspondingly rise. This would contribute to additional transport of sediments in the river systems originating in mountain regions. Other trigger mechanisms for rockfalls are linked to pressure-release joints following deglaciation (Bjerrum and Jfrstad 1968); such rockfalls may be observed decades after the deglaciation itself, emphasising the long time-lags involved. Freeze-thaw processes are also very important (Rapp 1960), and several authors have reported possible links between rockfall activity and freeze-thaw mechanisms linked to climate change (Senarclens-Grancy 1958; Heuberger 1966).

A further mechanism that would be responsible for decreased slope stability in a warmer climate is the reduced cohesion of the soil through permafrost degradation (Haeberli et al. 1990). With the melting of the present permafrost zones at high mountain elevations, rock and mudslide events can be expected to increase in number and possibly in severity. This will certainly have a number of economic consequences for communities, where the costs of repair to damaged communications infrastructure and buildings will rise in proportion to the number of landslide events.

4.3
Case Studies

The analysis of climate variability may be carried out through long-term changes in the territory and human society, as well as, through the detection of medium-short-term meteo-climatic scenarios, following the above suggested methodological approach.

Long-term climatic trends may be carried out, for instance, through the reconstruction of landsliding or flooding inventories as environmental indexes of meteo-climatic scenarios of a certain territory, while the reconstruction of medium-short-term climatic trend can be performed through the analysis of some meteorological parameters as precipitation and temperature.

4.3.1
Landslides

Climatic changes have largely influenced human activities and the features of the territory through the occurrence, for instance, of severe extreme events. Landslides occurrence, in most cases, are triggered by very intense or prolonged rainfalls; in such a way landslides induced precipitation have been investigated world-wide, even if the mechanism of transferring the effective precipitation to pore pressure is very complex and still to be exhaustively investigated and understood. The above process is also affected by climate changing due to the variability of amount and pattern of precipitation in long-time trends.

The correlation of meteorological data and release of landslides and the consequential attempt to detect the triggering threshold may be performed through: the examination of the relationship between types of slope failures and different climatic conditions; the investigation of landslide triggered by strong meteorological events; analysis of the relative importance of the different meteorological factors by means of statistical methods; evaluation of landslide prediction methods based on climatic factors.

In this context, it may be useful to understand the relationship between meteorological parameters and landslide triggering in the light of climate change, to report the main result of the European project MeFISSt that has analysed different case-studies in several test-areas on the basis of Varnes classification (1978) adopted as conceptual approach, in order to define the different landslides typologies.

The principal outlines of the researches on correlation between landslides and meteoclimatic aspect in the several selected area of the project can be briefly resumed as follows:

- Landslide inventories highlight that more than 50% of events strictly depends on meteorological factors, especially heavy and prolonged rainfalls;
- Different behaviours with respect to meteorological parameters have to be expected for different landslides typologies: superficial landslides as debris flows, mudflows, earthflows are triggered by very intense rainfalls whose threshold values may be generally expressed in terms of intensity with respect to the mean annual precipitation value;
- Deeper landslide triggering depend on an anomalous precipitation scenario, respect to the average precipitation, which may vary, following the time of water diffusion in the subsoil, from 3 months to 2–3 years;
- Rockfalls or rock slopes, seem to depend on a more complex mechanism, especially when resting over a plastic clayey bedrock, where precipitation has smaller effect and temperature may play an important role on instability triggering.

Climate change, considering the above mentioned case-studies, which can be as quite exhaustive of different landsliding mechanisms in Europe with respect to different climatic areas, can affect slope instability following landslide typologies and, consequently, following triggering mechanisms linked to the transferring of the effective precipitation into pore pressure. In such a way, climate variability may play an essential role changing the characters of landsliding with respect to the variation of long-term climatic scenarios in a certain territory and/or varying the triggering thresholds of landsliding in medium/short-term meteoclimatic scenarios, as well as, influencing the frequency of landslide occurrences.

4.3.2
Floods and Droughts

In the ambit of the realisation of the data bank EVA (*Environmental Events*) from ENEA, where over 2 500 different historical sources containing information on extreme events occurrence since A.D. 1000 have been retrieved, interpreted and catalogued, a long-term reconstruction of floods and droughts has been carried out with particular reference to climatic conditions and solar activity (ENEA, SGA 1987) (Fig. 4.5).

These events cluster in the 16th century, especially around 1500, suffering a drastic reduction for the two following centuries and then showing a high frequency from the half of 19th century up to 1940.

Along the temporal scale from 1000 up to date, all the data are obviously conditioned by historical, morphological and human factors even if the main "engine" can be considered the meteo-climatic variability.

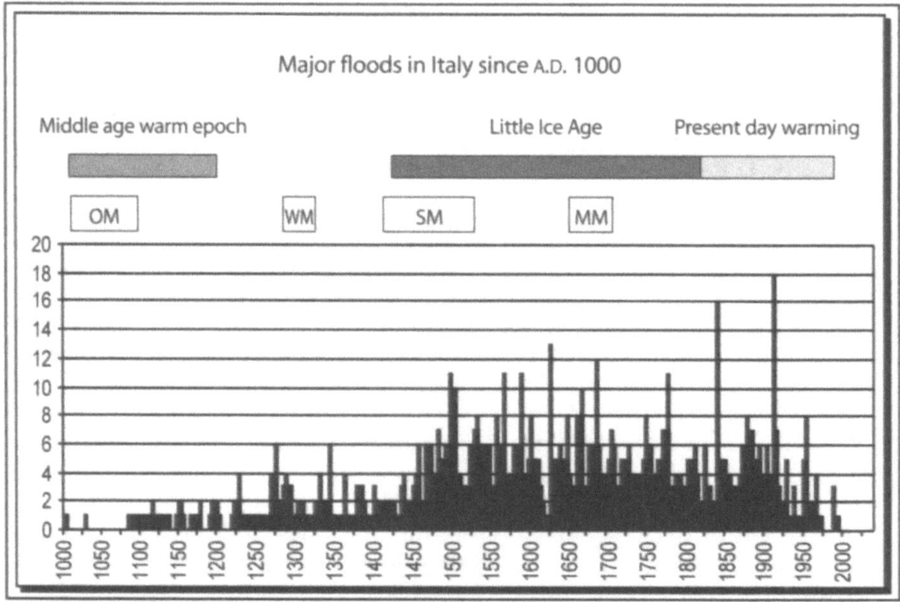

Fig. 4.5. Distribution of major floods occurred in Italy since A.D. 1000 compared with main climatic conditions and solar activity

Since A.D. 1000 up to date, the following climatic epochs and the solar activity minima may be detected: the Early Middle Age Warm Epoch (1000–1200), the Little Ice Age (1430–1850) and the Present-Day Warming (1850–today); the Oort Minimum (OM) 1010–1090, the Wolf Minimum (WM) 1282–1342, the Spörer Minimum (SM) 1416–1534 and the Maunder Minimum (MM) 1645–1714.

As regards to the correlation with the solar activity any consideration can be omitted about the OM, because of the lack of enough documentation. During the WM the minimum of the solar activity coincides with a minimum of floods fluctuations, in coincidence with a detected cooling of climatic conditions registered especially in Northern and Central Europe, but probably to be extended at a planetary scale (Pinna 1988); later, two main peaks can be detected at the end of 1400, at the middle of SM; hence, several peaks are regularly distributed during the 17th and 18th centuries. Data from 20th century may be explained both from a climatic and historical point of view: the period between 1880–1920 is characterised as the most humid in most long-term meteorological series available for Italy and information on flooding is doubtless more abundant and careful with respect to the past. From the second half of the century on, meteoric defluxes are influenced by the numerous hydraulic works realised in the Italian territory for hydro-electric purposes, as well as, to prevent large mountainous areas from erosion.

In conclusion, as drawn from other researches performed for some major basins of the Italian territory (Camuffo and Enzi 1994; Delmonaco et al. 1997) it seems that the climatic anomalies, and so flooding, are not due to changes in solar activities that often have been preceded by hydrological phenomena.

The time series of floods and droughts have also been analysed through simple techniques in order to find some important periodicities, likely correlable to climate cycles. The utilisation of the Fast Fourier Transform (FFT) estimates of power spectra has been applied for the temporal series of the extreme events (Fig. 4.6a,b).

The spectral estimate for floods shows important periodicities of 10, 13, 21 years and as well weaker long-term periodicities (32, 38, 53, 68 years). The same analysis for droughts shows periodicities of 19, 34, 48 and 95 years; the periodicity of 19 years for both seems to be the most representative as confirmed by the cross spectrum analysis (Fig. 4.7), while all the above mentioned return-time of events can be considered reliable as inferred by the analysis of the coherence (Fig. 4.8a). The 19-year cycle can be

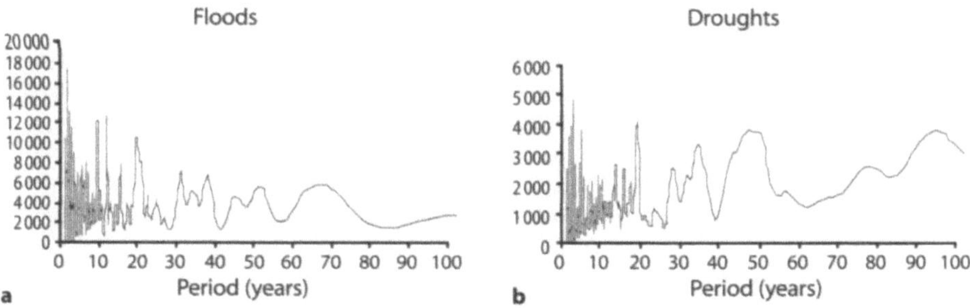

Fig. 4.6. Power spectra estimates applied for temporal series of **a** floods and **b** droughts occurred in Italy since A.D. 1000

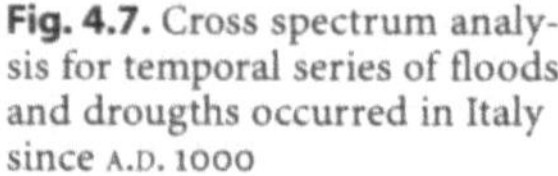

Fig. 4.7. Cross spectrum analysis for temporal series of floods and drougths occurred in Italy since A.D. 1000

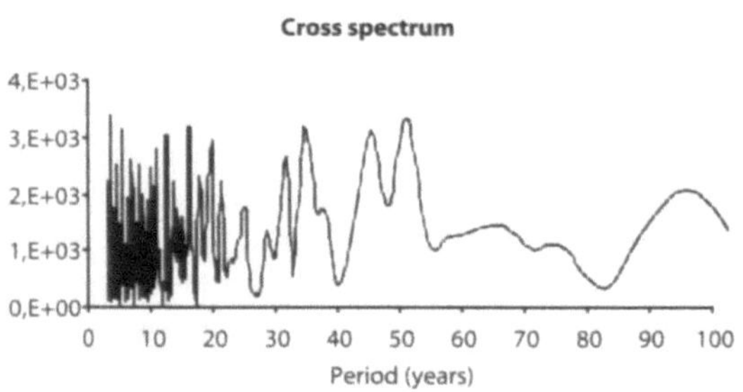

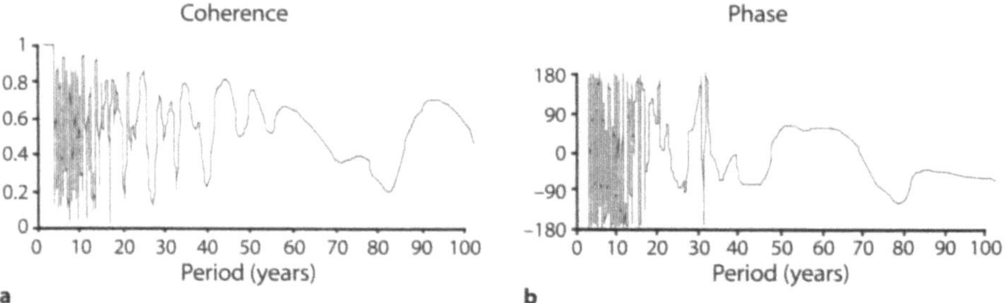

Fig. 4.8. Analysis of **a** coherence and **b** phase for temporal series of floods and drougths occurred in Italy since A.D. 1000

interpreted as the influence of the 18.6-year lunar nodal forcing that has been detected for other floods and drought cycles in Western North America, India, China, South America and Africa (Hugget 1991). The analysis of the phase (Fig. 4.8b) shows a substantial prevalence of opposite occurrence of floods and droughts even if sometime there is a slight phase difference or even a concordance as for long-time periodicities. The physical significance of the last situation may be interpreted as occurrence of floods events, for instance, occurred just after a drought period due to intense precipitation, which can be considered as quite frequent in some recent occurrence of Italian floods (i.e. in Piedmont, 4–6th November 1994 and Versilia, 19th June 1996).

Moreover, the detection of the same main time recurrence of 19 years for floods and droughts in opposite phase seems to confirm that both are connected to the same meteo-climatic scenario where precipitation assumes a fundamental role not only for floods, but even for droughts which seems to be a direct consequence of a hydraulic deficit rather than a temperature increasing.

4.3.3
Climatic Change and Impacts in the Po River Basin Between 1450 and 1650

A qualitative analysis on climate change and impacts in the Po River basin occurred between the 15th and 17th century, has been undertaken, mainly consulting historical sources. The Po River and its tributaries were affected by continuous flooding which represents the most relevant effect of the climatic change occurred during the middle of the 16th century at the start of the so called "Little Ice Age". In spite of the lack of instrumental data on meteorological parameters, the collection and analysis of historical data on climatic effects, stored in the data/bank EVA of ENEA, can be considered as a reliable indicator of the climatic change which affected both the natural environment and consequently the socio-economic aspects, in a well detected geographical and historical context (Fig. 4.4).

The analysis between natural phenomena and the anthropogenic system of the territory has been carried out considering the following elements:

- The mouth of the Po River is the final portion of one of the most articulate basin of Europe from a hydraulic and geological point of view.
- This area has been one of the most populated in spite of a certain difficulty of its agricultural development, essentially due to historical reasons.
- From the middle of 16th to the middle of 17th century, the agricultural system shows typical pre-industrial characters, that is to say a close system as concerning the possibility to procure energy from the sun or from biological sources (animals, wood, manure).
- The extreme fragility of a system totally dependent from weather contributes to magnify the negative effects of a climatic change, even lightly different from the average condition.
- The most critical impact of climate change on the agricultural system in the lower Po River basin, was caused by the increase of precipitation coupled with the feature of the basin itself. An increase of intensity and duration of rainfalls during periods where the Po River reached high levels, in autumn from Apennine tributaries and late spring from Alpine tributaries, could easily have collapsed the hydrographic system in the lower portion of the river, provoking large flooding in the lands devoted to agricultural practise.
- The loss of harvests occurred in concomitance of abnormal rainfall precipitation in spring, summer and autumn, as well as, by the effects of floods along the main stream of the Po River especially during the period 1550–1650.
- The available data confirm that flooding was also influenced by the agricultural exploitation of the hilly slopes and deforestation which preceded it, occurred almost everywhere during 16th century, due to the increasing population.
- The research of new land devoted to agriculture involved especially plans where reclamation provoked the shrinking of areas occupied by marshes and valleys.
- In order to contrast the climatic and meteorological conditions, towns and rural communities carried out intervention to regulate and control the hydraulic system, with artificial channels, embankments and dikes with a continuous raise of the base level of rivers.

The cold wave and heavy rainfalls starting around 1560, caused severe flooding events in the Po River and, as a direct consequence, many economic losses. The increase of precipitation is the predominant feature in this period, even if all the contemporary sources have recorded brief drought events: i.e. from May to November 1559; from February to October 1562; from May to September 1563. Very often these periods were followed by years characterised by intense spring and autumn rainfalls, which caused severe flooding.

The whole of Northern Italy was affected by an intense warm wave in the summer of 1564. Warm and humid sequences seem to characterise years 1571–1572, where a harmful invasion of grasshoppers occurred.

From 1576–1590, all the sources point out a period of frequent precipitation, flooding and cold wave, with a general decrease of the average seasonal temperature.

Summer rainfall, as known, are the most harmful, especially in the pre-industrial age, since they caused mould and parasites inducing a severe decrease of crop production.

In 1594, 1601 and 1660 heavy snowfalls were recorded in April and September, rainfalls and hail in the summer months.

A new and prolonged drought was recorded in 1611, followed by wet summers that triggered landslides and autumn floods.

The crisis of slope and hydraulic systems is well detected by the great deal of floods of the northern major rivers as; Po, Brenta, Reno, Adige, Secchia, Piave, Lambro, Crostolo, Ronco and their minor tributaries. Many inundations can be considered as a direct consequence of some important modification of the hydrographic features by human intervention in order to contrast the increasing precipitation and flooding, as the diversion of important tributaries into the main stream of the Po River (Indice, Santerno, Lamone and Reno Rivers) that caused important disasters in 1544, or the cutting of Porto Viro (1600–1604) carried out to prevent the Venetian Lagoon from the accumulation of solid load.

The floods reported by the analysed sources for the period 1450–1630, involved large portions of the Po River basin causing severe disasters; a detailed analysis of these aspects has pointed out the increasing of these phenomena during the period 1560–1590 with respect to the past (Fig. 4.9).

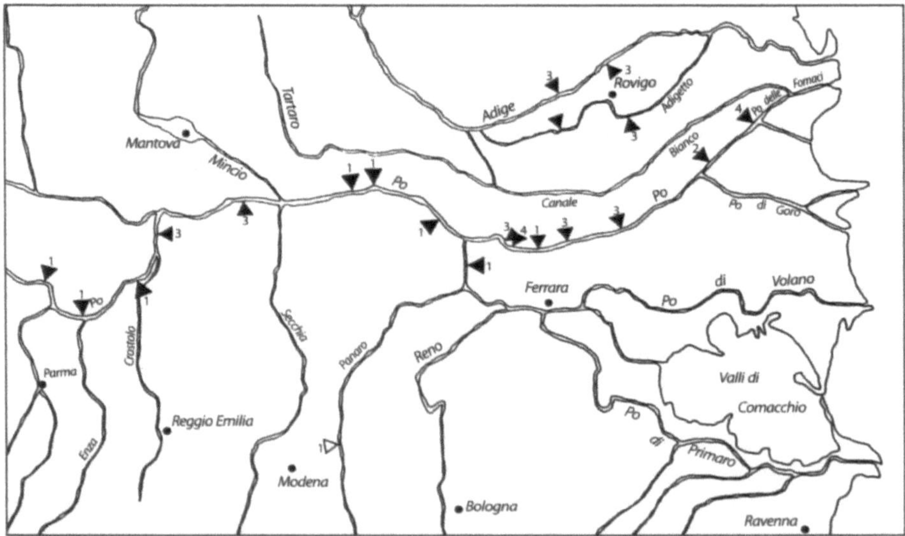

Fig. 4.9. Map of Po River and its tributaries floods during the decade 1621–1631: severe floods are indicated with black triangles while the number indicates the season (*1* spring, *2* summer, *3* autumn, *4* winter) (ENEA-SGA 1987)

4.4
Conclusions

The link between climate change and occurrence of extreme events represents a very debated topic especially since the 'greenhouse effect' detection and in general the anthropogenic activity are to be investigated in order to understand whether human influence on climate can be considered a cause or simply a factor of climate variability. The response is still far likely to be provided by scientists since the climate's future trend is a very difficult topic to be assessed due to the complexity of the Earth's system and the great number of variables which may influence the atmospheric circulation system. In addition, the methods in the assessment of maximum discharge expected for a water course or time returns of extreme meteorological events are based, as a rule, on historical instrumental records that, for most territories, date back to a time window which cannot be exhaustive of long-term trends (i.e. for precipitation and temperature almost 150–200 years of observation). The historical outline of the case studies mentioned above, highlights the importance of the memorialistic tradition (archives, chronicles, scientific literature) produced at the time of the events. The lack of numerical parametrization of descriptive information and data hampers, at present the applicative use: their importance is fundamental for the correct reconstruction of the meteo-climatic scenarios that lead to the occurrence of events.

Otherwise, the value and the reliability of meteo-climatic forecastings and the appraisal of their impact on the territory strictly depend on the degree of acquired knowledge of meteo-climatic changes which occurred in the past and of the natural and human reactions to climatic fluctuations. For this reason, there is a clear need to extend our knowledge as far as possible in the past, regarding both climatic features and their impact on the natural and anthropized environments, in a so-called 'historical monitoring of the territory', since the lack of understanding of climatic trends may compromise the reconstruction of numerical modelling for planning and management of the territory.

In particular, investigation on short-term impacts of meteo-climatic factors on landslides shows that flows (i.e. mud-flows, debris-flows) are mainly triggered by extreme precipitation occurring in the hours prior to the events; slides are often triggered by extreme rainfall due to the increase of water tables as a consequence of cumulated precipitation during previous months; falls seem to be linked to meteo-climatic factors only when triggered by frost-thawn cycles and, in some cases, for thermo-clastic mechanisms. Other parameters play a role in landslide triggering but their contribution seems to be relevant only when meteo-climatic factors do not exhibit a fundamental role (i.e. falls).

Medium-term evolution of floods and droughts investigated since A.D. 1000 for some major basins of the Italian territory shows a scarce correlation with solar anomalies; Fast Fourier Transform (FFT) estimates of power spectra for floods demonstrates periodicities of 10, 13, 21 years and, as well weaker long-term periodicities (32, 38, 53, 68 years), while the same technique applied for droughts shows periodicities of 19, 34, 48 and 95 years; the periodicity of 19 years for both seems to be the most representative and may be correlated with the 18.6-year lunar nodal forcing whose influence has been detected in other floods and drought cycles all over the world. The same periodicity for floods and droughts and their occurrence in opposite phases seems to confirm that both are connected to the same meteo-climatic 'engine' so that drought events are a direct consequence of precipitation affluxes rather than of temperature fluctuations.

Clearly, the impact of climate change has to be investigated in the contemporary socio-economic context in order to reconstruct not only the occurrence of extreme events but also the contribution of anthropogenic system (i.e. land-use, deforestation, artificial change on hydraulic streams) to the time/geographic events distribution: this is in order to avoid false interpretation in terms of climate impact instead of human interference on hydrological cycle. A model of relationships among different socio-economic and meteo-climatic factors has been developed for the Po Plain in the period 1450–1650 showing the complexity of the different elements and then the need of an integrated historical, socio-economic, geological and climatological approach.

References

Aguado E et al. (1992) Changes in the timing of runoff from West Coast streams and their relationships to climatic influences. Swiss Climate Abstracts, special issue, International Conference on Mountain Environments in Changing Climates, p 15

Bandis SC, Delmonaco G, DE Lotto P, D'epifanio A, Dutto F, Ferrer M, Frassoni A, Margottini C, Mortara G, Palandri M, Sandersen F, Serafini S, Stemberk J, Trocciola A, Anselmi B, Crovato C (1996) Meteorological events and natural disasters: An appraisal of the Piemonte (North-Italy) case history of 4–6 November 1994 by a CEC field mission. Casale R, Margottini C (eds), Roma

Barnola JM, Raynaud D, Korotkevich YS, Lorius C (1987) Vostok ice core provides 160 000-year record of atmospheric CO_2. Nature 329:408–414

Beniston M (ed) (1994) Mountain environments in changing climates. Routledge Publishing Company, London New York, 492 pp

Beniston M, Ohmura A, Rotach M, Tschuck P, Wild M, Marinucci MR (1995) Simulation of climate trends over the alpine region: Development of a physically-based modeling system for application to regional studies of current and future climate. Final Scientific Report N° 4031–33250 to the Swiss National Science Foundation, Bern Switzerland, 200 pp

Bjerrum L, Jfrstad F (1968) Stability of rock slopes in Norway. Norwegian Geotechnical Institute Publication 79:1–11

Bradley RS (1994) Reconstructions of climate from A.D. 1000 to the present. In: Speranza A, Tibaldi S, Fantechi R (eds) Global Change, Proceedings of the first Demetra meeting held at Chianciano Terme, Italy from 28 to 31 October 1991, European Commission EUR 15158 EN, pp 123–137

Bradley RS, Jones PD (eds) (1992) Climate since A.D. 1500. Routledge, London

Bultot F, Gellens D, Schädler B, Spreafico M (1992) Impact of climatic change, induced by the doubling of the atmospheric CO_2 concentration, on snow cover characteristics-case of the Broye drainage basin in Switzerland. Swiss Climate Abstracts, special issue, International Conference on Mountain Environments in Changing Climates, p 19

Camuffo D, Enzi S (1994) Climatic features during the Spörer and Maunder Minima. Pälaoklimaforschung/Palaeclimate Research, special issue 8, Pälaoklimaforschung

Chappellaz J, Barnola JM, Raynaud D, Korotkevich YS, Lorius C (1990) Ice-core record of atmospheric methane over the past 160 000 years. Nature 345:127–131

Davis MB (1986) Climatic instability, time lags, and community disequilibrium. In: Diamond J, Case TJ (eds) Community Ecology. Harper and Row, New York, pp 269–284

Delmonaco G, Margottini C, Trocciola A (1997) Non-stationarity of hydroclimatic data: The case study of the Tiber river basin.

ENEA, SGA (1987) Raccolta, analisi, elaborazione ed interpretazione di informazioni storiche sugli effetti prodotti da eventi eccezionali in Italia dall'anno 1000 al 1985. In: Progetto GIANO. Bologna 1987. Internal ReportENEA and CONSORZIO CIVITA (1995) CEC Project MeFISST. Final Report,

Eybergen J, Imeson F (1989) Geomorphological processes and climate change. Catena 16:307–319

Folland CK, Karl TR, Vinnikov KYA(1990) Observed climate variations and change. Climate Change: The IPCC Scientific Assessment, Houghton JT, Jenkis GJ, Ephramus JJ (eds), pp 194–238

Frenzel B (ed), Pfister C, Glaser B (co-eds) (1992) European climate reconstructed from documentary data: methods and results. Gustav Fischer Verlag, Stuttgart Jena New York

Genthon C, Barnola JM, Raynaud D, Lorius C, Jouzel J, Barkov NI, Korotkevich YS, Kotlyakov VM (1987) Vostok ice core: Climatic response to CO_2 and orbital forcing changes over the last climatic cycle. Nature 329:414–418

Giorgi F, Marinucci MR, Visconti G (1990) Use of a limited area model nested in a general circulation model for regional climate simulations over Europe. J Geophys Res 95(18):413–431

Giorgi F, Marinucci MR, Visconti G (1992) A $2 \times CO_2$ climate change scenario over Europe generated using a limited area model nested in a general circulation model, II. J Geophys Res 97:10011–10028

Giorgi F, Mearns LO (1991) Approaches to the simulation of regional climate change: A review. Rev Geophysics 29:191–216

Glantz MH (ed) (1988) Societal responses to regional climatic change. Westview Press, Boulder CO

Govi M (1990) Conférence spéciale: mouvements de masse récents et anciens dans les Alpes italiennes. Proc. Fifth Symposium on Landslides, Lausanne 3:1509–1514

Graumlich LJ (1993) A 1000-year record of temperature and precipitation in the Sierra Nevada. Quat Res 39:249–255

Gyalistras D, von Storch H, Fischlin A, Beniston M (1994) Linking GCM-simulated climatic changes to ecosystem models: case studies of statistical downscaling in the Alps. Clim Res 4:167–189

Haeberli W, Muller P, Alean J, Bösch H (1990) Glacier changes following the Little Ice Age. A survey of the international data base and its perspectives. In: Oerlemans J (ed.) Glacier Fluctuations and Climate. D. Reidel Publishing Company, London New York, pp 180–203

Heuberger H (1966) Gletschergeschichtliche Untersuchungen in den Zentralalpen zwischen Sellrain und Ötztal. Wiss. Alpenvereinhefte 20:126

Houghton JT, Jenkins GJ, Ephraums JJ (eds) (1990) Climate change. The IPCC Scientific Assessment. Cambridge University Press, Cambridge UK, 365 pp

Hugget J (1991) Climate, earth processes and heart history. Springer-Verlag, Berlin Heidelberg New York, 282 pp

IPCC WGI (1995) Climate change 1995 – The science of climate change. The Second Assessment Report of the Inter-Governmental Panel on Climate Change. Houghton JT, Meira Filho LG, Callander BA, Harris N, Kattenberg A, Maskell K (eds) Cambridge University Press, New York USA, 572 pp

IPCC WGII (1995) Climate Change 1995 – Impacts, adaptations and mitigations of climate change: scientific-technical analyses: The second assessment report of the inter-governmental panel on climate change. Watson RT, Zinyowera MC, Moss RH (eds) Cambridge University Press, New York, 880 pp

Kates RW, Ausubel JH, Berberian M (eds) (1985) Climate impact assessment. SCOPE 27, Wiley & Sons, Chichester UK, 78 pp

La Marche VC Jr (1973) Holocene climatic variations inferred from treeline fluctuations in the White Mountains, California. Quat Res 3:632–660

Lamb HH (1965) The early Medieval warm epoch and its sequel. Palaeogeogr Palaeoclimatol Palaeoecol 1, pp 13–37

Leavesley GH (1994) Modeling the effects of climate change on water resources-a review. In: Frederick KD, Rosemberg N (eds) Assessing the impacts of climatic change on natural resource systems, Kluver Academic Publishers, Dordrecht, Netherlands, pp 179–208

Lins H et al. (1990) Hydrology and water resources. In: McG Tegart WJ, Sheldon GW, Griffiths DC (eds) Climate Change: The IPCC impacts assessment, Australian Government Publishing Service Canberra, Australia

Luckman BH (1994) Using multiple high-resolution proxy climate records to reconstruct natural climate variability: An example from the Canadian Rockies. In: Beniston M (ed) Mountain environments in changing climates. Routledge Publishing Company, London New York, pp 49–52

Margottini C (1994) Natural and technological catastrophies and policy options: a review of some Italian experiences. World Conference on Natural Disasters Reduction. Yokohama Japan 23–27 May 1994

Margottini C (in press) Fenomeni evolutivi e catastrofi naturali. In: Le Scienze dell'Ambiente. RCS

Marinucci MR, Giorgi F (1992) A $2 \times CO_2$ climate change scenario over Europe generated using a limited area model nested in a general circulation model, I: Present day simulation. J Geophys Res 97(9)9989–10009

Marinucci MR, Giorgi F, Beniston M, Wild M, Tschuck P, Bernasconi A (1995) High resolution simulations of January and July climate over the Western Alpine region with a nested regional modeling system. Theor Appl Clim

Martin E (1992) Sensitivity of the French Alps' snow cover to the variation of climatic parameters. Swiss Climate Abstracts, special issue, International conference on mountain environments in changing climates, pp 23–24

Nash LL, Gleick PH (1991) Sensitivity of streamflow in the Colorado Basin to climatic changes. Hydrology 125:221–241

Oerlemans J (ed) (1989) Glacier fluctuations and climate change. Kluver Academic Publishers, Dordrecht, Netherlands

Parry ML et al. (1992) The potential socio-economic effects of climate change in south-east Asia. United Nations Environment Programme, Nairobi Kenia

Pinna M (1988) Il clima nell'Alto Medioevo conoscenze attuali e prospettive di ricerca, L'ambiente vegetale nell'Alto Medio Evo, 'L'ambiente vegetale nell'Alto Medio Evo', XXXVIII, Sett. di St. del Centro It. di Studi sull'AME

Rapp A (1960) Recent developments on mountain slopes in Kärkevagge and surroundings. Geografiska Annaler 42:1–158

Riebsame WE (1989) Assessing the social implication of climate fluctuations. United Nations Environment Programme, Nairobi Kenia

Rupke J, Boer MM (eds) (1989) Landscape ecological impact of climatic change on alpine regions with emphasis on the Alps. Discussion report prepared for European conference on landscape ecological impact of climate change, Agricultural University of Wageningen, Utrecht and Amsterdam Netherlands

Schuurmans CJE (1981) Climate of the last 1000 years. In: Berger A (ed) Climatic variations and variability: Facts and theories, D. Reidel Publishing Company, Dordrecht Boston London, pp 245–258

Schweingruber FH (1988) A new dendroclimatic network for Western North America. Dendrochronologia 6:171–180

Senarclens-Grancy W (1958) Zur Glazialgeologie des Ötztales und seiner Umgebung. Mittl. Geol. Ges. Wien 49:257–314

Slaymaker O (1990) Climate change and erosion processes in mountain regions of Western Canada. Mountain Research and Development 10:171–182

Smith EA, Voner Haar TH, Hickey JR, Maschhoff R (1983) The nature of the short period fluctuations in solar irradiance received by the Earth, Clim Change 5:211–235

Street RB, Melnikov PI (1990) Seasonal snow, cover, ice and permafrost. In: McG Tegart WJ, Sheldon GW, Griffiths DC (eds) Climate change: The IPCC impacts assessment. Australian Government Publishing Service, Canberra Australia

Tiedemann H (1992) Earthquakes and volcanic eruptions: A handbook on risk assessment. Swiss Reinsurance Company, Zurich, 952 pp

Varnes DJ (1978) Slope movements type and processes. In: Schuster RL, Krizek RJ (eds) Landslides analysis and control. Transp Research Nat Sc, Special Report 176:11–33

Vellinga P, Dorland K, Olsthoorm X, Tol R (1995) Climate change and extreme events: Droughts and floods. Change 26, The Netherlands

Webb III T, Kutzbach JE, Street-Perrot FA (1985) 20 000 years of global climate change: Paleoclimatic research plan. In: Malone TF, Roederer JG (eds) Global change. Cambridge University Press, Cambridge UK, pp 182–218

Part II
Tools – Methodology

Chapter 5

Dendrogeomorphology in Landslide Analysis

R. Fantucci

5.1
Introduction

Alestalo (1971) first outlined the basic principles of dendrogeomorphology. Basically geomorphic and geologic processes affect trees in a variety of ways that can be determined and dated through tree-ring analysis. Geomorphic stress can induce growth anomalies as suppression and release (sudden growth decrease or increase) (Shroder 1978; Shroder 1980). Whereas, climatic factors can also induce ring variations similar to those of geomorphic processes (Fritts 1971), it is important to exclude the climatic influence on the anomalies found in trees, affected by a geomorphic event, by control trees living in an undisturbed zone, close to the study area.

5.2
Kind of Investigable Landslides and Limitations

Dendrogeomorphology has been widely applied in the study of different kinds of mass movements in many countries. It was used to investigate almost all kinds of landslides: debris flow (Claugue and Souther 1982; Hupp 1984; Strunk 1992; Van Asch and Van Steijn 1991); mud flow (Braam et al. 1987; Fantucci and McCord 1995; Lionel and Jackson Jr 1977); earth flow (Bovis and Jones 1992); rock flow (Fantucci and Sorriso-Valvo 1997); rotational slides (Begin and Filion 1988; Burn and Friele 1989; Jibson and Keefer 1988; Terasme 1975); translational slides (Aipassa and Shimizu 1988; Jibson and Keefer 1988; Orombelli and Gnaccolini 1980); falls (Bednarz 1986; Moss and Rosenfeld 1978; Porter and Orombelli 1980), as well as, creeping (Denneler and Schweingruber 1993; Parizek and Woodruff 1957).

The limitations to the use of dendrogeomorphology are related to the presence of suitable tree species to dendrochronological analysis. Trees useful to dendrochronological analysis are sensitive species, which produce distinct high frequency growth variability with ring growth pattern well related to one or more casual factors (as temperature and rainfall). Tree complacent species produce, instead, always the same kind of growth and are less useful than sensitive species. Another limit of this kind of research is the time span investigation related to the age of trees sampled (usually the last few centuries).

5.3
Effects of Landslide on Trees

Landslides which occur on a tree covered slope can produce various events such as inclination of trees, shear of rootwood and stemwood, corrasion scars or bark removal, burial of stemwood, exposure of rootwood, inundation and nudation (production of

bare ground). Any one of these events may cause responses such as production of reaction wood, suppression of growth, release of growth, termination of growth, sprout-
ing and succession (Shroder 1978).

5.3.1
Trees' Inclination

One of the most common effects on trees due to landslide activity is stem tilting, which
will be differently oriented according to the kind and part of the landslide trees come from.
For example, in rotational slides, it is common for trees to have a radial tilting (Fig. 5.1).

Tilted trees tend to recover their straight geotropic growth developing an S shape
along the trunk (Fig. 5.2). First the upper part of the inclined tree begins to lean in the
opposite direction, past the vertical line in its recovery process of vertical growth and
then, at the end, will grow vertical again. A tree could be affected by more than one
tilting showing a complex bending form in the trunk, moreover a tilted form of the
tree can induce new sprouts to grow, helping the tree in its reaction.

The trees affected by stem tilting usually produce a "reaction wood" that is a growth
response which strengthens the tilted tree and may bring it back to a vertical position;
deciduous trees produce "tension wood" (grey or yellowish, long dense cells) on the
upper side of a tilted stem, while conifers produce "compression wood" (reddish and
yellowish brown, short, thick-walled dense cells) on the underside of inclined stem
(Shroder 1980). Although the year in which the onset of reaction-wood growth begins,
usually records the year of tilting, in some cases the onset is delayed a few years if the
tilting is severe (Carrara 1979). Many authors through the presence of reaction wood

Fig. 5.1. Radial tilting of trees from a rotational landslide in the Umbria district (Italy)

Fig. 5.2. Different trunks form developed by one or more tilting; *1–3* Gradual development of an S shape; *4* Two tilting in different directions; *5* Pistol butted form due to a severe inclination in a young tree; *6–7* Gradual continuous and extreme tilting; *8* Tree form resulting from successive opposite tilting; (From Braam et al. 1987)

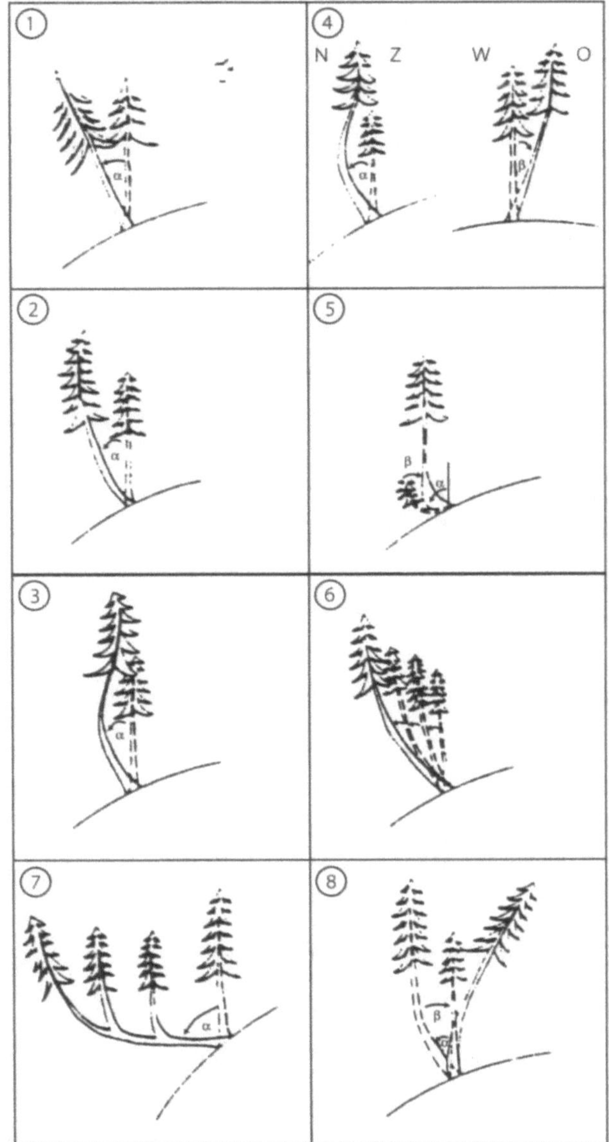

dated landslides movement events (Begin and Filion 1988; Claugue and Souther 1982; Denneler and Schweingruber 1993; Hupp 1984; Orombelli and Gnaccolini 1980; Shroder 1978).

Other common effects due to stem tilting on tree growth, evidenced by many authors, are the abrupt growth decrease (suppression) (Denneler, Schweingruber 1993; Fantucci and McCord 1995; Hupp 1984; Orombelli and Gnaccolini 1980; Shroder 1978; Strunk 1992; Terasme 1975) and the eccentricity variation (Braam et al. 1987).

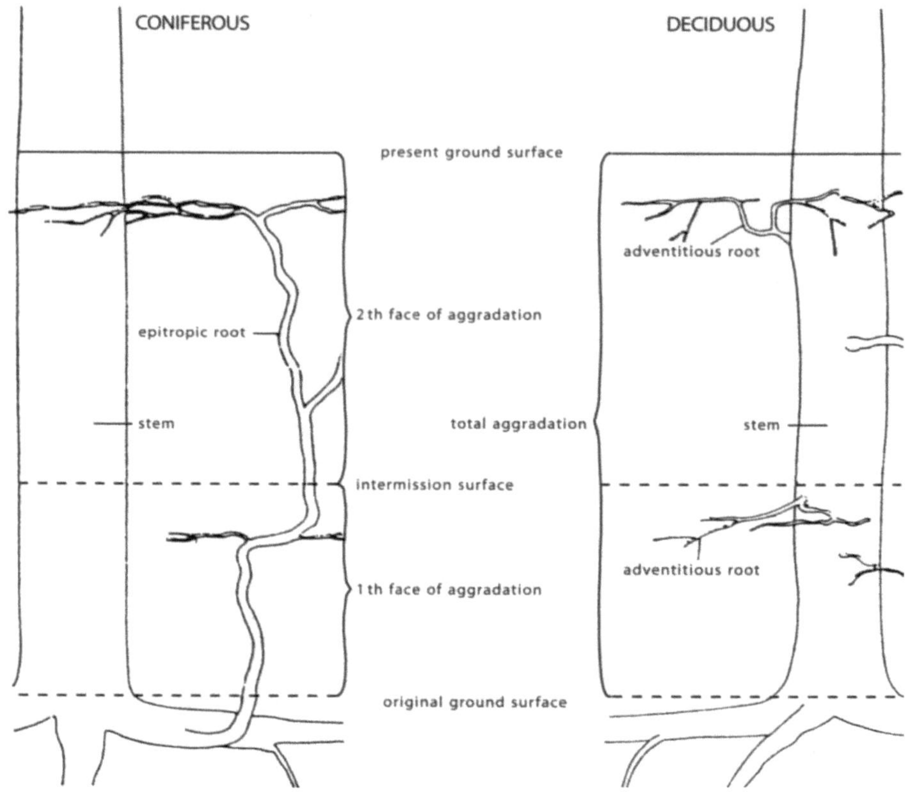

Fig. 5.3. Adventitious or epitropic roots (From Alestalo 1971)

5.3.2
Effects on the Root System

A Landslide can induce alteration also to the root system; in fact, the root system morphology can change to the effect of ground erosion which produces the formation of buttress roots, common in trees living close to the landslide crown (La Marche Jr 1963).

When there is a partial stem burial due to debris aggradation as in trees affected by debris flow deposit, they can develop adventitious roots (deciduous) or epitropic roots (conifers) (Fig. 5.3) (Alestalo 1971). These could be dated through the root ring analysis, even if this kind of analysis is still at the beginning.

5.3.3
Corrasion Scar

The rolling debris of some landslides types can provoke damages on the tree stem with their impact. They can brake the stem or branches of trees, as well as, produce corrasion scar. The kind of scar indicates the dimension of the rock (or debris) fragment that

Fig. 5.4. Stem scar in oak tree produced by a rock fall in Civita di Bagnoregio (VT)-Central Italy

crashed on the tree. The height of the corrasion scar is important because it marks the level at which the material was moving downslope (Fig. 5.4). The tree damaged will produce a callous margin that tends to cover the scar with time. These scars can be dated through wood wedge, multiple corings or a complete cross section of the stem. Dating tree scars can be very useful in dating debris flow landslides, as well as, rock fall.

5.3.4
Nudation

The Landslide process often exposes bare, inorganic surfaces that will become hospitable to plant growth. The trees' community will show an evolution from an earlier stage with a pioneer community to a climax community. The analysis of this vegetation community will help to know the age of the geomorphic event, giving an estimation of the minimum age of it (Aipassa and Shimizu 1988; Porter and Orombelli 1980).

5.4
Examples of Dendrogeomorphological Landslide Analysis in Italy

Three landslide areas, two in Central Italy (Mt. Rufeno Natural Park and Civita di Bagnoregio – Latium region) and one in South Italy (Lago – Calabria region) have been recently investigated through dendrogeomorphologic analysis to assess the activation or reactivation dating in the last centuries.

5.4.1
Sampling and Methodology

Sampling trees for this kind of research cannot be a random sampling but is important to collect those trees more affected by the landslide movement which show particular signs: stem tilting, corrasion scars, root exposure, partial burial of the stemwood. It is better to take samples with a non destructive method through the "Swedish incremental borer" by extracting a pencil-thin core of wood without causing harm to the tree.

In recent researches, for each one sampled tree two cores were usually collected: one on the side opposite to the direction of tilting and another orthogonal to or facing to the first one. In the field work, the following data for each sampled tree were recorded: the species, its position on the topographic map, the direction and degree of the stem tilting, the height and orientation of the samples.

Once transported to the laboratory, cores are allowed to dry, then are glue mounted on grooved sticks and sanded to a high degree of polish through progressively finer grades of sand paper.

All the samples were subject to the standard procedures used in dendrochronological researches such as; skeleton plots, cross dating, ring width measurements with a micrometer to build growth curves (Stokes and Smiley 1968). The ring width measurements were then checked with the program CHOFECHA (Holmes 1983).

A very important method used in these researches was the "visual growth analysis" introduced by Schweingruber (1990), to identify suppression or release in the cores (Fig. 5.5).

The visual growth analysis was used to identify the disturbances related to landslide movement that trees have recorded in their growth, mostly suppression kind, due to stem tilting or disturbances in the root system. The control trees living in the surrounding stable areas did not show signs of growth anomalies.

An example of the appearance of cores with and without growth anomalies are shown in (Fig. 5.6a,b) with oak samples (*Quercus cerris* L.) from the Mt. Rufeno Natural Park area – Central Italy.

The visual growth analysis was used to date the year of the events of the landslide movement activation or reactivation. A further investigation on the growth anomalies

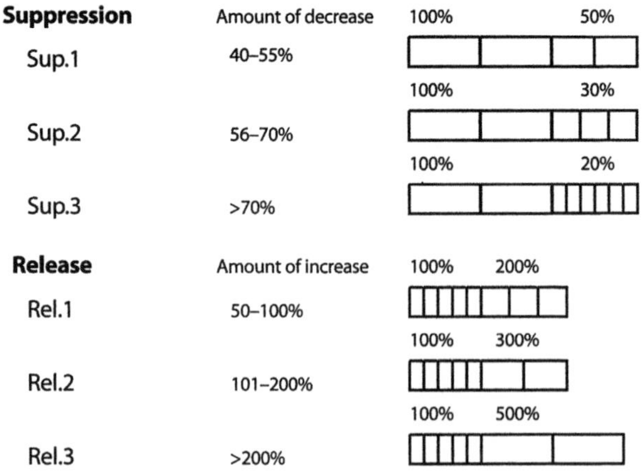

Fig. 5.5. Visual growth analysis: sudden growth decrease (suppression) or increase (release) divided in classes of intensity (Adapted from Schweingruber et al. 1990)

Suppression	Amount of decrease
Sup.1	40–55%
Sup.2	56–70%
Sup.3	>70%

Release	Amount of increase
Rel.1	50–100%
Rel.2	101–200%
Rel.3	>200%

Fig. 5.6. Oak tree samples (*Quercus cerris* L. 10 ×) from the Mt. Rufeno Natural Park area – Central Italy: **a** sample from the landslide area with strong suppression (more than 70% growth decrease); **b** sample from the control zone (stable area) without growth anomalies (Photographs by M Vollaro)

was carried out in the Mt. Rufeno Natural Park research to get a more precise (seasonal) dating of the 1963 event, recorded in more than 70% of samples from the landslide area.

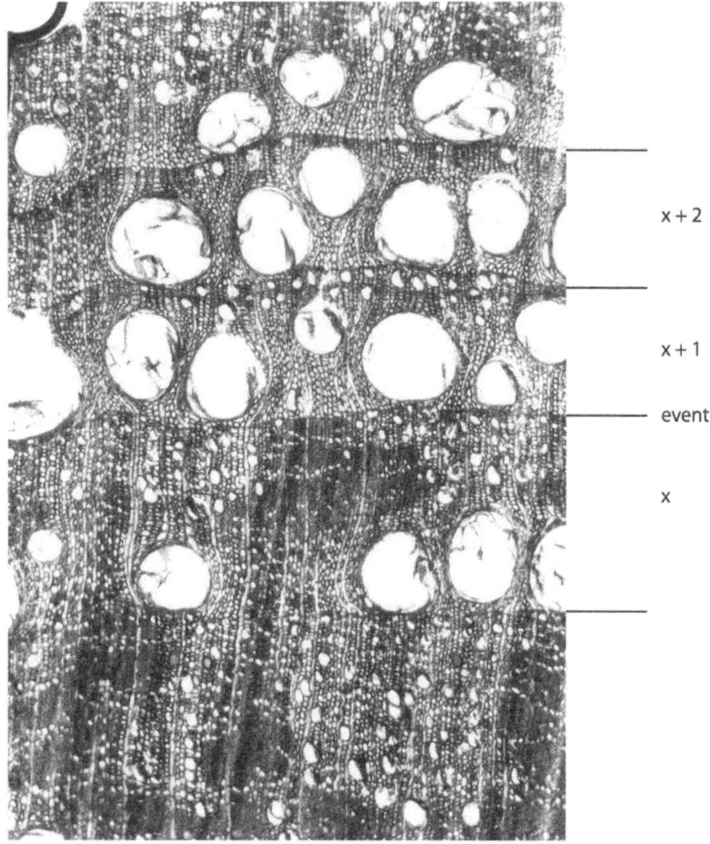

Quercus cerris L. 40×

Fig. 5.7. Microphotograph of a core (*Quercus cerris* L. 40 ×), representative of the 1963 anomaly, which dated the event in 1962 autumn. The ring structure after the event ($x + 1, x + 2, \ldots$) are abnormal because smaller than the ring previous the event (x) and the thick fibres are missing (darker part in the rings) (Photograph by F.H. Schweingruber)

The microphotographs (Fig. 5.7) made on disturbed samples evidenced that the disturbance on trees occurred during the rest season of 1992 (autumn-winter), which was well correlated with an extremely wet autumn, documented as a period of historical regional scale landslide crisis (Fantucci and McCord 1995).

5.4.2
Temporal Analysis

Temporal analysis is a tool to identify and date the activation and/or reactivation of landslide events. The results of visual growth analysis, made on the samples previously correctly dated by standard dendrochronological procedures, collected in landslide area, were utilised to calculate the index number I_t (%) for each class of anomaly, whose

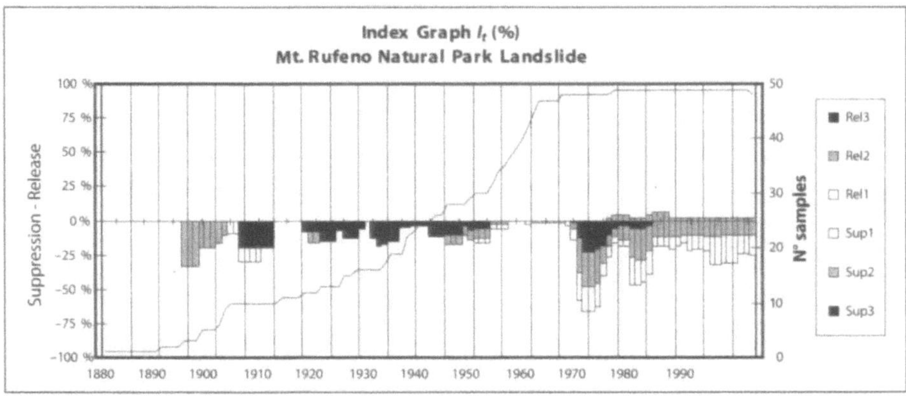

Fig. 5.8. Index graph I_t (%) of the landslide in Mt. Rufeno Natural Park. Negative peaks correspond to reactivation movement events (Adapted from Fantucci and McCord 1995)

result were summed up as total negative (suppression) and positive (release) anomalies at year t:

$$I_t = \left(\sum_{i=1}^{t} Ri \right) / \left(\sum_{i=1}^{t} Ai \right) \times 100 \;,$$

where:

- R_i = cores that have shown the anomalies in year t;
- A_i = total number of cores sampled in year t (From Fantucci and McCord 1995).

An index (I_t) graph related to the Mt. Rufeno Natural Park landslide area is shown in Fig. 5.8, where it is possible to identify the main landslide reactivation events with the suppression growth anomalies (negative peak values). In this site, the main reactivation events occurred in 1894, 1903, 1916, 1922, 1928, 1937, 1940, 1945, 1963, 1972 and 1986, with an increase of landslide activity from the 60s and two periods not affected by disturbances between 1909–1914 and 1948–1961 (Fig. 5.8).

The kind of information extracted from the temporal analysis permits an evaluation of the activity state of a landslide at present, as well as in the past; likewise, temporal analysis data are useful to analyse the landslide frequency recurrence in a study area (Strunk 1992).

5.4.3
Spatial Analysis

This analysis allows us to define the stable from the unstable areas where the stability of a slope is not defined; furthermore it is useful in rebuilding the history of landslide reactivation.

Dividing the examined time span in intervals and plotting on spatial-temporal maps the trees, living in that period, affected and not affected by growth anomalies, allows

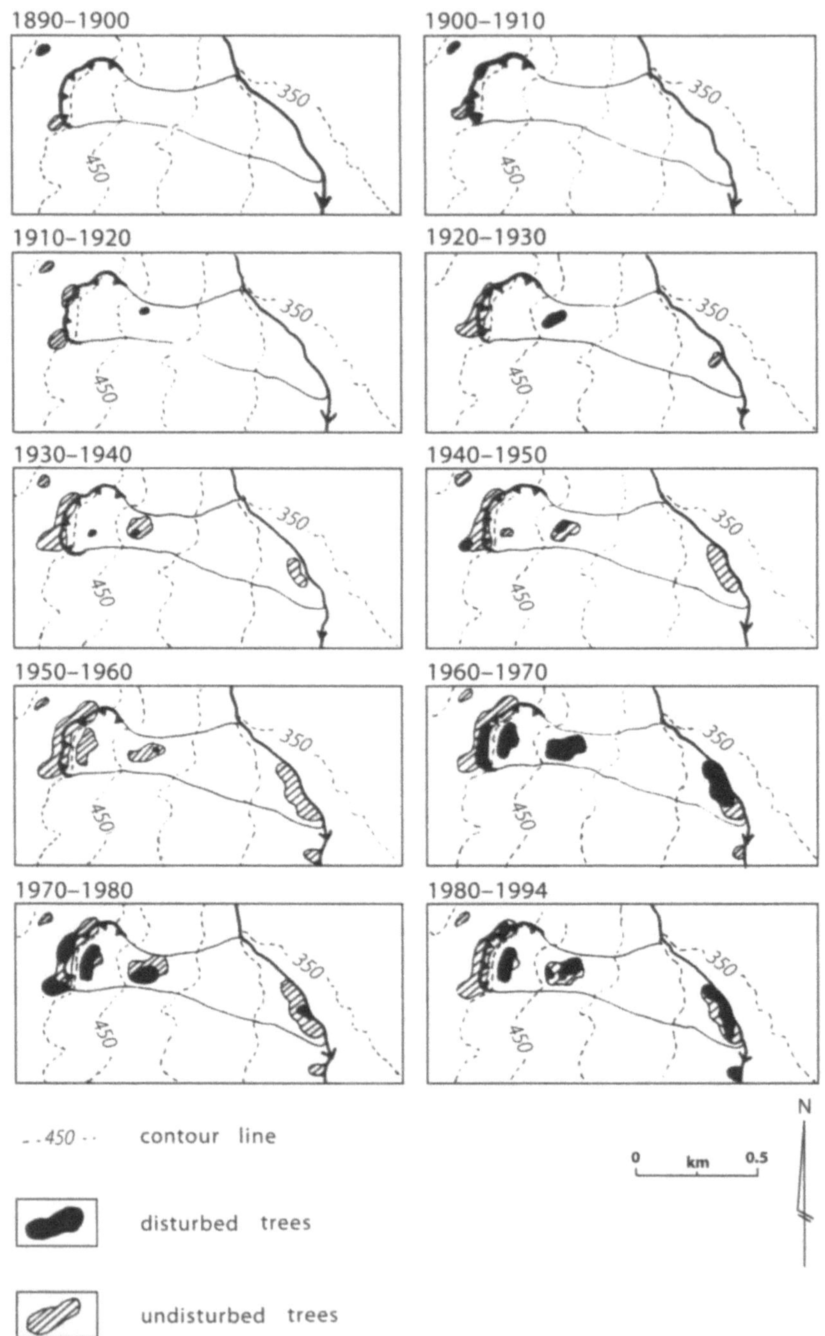

Fig. 5.9. Landslide spatial-temporal maps between 1890–1994 in the Mt. Rufeno Natural Park (From Fantucci and McCord 1995)

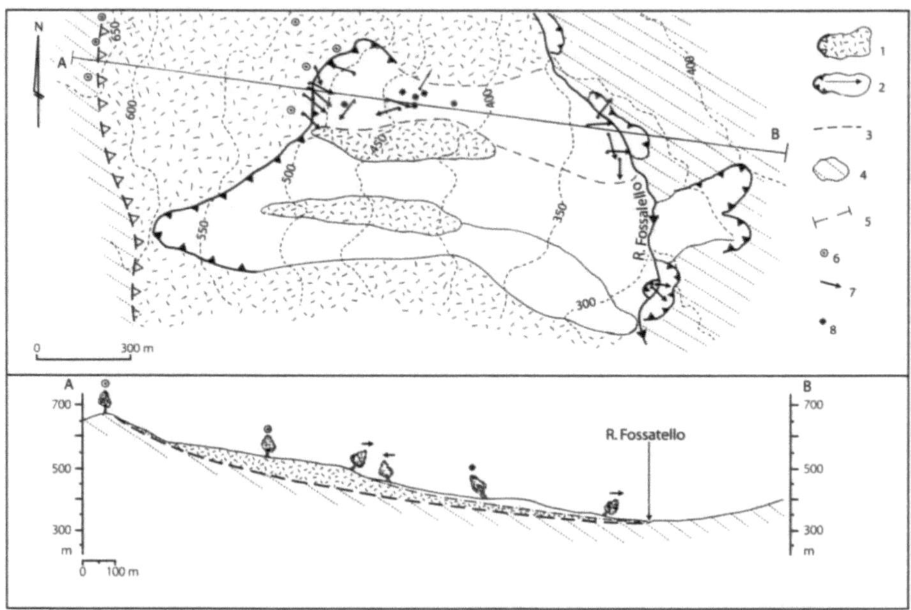

Fig. 5.10. Tilting map of trees and dendrogeomorphological section of Mt. Rufeno Natural Park landslide. *1* Old inactive landslide; *2* Active landslides; *3* Landslide study border; *4* Clayey flysh basement; *5* Dendrogeomorphological section; *6* Vertical trees; *7* Simply tilted trees; *8* Complex tilted trees; (From Fantucci 1997)

for distinction of the stasis and active periods of the landslide (Fig. 5.9). In a complex landslide this can permit to reconstruct the sequence of movements occurred (Shroder 1978).

Besides spatial-temporal maps, the map of tilting of trees sampled and dendrogeomorphological sections can be also useful to define the different landslide parts and the more disturbed ones using only the morphological aspect of the sampled trees (Fig. 5.10). Here is represented the Mt. Rufeno Natural Park landslide investigated with the sampled trees and their direction of tilting. In the active landslide investigated, trees are all tilted, sometimes more than once (complex tilting), while trees living in the old landslide deposit and in the stable areas did not show tilting but grow straight (Fantucci 1997).

5.4.4
Correlation Between Growth Anomalies Events and Geological Causes

The three landslide areas investigated gave different responses to the correlation between the geological causes and movement reactivations. In the first case, Civita di Bagnoregio landslide (a complex landslide in volcanic and clay sedimentary terrain) – Central Italy (Latium) the area was found stable during the time span investigated (1880–1994), in agreement with the historical data on landslide activity of the site. In the second case, Mt. Rufeno Natural Park landslide – Central Italy (Latium), investigated was a slide-flow type landslide on clayey flysh terrain. The growth anomalies recorded by trees were well correlated (75%) to particularly wet seasons, most of them occurred after a long dry period (Fantucci and McCord 1995), showing a strong relationship between rainfall and landslide reactivation. In the third case, Lago landslide in South Italy (Calabria), a large-scale deep-seated gravitational slope deformation of the sackung type was analysed, extended in low to medium grade metamorphics, represented essentially by ophiolite-bearing phyllite with metacalcarenite and quartzite banks, and by porphyroids. In this case the analysis proved to be of great help in determining the periods of the maximum instability between 1860 and 1895 and, through to a lesser extent, to find out the possible triggering causes of the slope mass movement. As a result, 80% of the strong earthquake dates (at least the 7th degree of Mercalli-Cancani-Sieberg (MCS) seismic intensity scale) corresponded with growth anomalies found in sampled trees while the correlation between rainfall and mass movement activity was weak (57%). This is probably due to the fact that the relationship, as the ca. 30 m deep, slow-moving sackung-type deformation should respond to much longer periods of increased rainfall; on the other hand, the continuous creeping of the sackung might irregularly interfere with the movement of shallower landslides, whose activities were reflected in the tree-ring growth records (Fantucci and Sorriso-Valvo 1997).

References

Aipassa M, Shimizu O (1988) Geomorphic process and natural revegetation on landslide scars in Teshio Experimental Forest. Hokkaido University. Res Bull of the College Experiment Forests, 45(3):691–716

Alestalo J (1971) Dendrochronological interpretation of geomorphic processes. Fennia 105:1–140

Bednarz Z (1986) An example of the application of the tree-ring chronology of the Dwarf Mountain Pine (Pinus Mugo var Mughus Zenari) for the dating of geomorphological processes in the Tatra Mts. Dendrochronologia 4:75–77

Begin C, Filion L (1988) Age of landslides along the Grande Rivière de la Baleine estuary, eastern coast of Hudson Bay, Québec Canada. Boreas 17:289–298

Bovis MJ, Jones P (1992) Holocene hystory of earthflow mass movements in South-Central Britisch Columbia: the influence of hydroclimatic changes. Can J Earth Sci 29:1746–1755

Braam RR, Weiss EEJ, Burroug PA (1987) Spatial and temporal analysis of mass movements using dendrochronology. Catena 14:573–584

Burn CR, Friele PA (1989) Geomorphology, vegetation succession, soil characteristics and permafrost in retrogressive thaw slumps near Mayo, Yukon Territory. Arctic 42:31–40

Carrara PE (1979) The determination of snow avalanche frequency through tree-ring analysis and historical records at Ophir, Colorado. Geological Society of America Bulletin, vol 90:773–780

Clague JJ, Souther JG (1982) The Dusty Creek landslide on Mount Cayely, British Columbia. Can J Earth Sci 19:524–539

Denneler B, Schweingruber FH (1993) Slow mass movement. A dendrogeomorphological study in Gams, Swiss Rhine valley. Dendrochronologia 11:55–67

Fantucci R (1997) La dendrogeomorfologia nello studio dei movimenti di versante: alcune recenti applicazioni in Italia (submitted to Geologia tecnica & ambientale)

Fantucci R, McCord A (1995) Reconstruction of landslide dynamic with dendrochronological methods. Dendrochronologia 13:33–48

Fantucci R, Sorriso Valvo M (1997) Dendrogeomorphological analysis of a landslide near Lago, Calabria Italy. (submitted to Geomorphology)

Fritts HC (1971) Dendroclimatology and dendroecology. Quaternary Research 1:419–449

Holmes RL (1983) Computer-assisted quality control in tree-ring dating and measurements. Tree-ring Bulletin 43:69–78

Hupp CR (1984) Dendrogeomorphic evidence of debris flow frequency and magnitude at Mount Shasta, California. Environ Geol Wat Sci 6(2):121–128

Jibson RW, Keefer D (1988) Landslide triggered by earthquakes in the Central Mississippi valley, Tennessee and Kentucky. US Geological Survey Professional Paper 1336–C

La Marche VC Jr (1963) Origin and geologic significance of buttress root of bristelcone pines, White Mountains, California. US Geological Survey Prof. Paper, 475–C, Art.98, C pp 148–149

Lionel E, Jackson Jr (1977) Dating and recurrence frequency of prehistoric mudflows near Big Sur, Monterey County, California. Jour. Research US Geol. Survey 5(1):17–32

Moss M L, Rosenfeld C R (1978) Morphology, mass wasting and forest ecology of a post-glacial re-entrant valley in the Niagara escarpment. Geografiska Annaler, 60A, pp 161–174

Orombelli G, Gnaccolini M (1980) La dendrocronologia come mezzo per la datazione di frane avvenute nel recente passato. Boll Soc Geol It 91:325–344

Parizek EJ, Woodruff JF (1957) Mass wasting and the deformation of trees. Am J Sci 25:63–70

Porter SC, Orombelli G (1980) Catastrophic rockfall of September 12, 1717 on the Italian flank of the Mont Blanc massif. Z. Geomorph.N.F. 24(2):200–218

Shroder JF (1978) Dendrogeomorphological analysis of mass movement on Table Cliffs Plateau, Utah. Quaternary Research 9:168–185

Shroder JF (1980) Dendrogeomorphology: review and new techniques in tree-ring dating. Progress in Physical Geography 4:161–185

Schweingruber FH, Eckstein D, Serre-Bachet F, Bräker OU (1990) Identification, presentation and interpretation of event years and pointer years in dendrochronology. Dendrochronologia 8:9–38

Stokes MA, Smiley TL (1968) An introduction to tree-ring dating. The University of Chicago Press, Chicago

Strunk H (1992) Reconstructing debris flow frequency in the southern Alps back to AD 1500 using dendrogeomorphologycal analysis. Erosion, Debris flows and environment in mountain regions (Proceedings of the Chengdu Symposium on Geomorphology, July 1992). IAHS Publ 209:299–307

Terasme J (1975) Dating of landslide in the Ottawa river valley by dendrochronology – a brief comment. Mass wasting Proceedings, 4th Guelph Symposium on Geomorphology, pp 153–158

Van Asch Th WJ, Van Steijn H (1991) Temporal patterns of mass movements in the French Alps. Catena 18:515–527

Chapter 6

Rainfall and Flow Forecasting Using Weather Radar

R.J. Moore

6.1
Introduction

Weather radar networks operated by national meteorological agencies are well serviced in the primary meteorological requirement of daily weather reporting and forecasting. Visual images of the spatial extent and propagation of storms provided by radar are now a familiar feature of such reporting on television. However, the need for quantitative estimates of rainfall to support applications in hydrology and water resources, especially flood forecasting, has not been so well serviced. Hydrologists and meteorologists have sought to improve the reliability of weather radar and to develop "radar hydrology" products characterised by greater resolution in space and time and improved quantitative accuracy.

Activity by this group of radar hydrologists is recorded in the proceedings to the symposium "Weather Radar and Flood Forecasting" at the University of Lancaster, UK, in 1985 (Collinge and Kirby 1987) and to the two symposia on "Hydrological Applications of Weather Radar", convened at the University of Salford, UK, in 1989 (Cluckie and Collier 1991) and at the University of Hannover, Germany, in 1992. The European Radar Hydrology Group, sponsored by the CEC, and now incorporating membership from 9 countries, also reported on their work at the workshop "Advances in Radar Hydrology" held at Lisbon, Portugal, in 1991.

The aim of this paper is to review developments in radar hydrology concerned with rainfall and flow forecasting. In scope, this review is restricted since it does not deal with advances in radar rainfall measurement that have led, in some instances, to the greater precision demanded by hydrologists. Reviews of radar rainfall measurement for hydrological application are provided by Joss and Waldvogel (1990) and Moore (1989). The use of radar for rainfall forecasting is considered first. Radar rainfall forecasts can provide the basis of an initial alert of possible flooding or form the input to flow forecasting models. The second part of the paper discusses the use of radar rainfall in flow forecasting, especially for flood warning and control. Applications of radar for storm hazard assessment, urban stormwater management and pollution hazard assessment are not considered.

6.2
Short-Term Precipitation Forecasting

6.2.1
Introduction

Reviews of radar-based short-term precipitation forecasting methods are available in Browning and Collier (1989) and Collier (1989). The replaying of weather radar images provides a strong visual impression of storm movement, raising the prospect of inferring the storm speed and direction and using this as the basis of forecasting. The majority of methods in operational use are based on an underlying advection model and differ in detail in terms of how the advection velocity is inferred and how the rainfall forecasts are subsequently generated.

6.2.2
Field Advection Methods

Arguably the most popular method is to take two consecutive radar images (typically 15 min apart) and to displace the first so as to get the best level of correspondence with the second. The optimal displacement provides an estimate of the storm speed and direction. This is then used to advect the second (most recent image) to form forecast images at future times. In early presentations of the method it was customary to use correlation as the measure of correspondence and consequently is often called the "correlation method". However, other correspondence measures are clearly possible and in general correlation as a measure is computationally expensive and not advised. Simpler measures such as those based on the sum of squared errors are usually more appropriate. Methods that initially reduce the field to a binary field, with ones assigned to radar pixels above some threshold rainfall intensity and zeroes below are not intrinsically different but result in merely another correspondence measure. It is probably best to refer to all such methods as *field advection methods* and to further classify them as of binary or real value type. Advection methods can also be classified in terms of whether they form the correspondence measure over a single region and advect this as a single entity or if they subdivide the region and infer localised velocities; in the latter case this can be seen to accommodate rotation of the storm in a possibly piecewise linear fashion.

An example of the operational implementation of the field advection method is provided by the IH Local Rainfall Forecasting System (Moore et al. 1991). This was one of two distinct developments aimed at making weather radar more useful for hydrological application. Its precursor was the IH Local Calibration System which combined radar and raingauge data in real-time to produce a rainfall field estimate that was, on average, more accurate than estimates based on each data source in isolation. The forecasting study also produced algorithms to pre-process radar data to automatically suppress anaprop and clutter. These radar preprocessing, calibration and forecasting procedures are now available as an integrated software product called HYRAD, an acronym for HYDrological RADar system (Moore 1993, 1994; Moore et al. 1994). The kernel to the system comprises the above procedures together with a post-processor to derive catchment average rainfall time series data (utilising digitised catchment boundary data) and an interface to the IH River Flow Forecasting System (Moore 1993). HYRAD incorporates real-time radar reception software, a radar data archiving facility and a Windows 3.1 radar display system incorporating hydrologically appropriate overlay information. The latter provides for client-server interaction between PCs and a host VAX or UNIX running the kernel, reception and archiving software. Figure 6.1 illustrates the system architecture of HYRAD together with an example weather radar display.

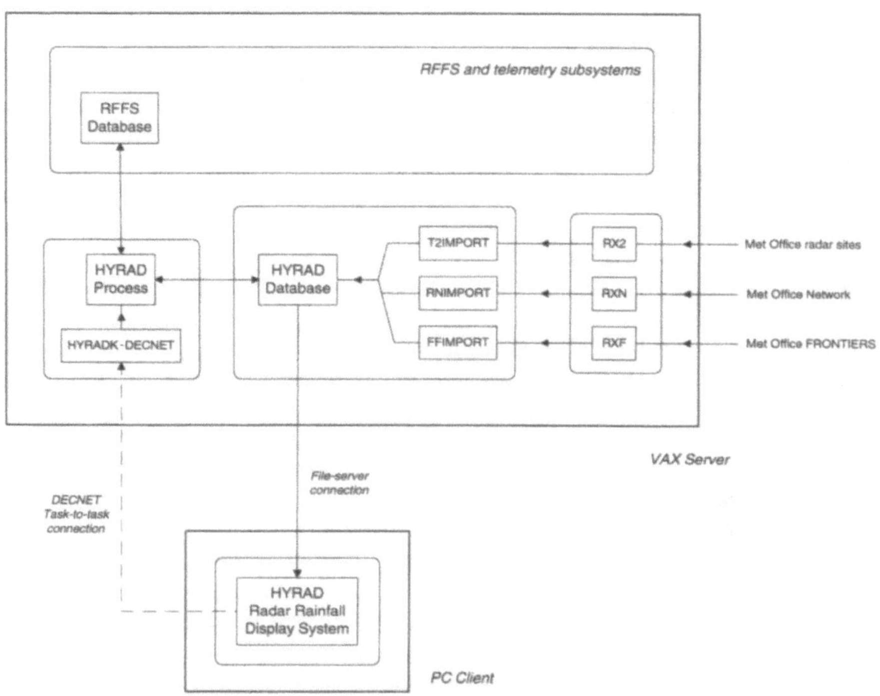

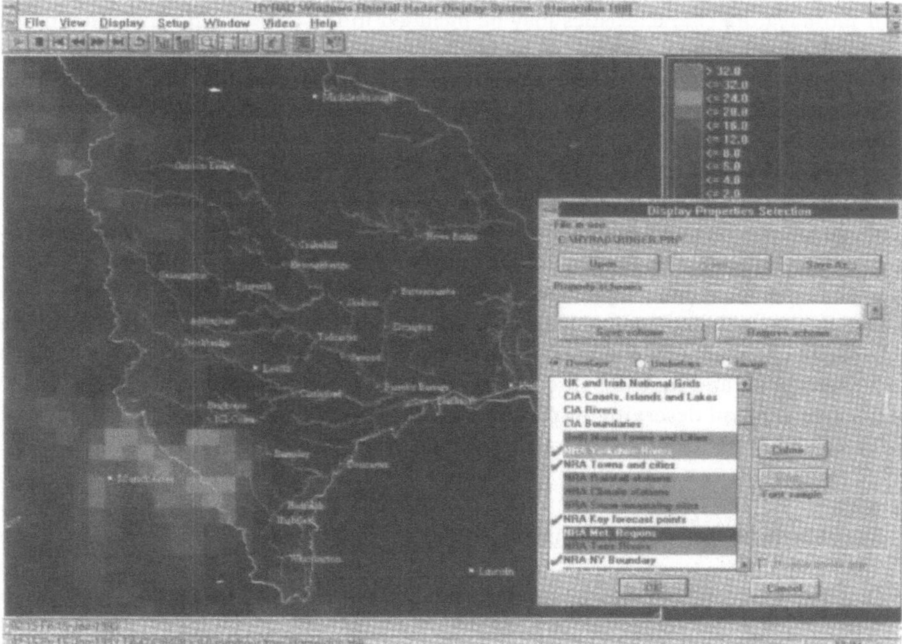

Fig. 6.1. The HYRAD system: system architecture and radar display

6.2.3
Feature Tracking Methods

A significantly different approach to radar rainfall forecasting is based on first delineating features in the rainfall field and attempting to track these features over successive radar images. Each feature is characterised by its own advection velocity in this approach. The main problem with the method is to reliably identify the same feature on successive images and to cope with the appearance of new features, the disappearance of old ones and the merging, fragmentation and change in size of existing ones.

A delineated feature is generally characterised by one or more quantitative *feature descriptors* such as the centre of gravity, mean intensity or size. These are compared with corresponding values in an adjacent radar image to identify corresponding features, the velocity of motion calculated and used as the basis of forecasting by extrapolation. *Feature delineation* may be achieved using an intensity threshold, a connectivity rule linking neighbouring pixels into a single feature or a clustering rule working outwards from a pixel with more intense rainfall. Clearly hybrid forms of these can be developed. Choices for feature characteristics include: size (number of pixels occupied by feature), mass (the average rainfall volume over the pixels), average intensity (mass divided by size), intensity variance, centroid, diameter, elongation and orientation and moments of inertia. The size and centroid of the feature are most commonly used.

Having chosen an appropriate set of feature characteristics the next step is to define a *matching rule* to establish a correspondence between features on successive images. The matching criterion might incorporate an assumption of a single, global, velocity vector or allow each feature to be unconstrained with regard to its velocity. One or more feature characteristics may be included in the chosen objective function for establishing a feature match.

The last step of *forecast construction* is generally based on similar principles assumed in the matching step, either assuming global or local velocities in advecting the features forward in time.

One example of such a feature tracking technique, in use for hydrological application, is the SCOUT II.o method (Einfalt et al. 1990). This includes procedures to recognise splitting and merging of features and to avoid feature fragmentation at the image boundary. Forecasts are used to support the real-time control of the sewer network in the Seine-Saint-Denis county near Paris, France. Control is in the form of two gate-controlled retention basins which are operated to minimise storm overflows in the sewer network whilst preserving water quality for recreational purposes in one of the reservoirs and limiting gate operations. Two recent developments of the feature tracking method deserving of mention are the contour method of Chen and Kavvas (1992) and the TITAN system for thunderstorm tracking of Dixon and Wiener (1993).

6.2.4
Man/Machine Interface Methods

Dissatisfaction with the fully automatic methods of rainfall forecasting, based on the inference of an advection velocity, led the UK Meteorological Office to investigate systems that introduced human judgement into the forecasting process (Carpenter and Browning 1984; Conway 1987). The concept was to combine radar and Meteosat display imagery with an experienced forecaster and to develop software tools to support the forecaster in using the radar imagery to make rainfall forecasts. The resulting man/machine rainfall forecasting system was called FRONTIERS.

A light pen touching a VDU screen is used by the operator to clean up an image, removing any anomalous pixels and truncating bright band affected pixels. The pen may also be used to point to features on successive images as a means of inferring storm velocities for an operator delineated set of storm clusters; tuning and verification of the inferred velocities is achieved through a Lagrangian replay technique in which a feature remains static for a correct choice of velocity. Additional techniques can be invoked to adjust for orographic enhancement effects on the radar field.

In the most recent implementation of FRONTIERS (Brown et al. 1994) movement of the radar image to construct forecasts is achieved using the 850 and 700 mb steering winds obtained from the mesoscale meteorological model. The system is operated 24 h a day and forecasts are constructed every half hour for 6 h ahead at intervals of 15 min on a 5 km grid for the whole of the UK. As a consequence the image used as the basis of forecasting incurs a transmission delay of the order of 40 min before it is available at flood warning centres; 30 min is lost through the time allowed for man/machine interaction and the remainder in the two-way transmission of data. This delay is particularly critical to hydrological applications, such as the provision of flash-flood warning to metropolitan areas.

Ongoing developments aim to speed up the forecasting procedure through greater automation, including consideration of expert system techniques (Conway 1989) and object programming methods. Progress on the new automated scheme called NIMROD has been reviewed by Ryall and Conway (1993): it is planned that it will replace FRONTIERS in 1995.

An operational assessment of the FRONTIERS system has recently been undertaken by the UK Meteorological Office in conjunction with the National Rivers Authority

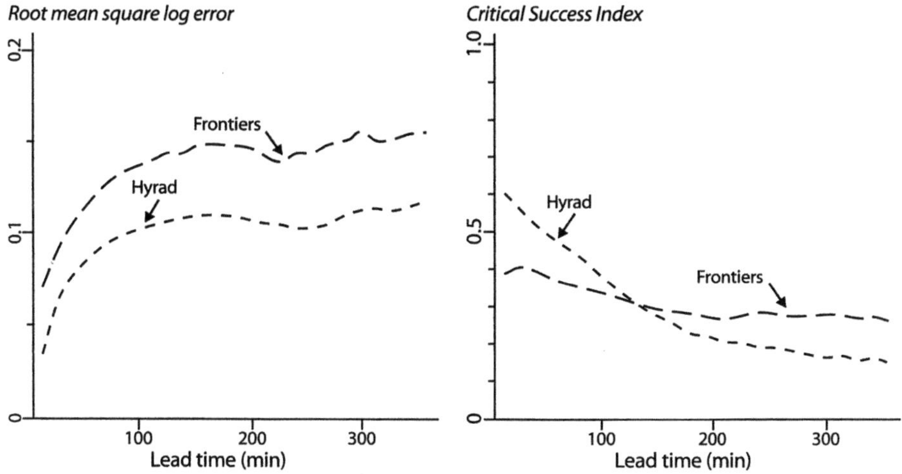

Fig. 6.2. Comparison of the performance of HYRAD and Frontiers forecasts of 15 minute rainfall totals (average over 14 events) over the Thames basin, UK

(NRA) in their Thames and north-west regions (NRA-Met Office 1992). The NRA also commissioned the Institute of Hydrology (IH) and the University of Salford to carry out independent assessments in these two regions, respectively. The IH assessment over London included consideration of their own advection forecasting method on which HYRAD is based. In terms of mean square error criterion the latter outperformed FRONTIERS at all lead times; using a pattern-based measure of performance, the critical success index, FRONTIERS provided some improvement at lead times beyond 2 h (Fig. 6.2). In many respects the two techniques should be seen as complementary. The HYRAD method provides high resolution (2 km square forecasts updated every 15 min) short-term regional forecasts, experiencing exhaustion effects beyond about 2 h on account of its use of a single radar. The coarser FRONTIERS product (5 km, 30 min update) provides national coverage from its use of the radar network data and full forecasts out to 6 h.

6.2.5
Numerical Weather Prediction Models

In contrast to the above inferential techniques for radar rainfall forecasting, some use is beginning to be made of radar in numerical weather prediction models. 4-D data assimilation procedures are being developed to accommodate radar, and other meteorological data, into these models to improve short-term forecasting of precipitation. The problem is perhaps better known to hydrologists as the state-updating problem where an adjustment to model states is made bearing in mind the relative uncertainties of the model state and the observations; in this case it might also be necessary to incorporate interpolation functions within the procedure and to ensure that the model remains numerically stable. Unfortunately, the current generation of mesoscale model (Golding 1990) represent storm dynamics on too coarse a grid to meet the hydrologists' needs – 16 km in the case of the UK Met. Office Mesoscale Model – with highly parameterised representations, for example, of convective cloud systems. Even if mesoscale models proved highly reliable at forecasting rainfall amounts their coarse grid scale would fail to meet the hydrological requirement for catchment scale rainfall, at least for the more spatially variable convective storms or for the smaller catchments of particular interest in urban areas. Disaggregation of mesoscale model rainfall to smaller scales provides one possible way forward. An interesting alternative is to pursue the physics-based approach at a smaller scale and a higher level of process representation. An extreme approach would be to employ one of the number of detailed cloud models (Smolarkiewicz and Clark 1985) currently being developed to support phenomenological studies of precipitation formation. However, in the development of an operational rainfall forecasting model it is important that the complexity of the model formulation is commensurate with the observation data available for assimilation into the model. This argument has led researchers to consider simple cloud model parameterisations which encapsulate the dominant dynamics affecting precipitation formation.

6.2.6
Simple Cloud Models

The coarse scale of rainfall forecasts made using mesoscale meteorological models is clearly not ideally suited to hydrological application. However, simple advection and feature tracking methods are known to be deficient under conditions of growth and decay, characteristic particularly of convective storm activity which can give rise to important flood-producing storms. These deficiencies have led to research into a hybrid approach in which a simple water balance model of a cloud column is advected to provide finer-resolution forecasts of rainfall fields (Kessler 1969; Georgakakos and Bras 1984; Lee and Georgakakos 1991). Recent progress has included the use of weather radar and satellite data into the formulation of these models (Seo and Smith 1992; French and Krajewski 1994). Weather radar data for multiple elevation scans are used to estimate the vertical integrated liquid content of the column which is used in the re-initialisation of the model's water balance in real-time. Satellite imagery in the infra-red waveband is used to estimate the temperature at the top of the column. Radar may also be used to infer the advection velocity, as outlined above, or alternatively be derived from a mesoscale meteorological model.

Clearly, the use of a simple water balance of the atmospheric column, with frequent state updating, in conjunction with radar inference of its advective movement, offers an attractive way forward for forecasting the development of storm systems. Since the model is essentially a simple dynamic water balance of the lower atmosphere it has much in common with the conceptual catchment water balance models familiar to hydrologists working in the land phase of the hydrological cycle. Using a rainfall model parameterisation commensurate with that of a catchment model clearly has much to commend it for the purposes of storm and flood forecasting. Depending on the resolution of the radar data the model is capable of representing rainfall fields for 1, 2 or 5 km grids and for time intervals of from 5–15 min, for example. It therefore, meets the hydrologist's requirement for forecasts at this fine resolution in space and time. An outline of one form of simple cloud model is given below.

An approximate water balance model of a cloud column can be represented as

$$\frac{dV_L}{dt} = w_o\,(Q_b - Q_t) - (W_t - w_o)\,M_b \quad . \tag{6.1}$$

This expresses the rate of change of the vertically integrated liquid water content of the cloud, dV_L / dt, as the difference between the inflow of water between the cloud base and cloud top and the outflow of water, in the form of rainfall from the cloud base. The inflow, $w_o(Q_b - Q_t)$, is the product of the updraft velocity (w_o) and the difference in the saturation water vapour density at the cloud bottom and top, Q_b and Q_t respectively. The outflow, $(W_t - w_o)M_b$, is the rainfall from the base of the cloud given by the product of the velocity of rainfall at the cloud base, $v_b = W_t - w_o$, and the rainwater content at the cloud base, M_b. Here W_t is the terminal free fall velocity of rain water in still air. For the assumptions underlying the derivation of this simplified dynamic cloud model the reader is referred to Georgakakos and Bras (1984), Seo and Smith (1992) and French and Krajewski (1994). This model may be used to represent the water mass balance of an atmospheric column defined as the grid square column above a weather radar pixel. In turn a two-dimensional model may be invoked by considering the model dynamics applied separately to each grid column across the weather radar grid. Advection is introduced by considering the model in a Lagrangian frame of reference, equivalent to moving the grid over the area of interest according to the radar-inferred storm trajectory.

Evaluation of the terms in Eq. 6.1 is possible given measurements at surface observing stations of temperature, humidity and pressure, from a satellite of infra-red derived cloud top temperature and from a weather radar of radar reflectivity at different elevations. The surface and satellite observations together with parcel theory (Wallace and Hobbs 1977) allow the pressure at the cloud top and in turn the saturation water vapour densities at the base and top of the cloud to be established. The updraft velocity w_o is considered to be controlled by the buoyancy of the air represented as an empirical function of temperature for which an effective value for the cloud is calculated at a prescribed level, again using parcel theory. Once w_o is known assumptions involving the drop size distribution and the terminal velocity of rain allow the velocity of rainfall at the cloud base (v_b) to be calculated. Finally, an expression for M_b in terms of V_L is required for the solution of Eq. 6.1. This is currently obtained by forming a regression relating $M_b H$ to V_L with the regression parameters re-estimated at each timestep; here H is the height of the cloud column. Multi-scan radar observations allow estimates of the vertically integrated liquid water content of the cloud column (V_L) to be derived. M_b is obtained from radar data as the lowest radar value in the vertical which is non-zero and H is the height range over which non-zero values occur. A simple deterministic dynamic prediction based on Eq. 6.1 is obtained by using the current radar-derived estimate of V_L at the forecast time origin and solving the nonlinear ODE over the forecast interval. For details of possible solution schemes see French and Krajewski (1994) and Bell et al. (1994). Forecasts of rainfall can be obtained if H is assumed constant (in a Lagrangian framework) over the forecast interval, along with surface temperature, humidity and pressure and cloud top temperature.

Preliminary results of the model applied to Southern England using data from the Wardon Hill weather radar in Dorset are reported by Bell et al. (1994). Six storm events from the period 8th to 12th June 1993 affected by a number of thunderstorms were forecast using the dynamic rainfall model. Whilst the model forecasts the development of frontal events with some success one hour ahead, it is less able to forecast the more rapid growth and decay of the two convective storms. One cause appears to be associated with the back-transformation of the vertically integrated liquid water content (V_L) into rain rate at the cloud base. This depends on the empirical equation linking V_L to rainfall rate and a drop size distribution parameter to which the results are sensitive. The only other parameter in the model, controlling the updraft velocity, also exerts a strong influence. Research is ongoing to improve the model in these and other areas under the NERC Special Topic HYREX (HYdrological Radar EXperiment) programme.

The general approach of using a simple dynamic rainfall model with frequent assimilation of real-time data from weather radar, satellite and surface weather stations has considerable appeal. It would also be straightforward to implement in practice either at regional flood forecasting centres or at a national meteorological forecasting centre. The National Rivers Authority is already taking a lead in the establishment of automatic weather stations linked via telemetry to regional centres and with the simple addition of an atmospheric pressure sensor would be well equipped to test a prototype operational system. The planned real-time dissemination of multi-scan radar data, in addition to the lowest scan, will make this possible. However, further work in a research context is required to develop and assess improved model formulations before operational implementation can be contemplated. Coping with the vagaries of radar data in a real-time implementation would clearly be an essential step towards providing robust forecasts. Use of full volume scan data from the new generation of UK Doppler weather radars should also bring a significant stimulus to progress in forecast accuracy from water balance rainfall models of this general type.

6.2.7
Limits of Predictability

In striving to improve the accuracy and precision of rainfall forecasts the natural limits imposed on predictability must be borne in mind. Each scale of motion in the earth's atmosphere is associated with an intrinsic finite range of predictability (Lorenz 1969). Whilst the largest scale motion may be predictable several weeks ahead and those at synoptic scales a few days in advance, the "cumulus-scale" motions of convective systems can be predicted for little more than an hour ahead. A revised estimate of 2 h is given by Lilly (1990) for motions with a length scale of 20 km, ignoring the effects of intermittency associated with convective storm activity. Even with detailed observations of the phenomenon in time and space supported by a commensurate high level of mechanistic model description these limits cannot be breached due to the chaotic nature of atmospheric motions (Lorenz 1993). However, the state-of-the-art of rainfall forecasting with real-time data assimilation has still much to progress before these limits can be said to have been approached. The utility to flood forecasting and warning of rainfall forecasts even half-an-hour ahead during convective storms, at spatial and temporal scales of 2 km and 15 min, makes research progress in this area particularly worthwhile.

6.3
Flow Forecasting

6.3.1
Rainfall Forecasts for Flood Forecasting

A study to investigate the utility of the radar-derived rainfall forecasts in the rainfall-runoff models used operationally for flood forecasting over London is reported in Moore et al. (1993). This work is somewhat of a rarity in providing a detailed assessment of radar rainfall measurements and forecasts, in an operational flood forecasting context, and utilising a variety of storm events, catchments and models.

A set of nine catchments in the Thames basin, ranging in size from 30–750 km^2, were used along with three rainfall-runoff models used operationally within the National Rivers Authority. A total of 30 storm events were employed, 13 for calibration and 17 for independent assessment. Flow forecasts were obtained, first assuming perfect foreknowledge of rain in the form of raingauge, uncalibrated radar and HYRAD calibrated radar estimates. Further flow forecasts were obtained emulating real-time conditions using rain forecasts obtained from HYRAD, from Frontiers and from a simple conditional Markov chain model used as a baseline. In addition to the usual R^2 evaluation statistic, skill scores were formulated which measure the success of forecasting the exceedence of flow thresholds. These skill scores are of particular relevance to the practical problem of when to issue warnings of differing severity to the public. It was found that use of radar rainfall forecasts consistently increases the probability of detection of an exceedence (warning), with recalibrated HYRAD forecasts performing best. When

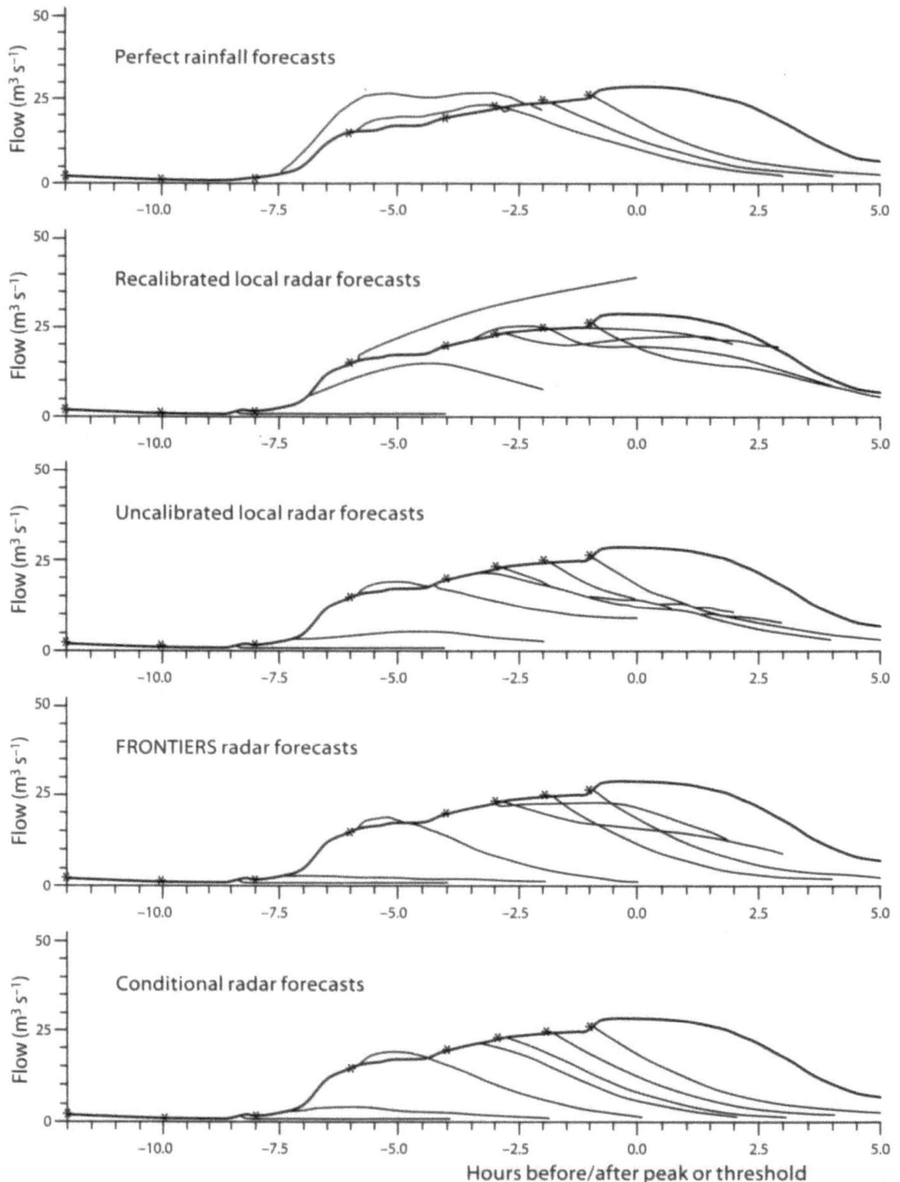

Fig. 6.3. Forecast hydrographs for the Silk Stream at Colindeep Lane made at different time origins before the peak at 05.00 a.m. 23rd September 1992. Flows computed using the PDM model using recalibrated data and forecast rainfall from the 5 sources indicated

event-pooled measures of performance are employed the value of rainfall forecasts, particularly for larger catchments, is perhaps less than is generally acknowledged. Persistence in the flow hydrograph for larger catchments increases the importance of model updating using observed flows, thereby reducing the value of rainfall forecasts, at least for smaller lead times. However, closer inspection of the skill score results demonstrate a more consistent improvement using radar rainfall forecasts, with a preference for HYRAD over Frontiers forecasts. Figure 6.3 illustrates a typical set of results obtained using five types of rainfall input to the PDM (Probability Distributed Moisture) model (Moore 1985; Institute of Hydrology 1992): perfect foreknowledge (locally recalibrated radar), locally recalibrated radar forecasts, uncalibrated radar forecasts, Frontiers forecasts and the simple Markov chain conditional raingauge forecasts. The best overall performance was obtained using the locally recalibrated radar forecasts obtained from HYRAD with the PDM rainfall-runoff model.

6.3.2
Distributed Radar Grid Square Flow Forecasting Models

The availability of radar rainfall distributed on a square grid has provided a stimulus to develop new rainfall-runoff models configured to make better use of data in this form. Previously, models relied on data from a sparse raingauge network and interpolation procedures were used to derive areal rainfall, often for complete river basins, for input to lumped rainfall-runoff models. More distributed formulations rarely demonstrated improved performance on account of the limitations imposed by the sparse point estimates of rainfall obtained from gauges. This constraint does not exist with weather radar data and the prospect for improvement utilising more distributed rainfall-runoff formulations clearly demanded investigation.

An early landmark in the development of grid-square rainfall-models was reported in the paper by Anderl, Attmannspacher and Schulz in 1976. They used data from a semi-automatic X-band weather radar at Hohenpeissenberg in Upper Bavaria along with flow data from two flow gauging stations in the vicinity draining basin areas of 34 and 56 km². An assessment of model performance demonstrated superior performance using weather radar data relative to the standard raingauge network (density of one gauge per 500 km²) and equal performance using a special dense raingauge network (one gauge per 25 km²). The rainfall-runoff model used was adapted for use with radar data from a previous model called Hyreun. Essentially the model is configured on a grid of 1 km square elements corresponding to the radar grid, and incorporates a runoff production function to generate excess runoff from each area element. Excess runoff from each grid is translated to the basin outlet, initially by pure advection, using a simple time delay depending on the distance between the area element and the outlet (that is using an isochrone principle), and then by storage routing using two linear reservoirs in parallel representing surface and subsurface translation routes. A limitation of the study was the X-band radar which only provided reliable measurements of rain when isolated single precipitation areas occurred, for which attenuation effects were weak. Only 3 or 4 such events were available for analysis.

More recent radar grid-square model formulations have been developed for real-time flood forecasting in Italy (Chander and Fattorelli 1991) and the UK (Moore and Bell 1992; Moore et al. 1994). Both models take as their starting point the translation of excess runoff from a grid square using a linear channel, equivalent to a time delay from each grid to the basin outlet. The discrete time formulation results in a double summation, the inner one representing the convolution of excess runoff for the grid with the time delay and the outer one adding the convoluted responses from each grid square to give the total basin runoff. The models differ primarily in the production function used, the Italian model utilising a Philip infiltration approach whereas the UK model, whilst incorporating an infiltration function, focuses on storage capacity controlled runoff. Results obtained for the Italian model on the 1 408 km² Bacciglione River, divided into 88 4 km square areas and using data from 11 raingauges, demonstrated superior performance to an equivalent lumped model when the flood-generating rainfall was not spatially uniform. However, no demonstration using radar data was made due to lack of available data at the time. The UK model, referred to as the IH Grid Model, is reviewed below.

The IH Grid Model provides a practical methodology for distributed rainfall-runoff modelling using grid-square weather radar data that is especially tailored for use in real-time flood forecasting. The model is configured so as to share the same grid as used by the weather radar, thereby exploiting to the full these distributed rainfall estimates. An intrinsic problem associated with models configured on a square grid is the potentially large number of model parameters which can be involved. If a different set of parameters is used for each grid square, the total number of parameters can become larger, even for basins of modest size, and there may be strong dependencies between sets. This problem of "over-parameterisation" is avoided by using measurements from a contour map or digital terrain model (DTM) of the basin together with simple linkage functions. These functions allow many model variables to be prescribed through a small number of regional parameters which can be optimised to obtain a good model fit.

The Grid Model has two main components. First, a runoff production function based on a saturation excess principle is used to represent the absorption of rainfall, by the soil/vegetation assemblage over the grid square area, and the generation of lateral water transfers to fast and slow response pathways. The second component is responsible for the translation of water along these pathways. The key linkage function used to define runoff production is a relationship between slope, as measured from the DTM, and absorption capacity. In the simple form of model, termed the SGM, the average slope within a grid square is related to the total absorption capacity of the square. Two basin-wide parameters, defining the upper limits of slope and absorption capacity establish the relationship across the basin. A probability-distributed variant of this model, the PGM, accounts for the frequency of occurrence of different slopes within the square, and hence for varying absorption capacity and runoff production. Further variants considered include a formulation based on a topographic index control on saturation (Beven and Kirkby 1979; Beven and Wood 1983) and the use of Integrated Air Capacity soil survey data as a surrogate for absorption capacity (Boorman et al. 1991). Impervious areas may also be introduced through the identification of urban areas, using a land cover classification derived from Landsat satellite imagery.

The "linkage function" used to define the translation component of the model is based on a relationship between distance to the basin outlet, as inferred from the DTM flow paths, and time-of-travel to the basin outlet. A velocity model is introduced to establish the link between these two quantities. The basic velocity model assumes two path-dependent velocities, one characteristic of hillslope pathways and the other of river pathways. This allows isochrones-lines joining points of equal time-of-travel to the basin outlet – to be derived automatically using the DTM (Fig. 6.4). Translation of water through a zone between two adjacent isochrone lines is represented as a discrete kine-

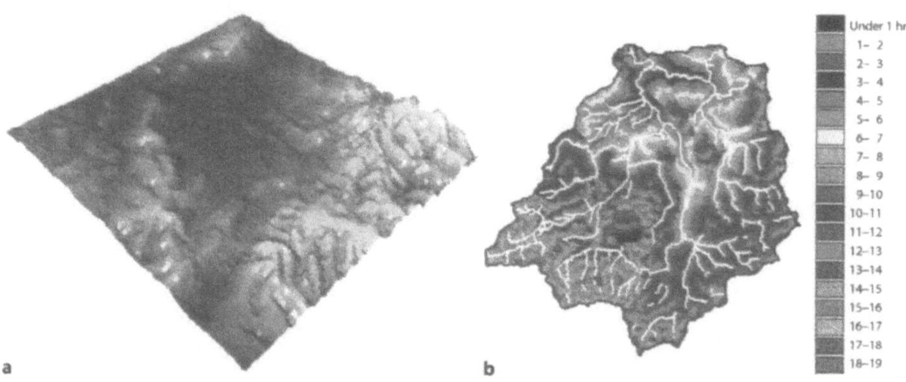

Fig. 6.4. Isochrone map derived from a digital terrain model: the Mole at Kinnersley Manor. **a** Digital Terrain Model; **b** Isochrone map

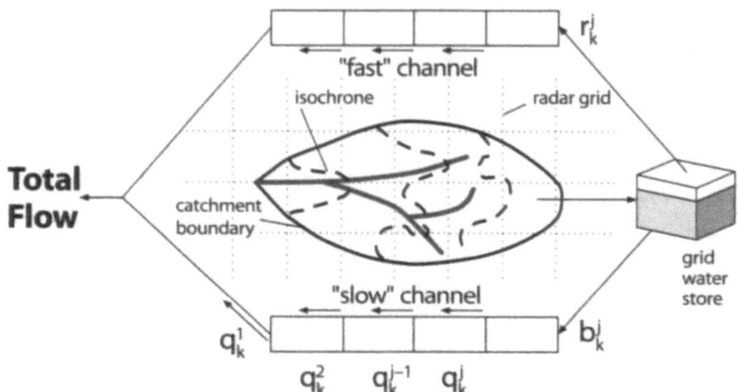

Fig. 6.5. The IH Grid Model

matic wave model, receiving lateral inflows from the areas of runoff production associated with the zone. A feature of this routing formulation is that, whilst the simple isochrone method is wholly advection-based, it also incorporates diffusion effects through its discrete space-time formulation. The approach also has significant computational speed advantages over schemes that use a convolution operation.

In practice, when calculating isochrones, a "Catchment Definition" algorithm developed for use with the DTM is first used to define flow path distances to the catchment outlet. Transfer of DTM path information to the distributed model then allows hillslope and channel velocities to be estimated as part of the overall model calibration process, yielding optimised isochrones and improved forecast performance. Consideration of alternative velocity models, including a form which employs slope as part of a Chezy-Manning relation, failed to provide improved performance. The final form of translation component adopted employs two kinematic routing cascades operating in parallel, one acting as a pathway for saturation excess runoff and the other for soil water drainage to groundwater (Fig. 6.5).

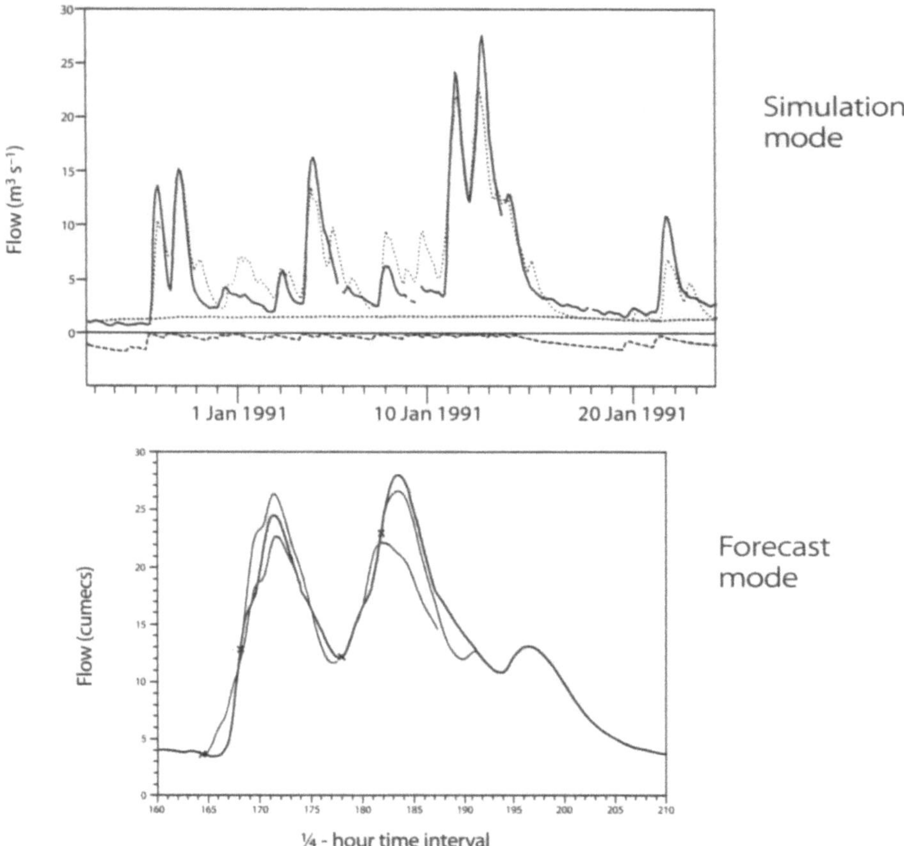

Fig. 6.6. Flow forecasts made using the IH Grid Model and weather radar data: the Mole at Kinnersley Manor

Four catchments, ranging in size from 100–275 km², were used for model development, calibration and assessment: one in the Thames basin, one in Northwest England, and the other two in South Wales. The results obtained show that when prevailing conditions make weather radar a good estimator of rainfall at ground level, significant model improvement is obtained by replacing data from a single raingauge by 2 km grid square radar data (Fig. 6.6). However, the possibilities of low-level enhancement of rainfall below the beam, and blockage effects, adversely affects the flow forecasts obtained for the more hilly catchments.

Whilst no one model variant consistently outperforms the rest, the Probability-distributed Grid Model (PGM) based on a power distribution of slope gave the best overall performance. Assessments using the Grid Model with areal average rainfall data as input and using a lumped rainfall-runoff model, the PDM, allowed the value of distributed data and a distributed model structure to be tested. The case for using either distributed data or a distributed model structure for flood forecasting is not overwhelming, at least when based on the storm periods considered for model assessment which are dominated by more widespread low pressure storm tracks. However, there is evidence that even for these storms the use of distributed forms of data and model can improve the accuracy of flood forecasts for larger catchments.

A broad conjecture, in part supported by the results of the assessment, is that a lumped rainfall-runoff model should suffice for widespread low pressure storms, particularly if the catchment is not hydrologically varied or too large. However, for storm structures whose magnitude is less than that of the basin for which flood forecasts are required, there is value in adopting a distributed approach. This is likely to be the case for convective storms which were absent from the storms used for assessment. A further potential advantage of distributed models is in their transferability to forecast flows for an ungauged catchment. Both good and bad results have been obtained and further work is needed to achieve a more resilient transfer.

For forecast updating in real time, a new state-updating form of the Grid Model has been developed and assessed against an error prediction technique. The method of state updating applies a smoothly varying adjustment to the water contents of the cascade of kinematic routing stores, the adjustment decreasing upstream, so as to achieve better agreement between modelled and observed flow at the basin outlet. This method of updating proves better than the persistence-based error predictor approach for the faster responding Welsh catchments although there is little difference overall. A particularly important practical result of the assessment is the excellent performance of both the lumped PDM and distributed Grid Model when applied to the Rhondda and Cynon catchments in South Wales. Whilst the forecasts are obtained assuming perfect foreknowledge of future rainfall, they do seem to indicate the significant potential value of using rainfall-runoff models as a basis for flood warning. Indeed, forecasts as good as these are rarely obtained from other UK catchments served by a single raingauge.

6.4
The Way Forward

There are many open questions concerning the way forward in the use of weather radar for rainfall and flow forecasting. In rainfall forecasting at smaller scales, do we follow a path of developing improved cloud physics parameterisations of increasing complexity? Experience with physics-based distributed catchment models in the hydrological sciences suggests that if forecasting is the goal, and not understanding, this may not be the right way. Simpler conceptual models which capture the dominant behaviours are often difficult to out-perform when calibrated to observed data. When the prospect of data assimilation, or updating, is possible, as it is in real-time forecasting, then the case for simpler conceptual models is arguably even stronger. Both these statements point towards a more interactive approach to conceptual rainfall model formulation in which atmospheric observation data are used to develop model structures appropriate to the scale of interest, to calibrate model parameters and to assess model performance.

Progress will be made as a result of real-time access to more atmospheric data. Establishment of an integrated network of telemetering automatic weather stations in the UK would seem a sensible way forward, and serve a variety of purposes. The NRA is taking regional initiatives in this regard but without atmospheric pressure sensors.

Use of full volume scan and Doppler wind data from the UK's new generation of radars provides further opportunities as does the prospect in the future of access to wind profiler data and improved satellite imagery. There remain problems associated with the measurement of rainfall by radar, particularly in conditions of bright band and orographic enhancement. Some of these problems may be resolved by considering rainfall measurement and forecasting as an integrated model-based problem tackled using an appropriate physical-conceptual representation with emphasis on the vertical dimension. Proper attention to appropriate data assimilation techniques which make best use of these real-time data sources should bring worthwhile dividends in improved rainfall forecast accuracy.

With regard to catchment models for flood forecasting it is still difficult in many situations to improve upon lumped conceptual models. Progress on distributed models will depend partly on improved radar measurement of rainfall, particularly in more hilly regions. Despite present difficulties, distributed models of simple conceptual form deserve continued attention as a means of forecasting for ungauged catchments and for coping with the variable catchment flood response observed when the storm scale is significantly less than the catchment scale.

Adopting an ensemble approach to flood forecasting, flood risk assessment and flood warning decision making in which alternative rainfall scenarios are employed has much to commend it and deserves further investigation. The Institute of Hydrology's River Flow Forecasting System already incorporates logical "What if?" switches which allow the user to "switch out" radar estimates of catchment rainfall in favour of raingauge estimates, when the latter are known to be unreliable, or when the sensitivity of flood forecasts to the alternative rainfall inputs is of interest. A more conventional ensemble flood forecast might derive from the use of ensemble rainfall forecasts as inputs to a catchment model. This might take the form of a dynamic rainfall model with radar data assimilation, whose initial conditions are stochastically perturbed, possibly exposing chaotic deterministic variability in some models beyond the limits of predictability.

Acknowledgements

The strategic support of the UK Ministry of Agriculture, Fisheries and Food Flood Protection Commission in this work is acknowledged along with the Commission for the European Communities under the EPOCH project "Weather radar and storm and flood hazard" and the Environment Programme project "Storms, floods and radar hydrology". Funding by the National Rivers Authority (now the Environment Agency) and by NERC under the HYREX Special Topic is also acknowledged. This paper was originally presented at the British Hydrological Society National Meeting on Hydrological Uses of Weather Radar, 9th January 1995, Institution of Civil Engineers, London.

References

Anderl B, Attmannspacher W, Schulz GA (1976) Accuracy of reservoir inflow forecasts based on radar rainfall measurements. Wat Resources Research 12(2):217–223

Bell VA, Carrington DS, Moore RJ (1994) Rainfall forecasting using a simple advected cloud model with weather radar, satellite infra-red and surface weather observations: An initial appraisal under UK conditions. Institute of Hydrology

Beven KJ, Kirkby MJ (1979) A physically based, variable contributing area model of basin hydrology. Hydrol Sci Bull 24(1):43–69

Beven KJ, Wood EF (1983) Catchment geomorphology and the dynamics of runoff contributing areas. J Hydrology 65:139–158

Boorman DB, Hollis J, Lilly A (1991) The production of the hydrology of soil types (HOST) data set. BHS 3rd National Symp., Southampton, 6.7–6.13, British Hydrological Society

Brown R, Newcomb PD, Cheung-Lee J, Ryall G (1994) Development and evaluation of the forecast step of the FRONTIERS short-term precipitation forecasting system. J Hydrol 158:79–105

Browning KA, Collier CG (1989) Nowcasting of precipitation systems. Rev Geoph 27(3):345–370

Carpenter KM, Browning KA (1984) Progress with a system for nowcasting rain. Preprint vol Nowcasting II, 3–7 September, Norrkoping Sweden, ESA SP–208, pp 427–432

Chander S, Fattorelli S (1991) Adaptive grid-square-based geometrically distributed flood-forecasting model. In: Cluckie ID, Collier CG (eds) Hydrological applications of weather radar, Ellis Horwood, pp 424–439

Chen Z, Kavvas ML (1992) An automated method for representing, tracking and forecasting rainfall fields of severe storms by Doppler weather radars. J Hydrol 132:179–200

Cluckie ID, Collier CG (eds) (1991) Hydrological applications of weather radar. Ellis Horwood, 644 pp

Collier CG (1989) Weather radar forecasting. In: Weather radar and the water industry: Opportunities for the 1990s. British Hydrological Society Occasional Paper N° 2, pp 35–55

Collinge VK, Kirby C (eds) (1987) Weather radar and flood forecasting. J Wiley, 296 pp

Conway BJ (1987) FRONTIERS: an operational system for nowcasting precipitation. Proc. Symp. on Mesoscale Analysis and Forecasting, Vancouver, 17–19 August 1987, Special Pub. No. ESA SP-282, European Space Agency, pp 233–238

Conway BJ (1989) Expert systems and weather forecasting. Met Mag 118:23–30

Dixon M, Wiener G (1993) TITAN: Thunderstorm identification, tracking, analysis and nowcasting – a radar-based methodology. J Atmos Oceanic Technol 10(6):785–797

Einfalt T, Denoeux T, Jacquet G (1990) A radar rainfall forecasting method designed for hydrological purposes. J of Hydrology 114:229–244

French MN, Krajewski WF (1994) A model for real-time quantitative rainfall forecasting using remote sensing: 1. Formulation. Wat Resourc Res 30(4):1075–1083

Georgakakos KP, Bras KL (1984) A hydrologically useful station precipitation model, Part I: Formulation. Wat Resourc Res 20(11):1597–1610

Golding BW (1990). The Meteorological Office mesoscale model. Meteorol Mag 119:81–86

Institute of Hydrology (1992) PDM: A generalised rainfall-runoff model for real-time use. Developers Training Course, National Rivers Authority River Flow Forecasting System, Version 1.0, Institute of Hydrology, 26 pp

Joss J, Waldvogel A (1990) Precipitation measurement and hydrology. In: Atlas D (ed) Radar in Meteorology. Battan Memorial and 40th Anniversary Radar. Meteorology Conference, pp 577–606

Kessler E (1969) On the distribution and continuity of water substance in atmospheric circulation, Meteorol Monogr 10, 86 pp

Lee TH, Georgakakos KP (1991) A stochastic-dynamical model for short-term quantitative rainfall prediction. Iowa Inst. Hydr. Res. Report N° 349, University of Iowa, 247 pp

Lilly DK (1990) Numerical prediction of thunderstorms – has its time come? Quart J Roy Met Soc 116:779–798

Lorenz EN (1969) The predictability of a flow which possesses many scales of motion. Tellus, XXI 3:289–307

Lorenz EN (1993) The Essence of Chaos, UCL Press, London, 227 pp

Moore RJ (1985) The probability-distributed principle and runoff prediction at point and basin scales, Hydrol Sci J 30(2):273–297

Moore RJ (1989) Weather radar measurement of precipitation for hydrological application. In: Weather radar and the water industry: Opportunities for the 1990s. British Hydrological Society Occasional Paper 2:24–34, Institute of Hydrology

Moore RJ (1993) Real-time flood forecasting systems: Perspectives and prospects. British-Hungarian Workshop on Flood Defence, Budapest, 6–10 September 1993, 51 pp

Moore RJ (1994) Integrated systems for the hydrometeorological forecasting of floods. In: Verri G (ed) Proceedings of the International Scientific Conference EUROPROTECH. Part I: Science and Technology for the Reduction of Natural Risks, , 6–8 May 1993, CSIM, Udine Italy, pp 121–137

Moore RJ, Bell V (1992) A grid-square rainfall-runoff model for use with weather radar data. Proc. Int. Workshop, 11–13 November 1991, Lisbon, Portugal, Commission for the European Communities, 9 pp

Moore RJ, Hotchkiss DS, Jones DA, Black KB (1991) London weather radar local rainfall forecasting study: Final report. Contract Report to NRA Thames Region, Institute of Hydrology, 124 pp

Moore RJ, Austin R, Carrington DS (1993) Evaluation of FRONTIERS and Local Radar Rainfall Forecasts for use in Flood Forecasting Models. R and D Note 225, Research Contractor: Institute of Hydrology, National Rivers Authority, 156 pp

Moore RJ, Bell V, Roberts GA, Morris DG (1994) Development of distributed flood forecasting models using weather radar and digital terrain data. R and D Note 252, Research Contractor: Institute of Hydrology, National Rivers Authority, 144 pp

Moore RJ, Jones DA, Black KB, Austin RM, Carrington DS, Tinnion M, Akhondi A (1994) RFFS and HYRAD: Integrated systems for rainfall and river flow forecasting in real-time and their application in Yorkshire. In: Analytical techniques for the development and operations planning of water resource and supply systems. BHS National Meeting, University of Newcastle, 16 November 1994, BHS Occasional Paper N° 4, British Hydrological Society, 12 pp

NRA-Met Office (1992) Evaluation of FRONTIERS accumulation forecasts in the NRA Thames and North West Regions. Final Report, UK Met. Office, 90 pp

Ryall G, Conway BJ (1993) Automated precipitation nowcasting at the UK Met Office. First European Conference on Applications of Meteorology, 27 September–1 October 1993, University of Oxford UK, 4 pp

Seo DJ, Smith JA (1992) Radar-based short-term rainfall prediction. J Hydrol 131:341–367

Smolarkiewicz PK, Clark TL (1985) Numerical simulation of the evolution of a three dimensional field of cumulus clouds. Part I: Model description, comparison with observations and sensitivity studies. J Atmos Sci 42:502–522

Wallace JM, Hobbs PV (1977) Atmospheric science. An introductory survey. Academic, San Diego, Calif, 467 pp

Chapter 7

Numerical Modelling Techniques as Predictive Tools of Ground Instability

S.C. Bandis

7.1
Introduction

The engineering assessment of the stability of natural and man-made slopes as influenced by natural or induced changes to their natural environment can be aided by the application of analytical and numerical techniques.

The concepts and approaches of modelling geotechnical problems differ from those in deterministic fields of engineering. A model (analytical or numerical) generally represents a realistic simplification, rather than an imitation of reality, since geology and associated mechanical interactions can never be known in detail sufficient enough to allow unambiguous representations.

The various techniques available today for the stability assessment of soil or rock slopes are distinguished into two broad classes:

a *Analytical (or closed-form) solutions.* These are deterministic techniques providing a single answer for a set of input. The method most widely used is that of limit equilibrium, whereby force or/and moment equilibrium conditions are examined on the basis of statics. The commonly derived output is the Factor of Safety (Graham 1984).
b *Numerical (or approximate) methods of deformation analysis* (e.g. Gudehus 1977). These include mainly the Finite Element, Finite Difference and Boundary Element Methods. Advanced techniques for analysing discontinuous behaviour are also available (Distinct Element Method, DEM).

   The results are obtained in the form of stress-deformation histories, induced stress and displacement fields and overstress states (permissible over induced stresses).

7.2
Slope Stability Evaluation

The stability of a slope depends on its ability to sustain load increases or changes in environmental conditions, which may affect the geomaterials mechanically or chemically.

The analysis of slope stability may be implemented at two distinct stages:

- *Pre-failure analysis.* Applied to assess safety in a global sense to ensure that the slope will perform as intended.
- *Post-failure analysis.* To provide consistent information for landslide events. Post-failure, otherwise termed back-analyses, should be responsive to the totality of processes which have led to failure.

The fundamental requirements for a meaningful analytical or numerical modelling exercise should include the following steps of data collection/evaluation:

- Site characterization (geological and hydrogeological conditions);
- Ground water conditions (pore pressure model);
- Geotechnical parameters (strength, deformability, permeability);
- Primary stability mechanisms (kinematics or potential failure modes).

Overall appreciation of the above allows a realistic conceptualization of a slope problem into a form amenable to mathematical solution.

7.3
Limit Equilibrium Methods

The fundamental concepts of the Limiting Equilibrium Method for slope stability analysis are the following (Morgenstern 1995):

- Slip mechanism results in slope failure.
- Resisting forces required to equilibrate disturbing mechanism are found from static solution.
- The shear resistance required for equilibrium is compared with available shear strength in terms of the Factor of Safety.
- The mechanism corresponding to the lowest Factor of Safety is found by iteration.

There are several methods available, all of which are based on the above fundamental concepts but differ in the implementation. Of the best known are the following:

- *Ordinary Method of Slices:*
    - Suitable for analysis of circular slip;
    - Based on moment equilibrium;
    - Neglects force equilibrium.
- *Bishop's Modified Method:*
    - Suitable for circular slip;
    - Based on vertical force and moment equilibrium;
    - Neglects horizontal force equilibrium.
- *Force Equilibrium Methods:*
    - Suitable for any shape of failure surface;
    - Based on force equilibrium;
    - Neglects moment equilibrium.
- *Jambu's and Morgenstern-Price's Methods:*
    - Suitable for any shape of failure surface;
    - Based on force and moment equilibrium;
    - Transferred lateral (side) force locations can be varied.
- *Spencer's Method:*
    - Suitable for any shape of slip surface;
    - Based on force and moment equilibrium;
    - Sideforces are taken parallel to slip surface.

In reviewing these available methods one may pose the following questions:

- Which of these methods are accurate?
- Which are inaccurate?
- Which is the reference for assessing the accuracy or otherwise of the various methods?
- Which of the accurate methods can be applied most easily?

Prior to providing some general answers, it is appropriate to consider the common fundamental concept, i.e. the Factor of Safety (*SF*):

$$SF = \frac{shear\ strength\ of\ material}{shear\ stress\ required\ for\ equilibrium} \cdot$$

Limitations of the Limit Equilibrium Method are:

- The implicit assumptions of ductile stress-strain behaviour for the material.
- Most problems are statically indeterminate.
- *SF* is assumed to be constant along the slip surface.
- Computational accuracy may vary.

The following are some comments of relevance to the above limitations:

- The role of the *SF* is quite complex. One well-recognized role is to account for uncertainty and to act as a factor of ignorance with regard to the reliability of the input to the analysis (Morgenstern 1993).
    An additional (major) role is that it constitutes the empirical tool, whereby deformations are limited to tolerable amounts. Then the choice of a particular *SF* is influenced by the accumulated knowledge with a particular soil or rock mass.
- Stress-strain behaviour implicitly is inferred as DUCTILE.
    Unless strengths used in the analysis can be mobilized over a wide range of strains, there is no physical justification for inferring constant peak shear resistance along the full length of the slip surface.
- Computational Accuracy.
    - Ordinary Method of Slices.
        Inaccurate for effective stress analysis of flat slopes with high pore pressures, calculated *SF* generally very conservative.
    - Modified Bishop's generally O.K.
        Some numerical problems.
    - Force equilibrium methods affected by inclinations of side forces.
    - Methods satisfying all conditions of equilibrium are generally O.K.

Although it can be said that the Limit Equilibrium Methods of stability analysis are generally well developed and arithmetically sound, responsibility still rests with the analyst to select a suitable sliding mechanism and to identify appropriate shear strength parameters.

Table 7.1. Geotechnical engineering computer programs (Bond 1995)

Field of application	Name of program	Operating system	Developed by	Marketed by
Slope stability	SLIP	DOS	CADS	CADS
	Slope	DOS	Borin	Geosolve
	Slope	DOS	Oasys	Oasys/Integer
	Slope/W	Windows	Geo-slope International	Geo-slope International
	STABL5	DOS	Purdue University	
	STABLE	DOS	MZ Associates	MZ Associates
Retaining walls	FREW	DOS	Oasys	Oasys/Integer
	GRETA	DOS	Oasys	Oasys/Integer
	GWALL	DOS	Borin	Geosolve
	Kzero	DOS	Savoy Computing	Savoy Computing
	RETAIN	DOS	CADS	CADS
	ReWaRD	Windows	Geotechnical Consulting Group	British Steel
	SHEET	DOS	CADS	CADS
	STAWALL	DOS	Oasys	Oasys/Integer
	WALLAP	DOS	Borin	Geosolve
Reinforced slopes	NailSolver	DOS	Oxford Geotechnical Software	Oxford Geotechnical Software
	ReActiv	Windows	Geotechnical Consulting Group	Transport Research Laboratory
	Scanslope	DOS	Strata Technology	Strata Technology
	Slope	DOS	Borin	Geosolve
	TALREN	DOS	Terrasol	Terrasol
Pile design	ALP	DOS	Oasys	Oasys/Integer
	CAPWAP	DOS	Goble Rausche Likins	Goble Rausche Likins
	Cemset	DOS	Cementation	Cementation
	DEFPIG	DOS	Poulos	
	PC-MPILE	DOS	Rendolph/Mott MacDonald	Mott MacDonald
	PC-PGROUP	DOS	Banerjee & Driscoll/Mott MacDonald	Mott MacDonald
	PILSET	DOS	Oasys	Oasys/Integer
	Piglet	DOS	Randolph	University of Western Australia
	SIMPLE	DOS	Springman	Lynxvale
	SLAP	Speadsheet	Springman	Lynxvale
Settlement analysis	CLOG	DOS	Oasys	Oasys/Integer
	MINLIN	DOS	Oasys	Oasys/Integer
	ONEDIM	DOS	Dept of Transport	HMSO
	TWODIM	DOS	Dept of Transport	HMSO
	VDISP	DOS	Oasys	Oasys/Integer
Groundwater flow	CTRAN/W	Windows	Geo-Slope International	Geo-Slope International
	CVM	Windows	Oxford Geotechnica International	Oxford Geotechnica International
	SEEP	DOS	Oasys	Oasys/Integer
	SEEP/W	Windows	Geo-Slope International	Geo-Slope International
	SefCut	Windows	Oxford Geotechnica International	Oxford Geotechnica International
	SefDam	Windows	Oxford Geotechnica International	Oxford Geotechnica International
	SEFTRANS	DOS	Oxford Geotechnica International	Oxford Geotechnica International
	SefWeir	Windows	Oxford Geotechnica International	Oxford Geotechnica International
General analysis	CRISP-90	DOS	Cambridge University	(To be replaced by Sage CRISP)
	Sage CRISP	Windows	Cambridge Univ/Sage Engineering	Sage Engineering
	FLAC	DOS	Itasca	Mott MacDonald
	PLAXIS	DOS	PLAXIS BV	AA Balkema
	SAFE	DOS	Oasys	Oasys/Integer
	SIGMA/W	Windows	Geo-Slope International	Geo-Slope International

The use of computers for geotechnical design has become widespread over the past 10 years and has allowed more realistic analyses and more trial calculations than traditional hand methods, which generally leads to better designs.

However, the benefits obtained from the "computer revolution" are not without their costs. In particular, the opportunities for unsafe design through mistaken reliance on computer analysis are steadily increasing. These include:

- Users without adequate geotechnical engineering knowledge or training carrying out the analysis.
- Communication gaps between the designer, software developer and user.
- Programs used out of context.
- Inadequacies in the checking process.
- The user being unaware of limitations of the software.

Table 7.1 lists a number of commercially available software packages used in Geotechnical Engineering (mostly for soils).

7.4
Deformation Analyses

The advent of computers in geotechnical engineering has undoubtedly enabled significant developments in terms of analytical capabilities. The Finite Element Method is the first numerical method which has been applied to the study of complex geomechanical problems (Gudehus 1977).

Although numerical modelling techniques provide powerful tools for solving complex problems, there are many intrinsic uncertainties, which need to be taken into consideration in the process of model development and assessment of the results. Some key questions which should guide future research and code validations could be as follows:

1. How can one assess acceptable vs. unacceptable simplification when modelling a geotechnical problem, without influencing the results ?
2. Can deterministic analyses represent inherently stochastic systems of behaviour?
3. Are 2-D simulations of any use?
4. Is it necessary to reproduce the "correct" path of loading/unloading or will any path do that satisfies equilibrium, compatibility and strength criteria?
5. Are the boundaries taken care of properly in a numerical model of a infinite problem?
6. Is it possible to account for time-dependence?

These questions probably set the capabilities of modelling to the right perspective and underline the restrictive factors.

A key question concerning the reliability of the available numerical modelling techniques and corresponding software systems concerns their verification. Some thoughts on verification requirements are summarized below, where three levels A–C are perceived:

- *LEVEL A* verification should provide documentation:
  a Against concealed computational deficiencies, which might produce not readily recognizable errors in the results;

b Of the overall numerical efficiency of the software system (e.g. reproducibility of the results, sensitivity to input parameters, etc.);
c To verify the overall "sensibility" of the results obtained from the analyses and to check if the predictions on problems which can also be solved using traditional methods are quantitatively comparable.

- *LEVEL B* verification should involve comparisons with other Code(s) which is(are) longer established and more widely tested.
- *LEVEL C* verification should focus on Code validations against factual data from well-documented case studies.

7.5
Analytical Studies

7.5.1
Scope of Work

Extensive analyses were conducted on a relatively simple slope stability problem to evaluate the accuracy and, generally, the efficiency of several commercially available software, as well as to compare different methods. The slope stability analyses were of parametric nature and involved variations of the following factors:

a Slope angle: 30°, 45° and 60°.
b Material constants c and Φ. Combinations of c and Φ were tested to establish sets of parameters, which corresponded to limiting equilibrium. Φ was taken equal to 20°, 25° and 30° and c was varied according to slope angle from 10–220 kPa.
c The above analyses were carried out assuming dry and fully saturated conditions.
d The analyses were carried out according to several of the known limit equilibrium methods. Suitably adopted numerical models using the codes UDEC and FLAC were also developed and analyzed to compare predictions between analytical and numerical techniques.
e The influence of the third dimension was examined by using FLAC-3D.
f Specific analyses were conducted to compare results for predefined circular or multiplanar failure surfaces.

The results are presented in the form of summary tables of input and typical output plots. Explanatory notes are provided as appropriate.

7.5.2
Presentation of Results

An overview of all analytical cases can be obtained from Table 7.2, where the following are displayed:

- a = slope angle 30°, 45°, 60° for constant slope height of 50 m;
- D / W = dry/wet analytical cases;
- mod = analytical model;
- Φ = angles of frictional resistance (20°, 25°, 30°) which were combined with a required c for limiting equilibrium for the various slope geometries.

Table 7.2. Summary of analytical study in matrix form

a = Slope angle
D/W = Dry/wet (fully-saturated)
mod = Model N°

Software References

FLAC (Version 3.30)	Itasca Consulting Group, Inc. Minneapolis, USA
FLAC-3D 1.02	Itasca Consulting Group, Inc.
UDEC (Version 2.01)	Itasca Consulting Group, Inc.
SLOPE/W, v.2.07	Geoslope, Edmonton, Canada
SLOPE	Geoslope, UK
PLAXIS	

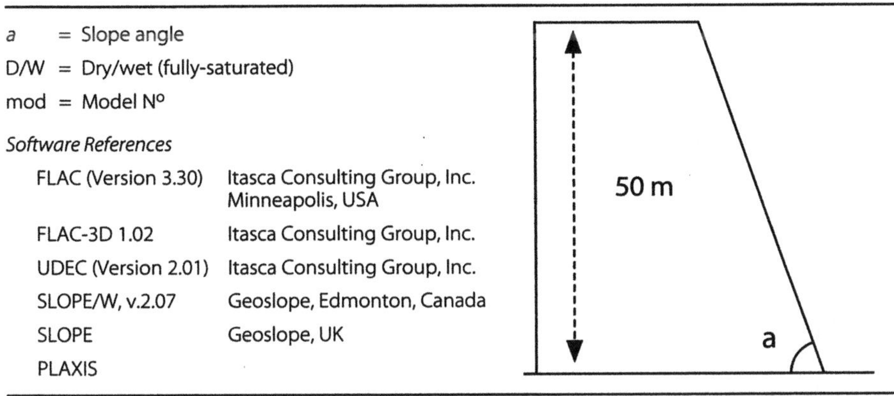

a	D/W	mod	Φ(°)	C	USED SOFTWARE
30	D	1.1	20	10 – 50	FLAC – SLOPE/W
30	D	1.2	25	10 – 50	FLAC – SLOPE/W
30	D	1.3	30	10 – 50	FLAC – SLOPE/W
30	W	1.1w	20	10 – 100	FLAC – SLOPE/W
30	W	1.2w	25	10 – 70	FLAC – SLOPE/W
30	W	1.3w	30	10 – 50	FLAC – SLOPE/W
45	D	2.1	20	10 – 100	FLAC – SLOPE/W
45	D	2.2	25	10 – 60	FLAC – SLOPE/W – PLAXIS
45	D	2.3	30	10 – 40	FLAC – SLOPE/W
45	W	2.1w	20	100 – 190	FLAC – SLOPE/W
45	W	2.2w	25	80 – 180	FLAC – SLOPE/W – PLAXIS
45	W	2.3w	30	100 – 160	FLAC – SLOPE/W
60	D	3.1	20	50 – 130	FLAC – FLAC3D – SLOPE/W
60	D	3.2	25	50 – 110	FLAC – SLOPE/W
60	D	3.3	30	10 – 100	FLAC – FLAC3D – SLOPE/W
60	W	3.1w	20	200 – 220	FLAC – SLOPE/W
60	W	3.2w	25	200 – 220	FLAC – SLOPE/W
60	W	3.3w	30	200 – 220	FLAC – SLOPE/W

The dry unit weight for all analytical cases was taken equal to 20 kN/m³ and c, Φ as discussed above. The numerical models were assigned the same c, Φ combinations (Mohr-Coulomb materials inferred throughout, and the elastic constants were set to $E = 30\,000$ kPa and $\nu = 0.3$.

The effect of the 3-D geometry and stress state were investigated using FLAC-3D and assuming planar, concave and convex slope geometries.

The influence of ground water on the calculated Safety Factor was examined with different programs and applying alternative methods.

Finally, a number a complex landslide problems were modelled using state-of-art numerical codes such as FLAC and UDEC.

7.6
Evaluation of Results

7.6.1
Comparisons Between Different Methods of Analysis

Different limiting equilibrium methods implemented in two software packages (Geoslope SLOPE/W and Geosolve SLOPE) were applied to the 50m slope with face inclinations 30°, 45° and 60°. The various methods are described in Table 7.3 and the results summarized in Table 7.4.

Table 7.3. Limiting equilibrium methods implemented in SLOPE/W by GEOSLOPE International and SLOPE by GEOSOLVE

Method	Force eq. D1[a]	D2[a]	Moment eq.	Assumptions
Ordinary	yes	no	yes	Interslice forces are neglected.
Bishop's Simplified	yes	no	yes	Resultant interslice forces are horizontal.
Janbu's Simplified	yes	yes	no	Interslice shear forces are account by an empirical correction factor.
Janbu's Generalized	yes	yes	[b]	Location of the interslice normal force is defined by an assumed line of thrust.
Spencer	yes	yes	yes	Resultant interslice forces are of constant slope throughout the sliding mass.
Morgenstern-Price	yes	yes	yes	Direction of resultant interslice forces is determined using an arbitrary function.
Corps of Engineers	yes	yes	yes	Direction of resultant interslice forces is equal to the average slope from the beginning to the end of the slip surface or parallel to the ground surface.
Lowe-karafiath	yes	yes	no	Direction of resultant interslice forces is equal to the average of the ground surface and the slope at the base of each slice.

[a] Any of two orthogonal directions can be selected for the summation of forces;
[b] Moment equilibrium is used to calculate interslice shear forces.

Table 7.4. Summary of results from comparative analytical studies. Numbers in columns are the calculated factors of safety for the combinations of c and Φ written under SLOPE 1, 2 and 3

| | | Geoslope SLOPE/W | | | | | | | | | | Geosolve SLOPE | | | | F.E.F. |
| | | Moment Equilibrium Factor | | | | Force Equilibrium Factor | | | | | | M.E.F. | | | | |
Phi	Coh	O	B	S	M-P	J	S	M-P	E1	E2	L-K	O	B	S	J(G)	J
SLOPE 1; a = 30																
A	10	0.918	0.959	0.958	0.958	0.913	0.959	0.959	0.961	0.965	0.959		0.945	0.942	0.942	
$\Phi = 20$	20	1.039	1.079	1.078	1.078	1.030	1.078	1.078	1.082	1.087	1.079		1.010	1.012	1.012	
	30											0.989				0.974
	40											1.061				1.043
B	0	0.949	0.967	0.966	0.949	0.949	0.958	0.967	0.949	0.967	0.965	0.990	1.120	1.117	1.117	0.987
$\Phi = 25$	10	1.142	1.190	1.189	1.189	1.137	1.188	1.188	1.193	1.193	1.188	1.062				1.050
C	0	1.176	1.197	1.196	1.196	1.175	1.196	1.196	1.197	1.197	1.195	1.226	1.386	1.384	1.383	1.222
$\Phi = 30$																
SLOPE 2; a = 45																
A	60	0.914	0.945	0.943	0.943	0.903	0.941	0.940	0.951	0.961	0.950	0.939	0.960	0.958	0.959	0.931
$\Phi = 20$	70	0.984	1.010	1.008	1.008	0.970	1.008	1.008	1.018	1.030	1.017	1.021	1.041	1.039	1.039	1.011
	80	1.048	1.074	1.072	1.072	1.028	1.073	1.072	1.084	1.097	1.083					1.092
B	40	0.882	0.925	0.921	0.920	0.874	0.921	0.920	0.930	0.935	0.922		0.932	0.929	0.929	
$\Phi = 25$	50	0.961	1.005	1.000	1.000	0.952	1.001	1.000	1.012	1.019	1.003	0.985	1.012	1.010	1.010	0.977
	60	1.039	1.084	1.079	1.079	1.030	1.082	1.081	1.095	1.100	1.086	1.066				1.057
C	20	0.856	0.898	0.892	0.892	0.852	0.887	0.885	0.895	0.900	0.891					
$\Phi = 30$	30	0.938	0.986	0.980	0.980	0.930	0.974	0.973	0.985	0.991	0.980	0.959	0.995	0.993	0.993	0.955
	40	1.018	1.070	1.064	1.064	1.009	1.057	1.056	1.070	1.077	1.065	1.041	1.076	1.073	1.073	1.035
SLOPE 3; a = 60																
A	90	0.896	0.894	0.896	0.895	0.899	0.901	0.904	0.913	0.917	0.908	0.982	0.987	0.986	0.986	0.982
$\Phi = 20$	100	0.978	0.957	0.958	0.957	0.962	0.963	0.966	1.010	0.981	0.970	1.057	1.061	1.061	1.061	1.058
	110	1.016	1.011	1.020	1.019	1.026	1.024	1.027	1.057	1.044	1.034					
B	80	0.918	0.925	0.926	0.907	0.907	0.927	0.930	0.935	0.944	0.939	0.993	1.000	0.999	0.999	0.992
$\Phi = 25$	90	0.987	0.988	0.989	0.977	0.977	0.990	0.990	1.009	1.015	1.004	1.068		1.074	1.074	1.067
	100	1.051	1.050	1.052	1.054	1.054	1.054	1.057	1.062	1.079	1.067					
C	70	0.934	0.957	0.957	0.956	0.927	0.951	0.953	0.964	0.968	0.961	1.011	1.022	1.020	1.020	1.009
$\Phi = 30$	80	1.004	1.025	1.025	1.025	0.991	1.022	1.025	1.025	1.039	1.030					
	90	1.073	1.088	1.089	1.088	1.060	1.089	1.092	1.109	1.107	1.098					

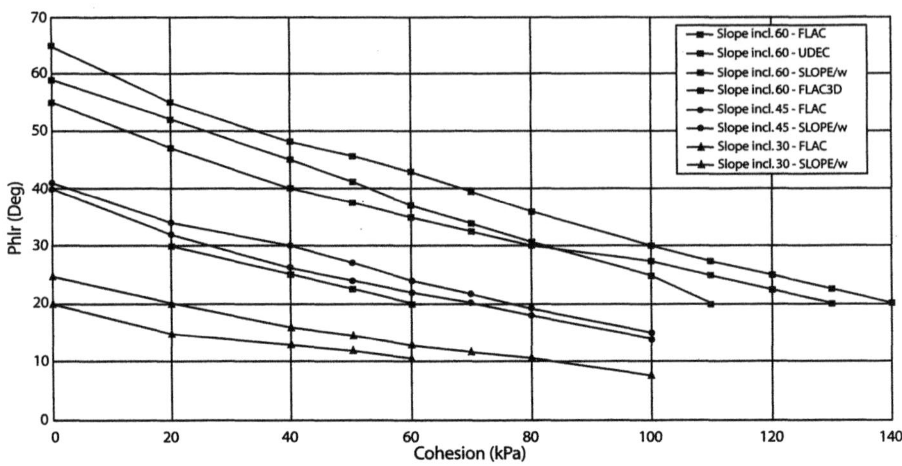

Fig. 7.1. Comparisons of predicted limiting equilibrium states of 30°, 45° and 60° slopes according to analytical solutions (Bishop) and numerical techniques

The comparisons demonstrate that for the type of problem analyzed, the softwares provided quite similar results. For the cases considered limiting equilibrium was attained for c values differing by 10 kPa to (rarely) 20 kPa.

A point of note is that a $SF = 1.08$ (≈ 1.1 which may be acceptable) may easily revert to 0.97 (which may be unacceptable) if a $c = 30$ kPa is used instead of $c = 20$ kPa. Such small differences in "c" value assessments are beyond the accuracy of measurement or assessment and exacerbate the problems of Safety Factor interpretation in practice. The effect is more pronounced in cases of flatter slopes.

7.6.2
Comparisons Between Different Softwares

The results from the application of four different softwares to the same slope problem are summarized in plotted form on Fig. 7.1.

The softwares were SLOPE/W, FLAC, FLAC 3D and UDEC. Some general remarks are as follows:

- The results follow expected trends of increasing values of Φ (required) for given c (or vice versa) to attain limiting equilibrium state as the slope angle increases.
- Limiting equilibrium analyses tended to be somewhat more conservative compared to FLAC.
- The 3-D analysis of FLAC-3D yielded very similar results with FLAC-2D.
- Predefinition of failure surfaces in a distinct element code such as UDEC appears to be a reasonable approach with significant advantages in terms of mechanism modelling.

A typical companion between the predicted failure zones from different approaches is given in Fig. 7.2 for one of the cases analyzed.

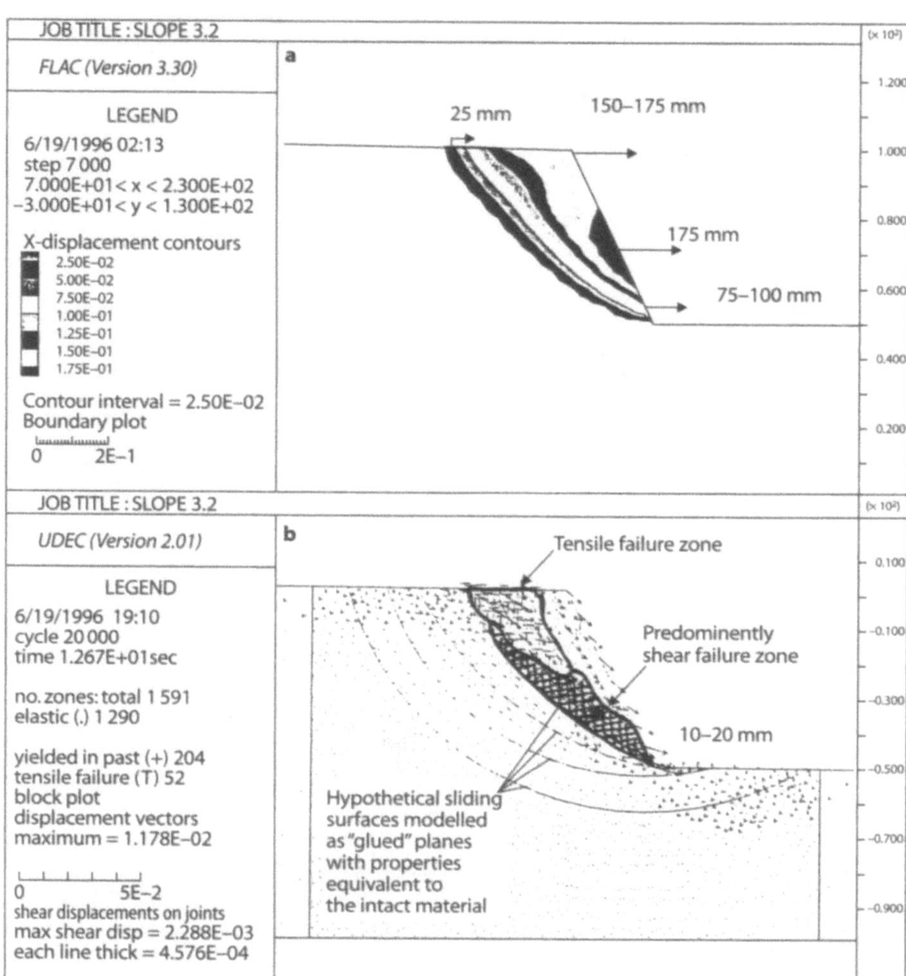

Fig. 7.2. Comparisons between the induced failure zones at limiting state of the 50 m slope modelled according to FLAC and UDEC; **a** illustrates the zonal distribution of horizontal displacements at the onset of failure (c = 0.10 MPa, Φ = 25°) in the FLAC model; **b** illustrates the zonal distribution of resultant displacements (vector plot) at the onset of failure (c = 0.12 MPa, Φ = 25°). The "pre-existing" circular sliding planes were modelled as "glued" fractures with properties of intact soil

A generally good agreement in apparent between the predictions from FLAC and UDEC. The displacements differ, although this is associated with the run-time of the two models.

7.6.3
Code Validations Against Case Studies

Two different case histories were modelled and analyzed using the UDEC software. One of the cases refers to a typical landslide of significant length and extent (Pieria landslide) in marl-sandstone successions in Northern Greece. The second case is an un-usual land instability phenomenon involving both soil and rock formations in the area of Civita di Bagnoregio, Central Italy.

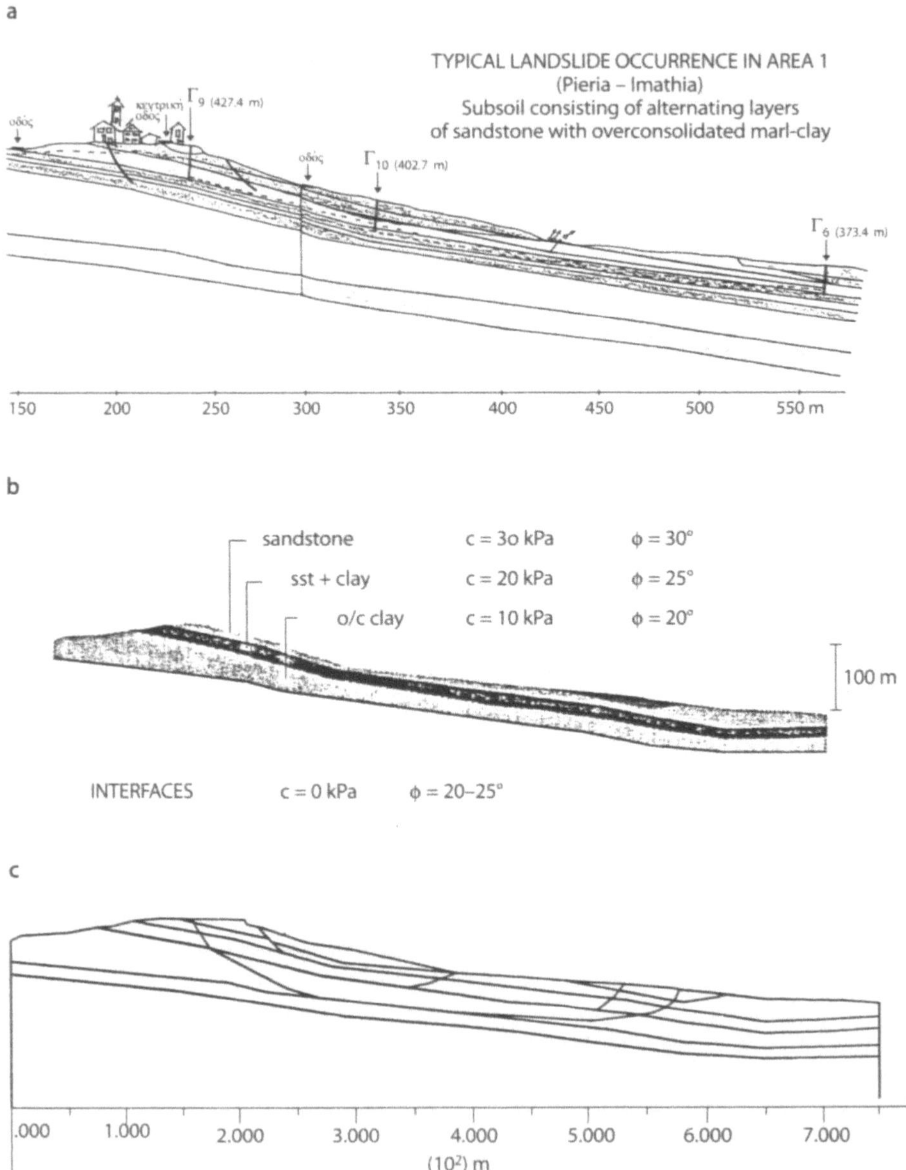

Fig. 7.3. a Typical landslide occurrence in the Pieria region; **b** UDEC model material and strata inter-face input data; **c** UDEC model block

7.6.4
The Pieria Landslide Case History

The landslide has been thoroughly examined in previous studies (Tsotsos and Anagnostopoulos 1992) and a considerable volume of geological and geotechnical data is available.

The ground profile in the area consists of alternating layers of sandstone and overconsolidated marl-clay of intermediate to high plasticity. From the study of borehole logs, the correlation between meteorological data and slide occurrences and the piezometer response the following were deduced regarding the principal mechanisms:

- Infiltrating water circulates through the relatively permeable, weathered sandstone, the partially broken marly silts and through the strata interfaces.
- The underlying layers of impermeable marly clay absorb water (20% montmorillonite), the highly plastic material looses its consistency and sliding commences under shear stresses induced by gravity or other loading.
- Similar phenomena but to a lesser extent were observed in the marly silts which may also become plastic under the influence of water.

A typical section is illustrated in Fig. 7.3 which also includes the inferred ground profile and interface-sliding (rupture) surfaces in a UDEC model representation.

In Fig. 7.4, the series of output plots demonstrates the key geotechnical factors that may trigger the landslide:

a Landslide under self-weight loading (dry): limited areas of shear overstress are predicted.
b Assumed distribution of pore pressures along the sandstone-marl interfaces.
c Significant shearing (up to 1 m dislocation) occurring in the upper and steeper zones due to water pressures.
d Deep seated shearing is also evident extending also to the downhill (known) limit of the slide.
e The extent of shear overstress developed further as the c, Φ parameters were reduced from $c = 0$, $\Phi = 20\text{--}25°$ to $c = 0$, $\Phi = 15°$.

The above responses are considered quite representative of the mechanics of the slide. It is of interest that analyses by other methods (Tsotsos and Anagnostopoulos 1992) indicated also that $SF < 1$ was obtained when $\Phi_r = 10\text{--}13°$ values (laboratory measurements) were used.

7.6.5
The Civita di Bagnoregio Case History

The ancient town of Civita di Bagnoregio is built on the top of a volcanic cliff, totally isolated from the neighbouring, surrounded by steep valleys formed by intensive erosion. A view of the cliff is shown in Fig. 7.5a. The cliff is composed of a top layer of jointed tuff underlain by a lithologically similar stratified formation. The tuffs rest on a suc-

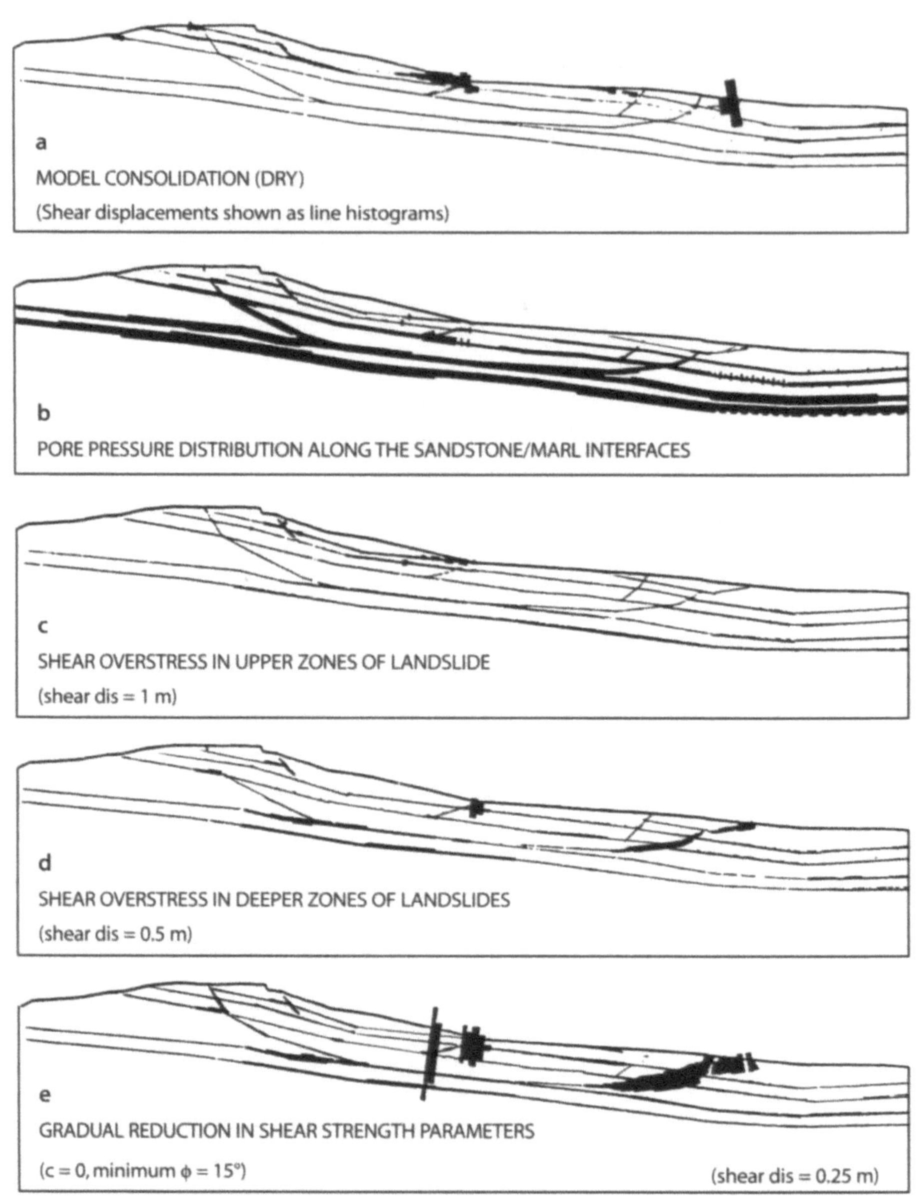

Fig. 7.4. Successive stages of landslide modelling indicate progressive development of shear overstressing due to self-weight loading (**a**), pore pressures (**b, c, d**) and strain-softening behaviour (**e**)

cession of clayey formations which form the valleys in the major area. A typical valley profile as reproduced from historical records in Fig. 7.5b indicates the rate of erosion during the last few centuries, that has marked the landscape evolution in the area.

Due to the particular geological and morphological features, the town of Civita di Bagnoregio has suffered from large and frequent slope instability phenomena (since 1450 according to historical records).

Deterioration of the clay subgrade has induced toppling, rotational failures and other complex gravitational phenomena of progressive nature. Consequently, a number of buildings has been totally destroyed while several buildings founded close to the cliff faces are endangered. The recently undertaken operation to restore the once called "dead town" has led to the design of permanent reinforcement measures for stabilization of the cliff and a pilot construction project is about to commence.

The numerical modelling studies described in the paper were undertaken to attempt a rigorous investigation of the mechanics of the progressive complex stability phenom-

Fig. 7.5 a. View of Civita di Bagnoregio and the topographic set up of the area

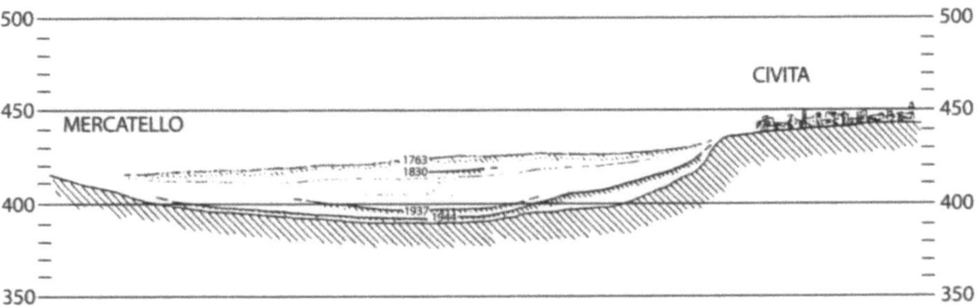

Fig. 7.5 b. Reconstruction of the process stages according to historical evidence

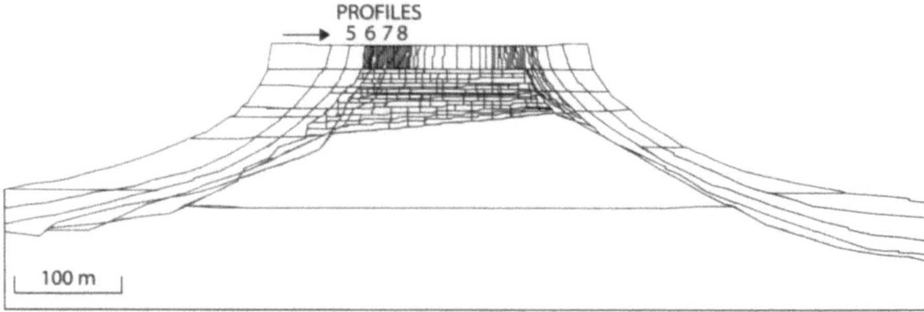

Fig. 7.5 c. Numerical representation (UDEC model) of ground profile corresponding to 1600. The geological set up and predominant rock mass structure are also represented

Fig. 7.5 d. Recent landslide failure involving the stratified tuff and the clayey substrata

Fig. 7.6. Modelling of successive stages of landslide development. Numerical analyses indicate that rock slope instability is related with significant build-up of differential (shear) stresses in the stratified tuff substrate. The increase in the shear stresses is the result of lateral unloading (stress relief) imposed by the erosion of the valley materials. Long-term overstressing of the materials in the outer slope zones and weathering effects both causing material weakening also appear as an additional contributing factor. Shear overstress development along potential failure planes and excessive shear strains (upper series of diagrams) may lead to progressive failure mechanisms (lower series of diagrams). The process is shown to apply repeatedly to several of the slope profiles numbered 5, 6 etc. Stress path within the stratified tuff associated with the simulated erosional processes (lateral unloading) as predicted from the UDEC model. A small decrease in cohesion apparently suffices to induce failure in the strata underlying the jointed (massive) tuff. Rock cliff failures may then develop due to loss of foundation reactions or excessive straining, promoting rotational failures of the rock columns (direct toppling may also occur depending on geometry)

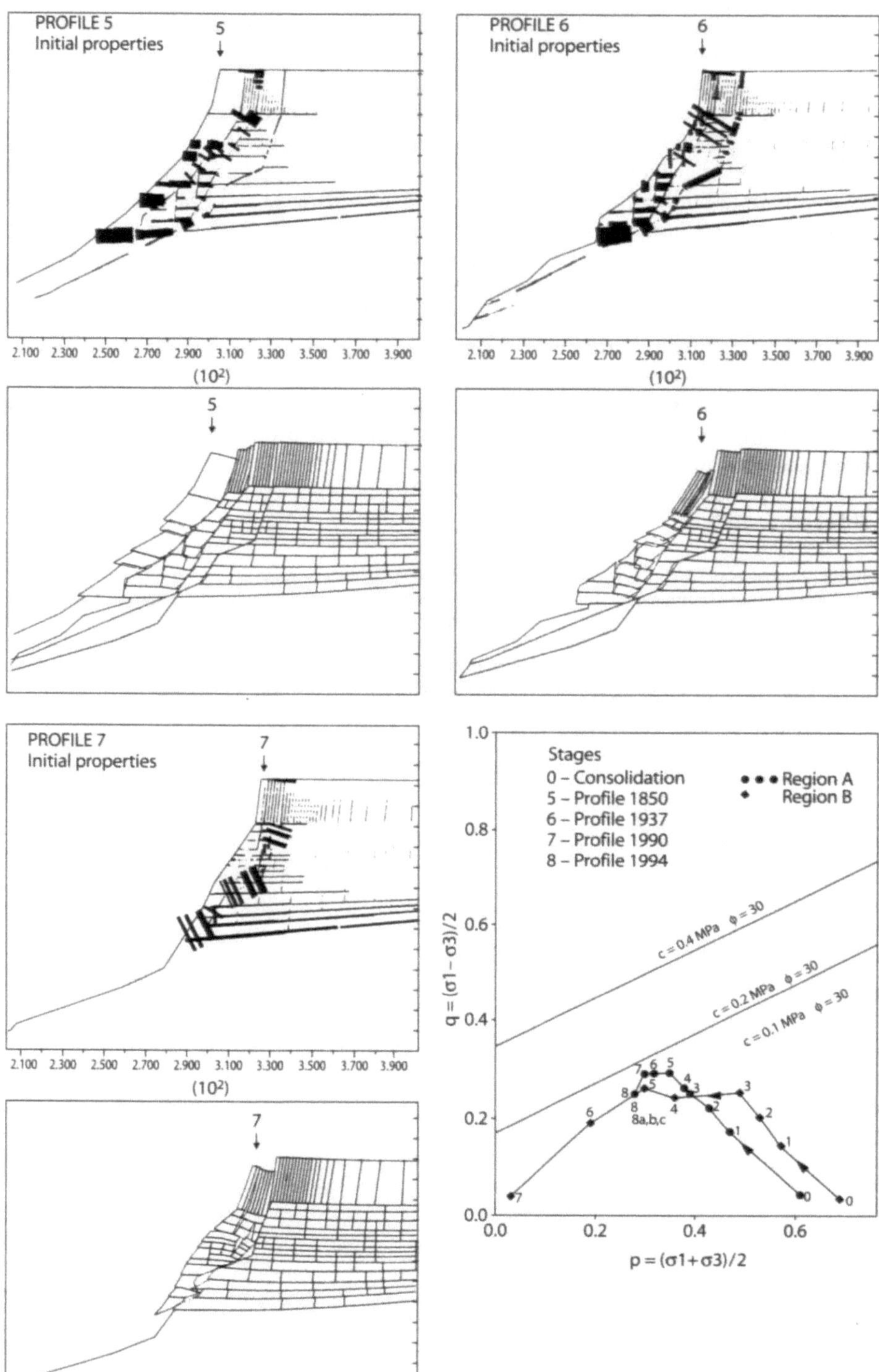
PROFILE 5
Initial properties
5
2.100 2.300 2.500 2.700 2.900 3.100 3.300 3.500 3.700 3.900
(10²)
PROFILE 6
Initial properties
6
2.100 2.300 2.500 2.700 2.900 3.100 3.300 3.500 3.700 3.900
(10²)
PROFILE 7
Initial properties
7
2.100 2.300 2.500 2.700 2.900 3.100 3.300 3.500 3.700 3.900
(10²)
Stages
0 – Consolidation
5 – Profile 1850
6 – Profile 1937
7 – Profile 1990
8 – Profile 1994
Region A
Region B
c = 0.4 MPa   φ = 30
c = 0.2 MPa   φ = 30
c = 0.1 MPa   φ = 30

ena. The numerical simulations involved "reconstruction" of the valley erosion and slope forming processes and monitoring of the associated (induced) changes in the stress regimes close to the critical zones. The practical implications of the back-analyses concerned the selection of type and the dimensioning of the reinforcement measures.

A 2-D model of a representative section was developed using the program UDEC (Universal Distinct Element Code), which is shown in Fig. 7.5c. A stepwise process of valley erosional unloading was implemented according to records and the induced principal stresses in the tuff formations were monitored at all stages of the simulated landscape development. The traces of known failure surfaces were introduced as fictitious cracks with properties equivalent to those of the tuff formations.

The series of numerical results in Fig. 7.6 indicate that failures probably initiated within the stratified tuff are due to a significant increase of the deviatoric stresses associated with the lateral unloading of the cliffs. Consequently, the gravitational falls of the top layer of tuff with columnar jointing were more of a rotational rather than of a toppling mode. Confirming evidence on the latter was provided from field observations and monitoring data.

References

Bond A (1995) Hard facts on software. Ground Engineering 28(8):14–16
Graham J (1984) Methods of stability analysis. In: Brudsen D, Prior DB (eds) Slope stability. pp 171–215
Gudehus G (ed) (1977) Finite elements in geomechanics. J Wiley & Sons, p 573
Morgenstern NR (1993) The evaluation of slope stability – A 25 year respective. Slopes and embankments.
Morgenstern NR (1995) The role of analysis in the evaluation of slope stability. Keynote Paper. Proc. 6th Int. Symp. on Landslides, Bell DH (ed), Chistchurch, N. Zealand, pp 1615–1629
Tsotsos S, Anagnostopoulos C (1992) Investigation of Landslides in the Pieria region. Proc. 6th Int. Symp. on Landslides, Bell DH (ed), pp 237–242

New Applications of Monitoring Networks, Multipurpose, Cost/Benefits, Management

M. Crespi

8.1
New Meteorology

Several human activities are characterised by a strong interaction with meteorological phenomena, particularly complex social-production-environmental systems are linked to this topic.

Rationalisation and decision making requirements need organisations and technologies which can offer good time-spatial resolution data and real-time information.

Resulting from this is a new local scale of meteorological service, an efficient and *multiuse* service, aimed at supporting strategic aspects of region life and work.

Modern meteorology must diversify its own products, to cover marketing areas where a demand rise is foreseen or foreseeable (Table 8.1).

Generally the low development achieved by these kinds of services leaves the possibility of growth.

Between local scale Modern Meteorology application fields, the more strategic are:

- Urban meteorology;
- Agrometeorology;
- Civil protection;
- Hydrology and territorial water defence;
- Environment and pollution defence;
- Tourism and sports;
- Planning and projecting productive activities;
- Services.

Concentrating the attention on urban problems, the supporting services in Fig. 8.1 could be identified.

Table 8.1. Meteorology applications

Traditional	Modern
Air navigation	Agrometeorology
Civil defence	Hydrology
	Environment
	Tourism
	Urban areas
	Others

Fig. 8.1. Urban meteorology applications

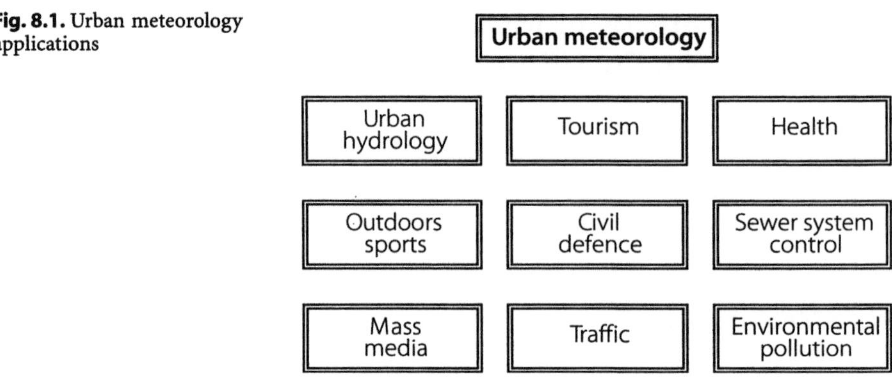

Table 8.2. Illustrative scale of meteorological information additional value

Value (example)	100	120	150	200
Additional value flux	Original data	Original data sale to privates	Original data elaboration	Elaborated data sale
Owner or thirds	Weather centre	Private	Private	Private or weather centre

Modern Meteorology products and services are very new and therefore, not yet well known. Consequently, demand must be stimulated with a good diffusion, following the experience already made.

Market problems and users type number knowledge is necessary because Mesoscale Weather forecasting monitoring systems require big investments.

These systems main characteristics are the availability of primary data (ground-stations data, forecasts) that could be emphasised by the Service (elaborating and integrating the information) or that could be diffused to the users as rough data.

In this way, the choice of know-how diffusion over the territory acquires great importance, to qualify and rise users standard (elaborated radar images diffusion). An example can be seen in Table 8.2.

8.2
Services Rate

National Services began charging for meteorological information, and more recently, followed by multiuse regional services. Particularly, the last ones, nevertheless, support environment or productive sectors, frequently had developed a wide number of users because of data and forecasts of high quality.

Consequently, a specific information demand rise took place, coming from: Universities, Research Centres, Enterprises, Professionals and Service Organisations.

This situation, which determines an increase of Organisations work, forces an adequate organisations procedures activation.

The result is services rating, aimed to emphasise high quality services and to obtain, at least a part of, the necessary money.

Information charge allows to:

- measure the demand and to know its satisfaction level,
- partially cover the expenses,
- better plan the investments, that must also be based on demand and
- develop information multiuse.

Particularly important are the possibilities opened by products charge on evaluating different types of information costs/benefits relationship.

Free of charge services are also considered to allow total benefits valuation (Table 8.3).

Considering that product rate is not able to cover system investments, but only a small part of its variable cost, there are two possibilities:

- A management strongly aimed at payable services, that covers costs but punishes service public role.
- A management privileging free costs information following a deep users analysis.

We think that, although public resources are limited, the services must choose the second hypothesis, mainly to maximise environmental benefits linked to the massive productive activities (agriculture, natural resources use, etc.) through free cost services.

This type of services benefits are greater than those that interest the privates; though it is necessary to individuate the right valuation methods.

Public organisations strategy towards information charge is summarised in two aspects:

- Users and products category definition;
- Rates definition.

Concerning the first item, the discriminating element becomes the relation *user/ product*, considering single and collective users benefits and resources used to satisfy the specific need.

Where collective benefit prevails the maximum demand cover is necessary, trough free information mainly with "territorial" (bulletins) products but also through special reports like high resolution data supply, specific elaboration, and so on.

Table 8.3. Example of benefits derived from an agrometeorological information service

Agrometeorological service benefits	
Economically quantifiable	Costs reduction Damage reduction Quality development Greater productivity (high profit crops)
Non economically quantifiable	Environment impact reduction Limited resources rationalization use Planning and policy Cultural development of farmers

In the case of Veneto Region this strategic role belongs to the agriculture, the environment, the civil protection and the territory defence.

The first, particularly, is supported because it is in a critical level from the economic-productive and environmental points of view.

Briefly, the scheme of users/products classes of a Weather monitoring multiuse Regional Service could be the following:

- Public and free of charge general territorial information;
- Free of charge specific strategic sectors information;
- Reasonably charged specific public information (mass media, research, experimentation, etc.).

The products supplying distribution, by users type, during the first Veneto Region experimentation period can be seen in Table 8.4.

The second strategic aspect for Public Organisation products supplier is determining prices. These could be fixed following two criteria:

- Based on service production *cost*;
- Based on *marketing*, that is based on the money that users are willing to pay for the services.

Nevertheless, the second criteria seems the correct one, because fees are proportional to the real benefit of each users category, the possibility of applying it seems complex, because the difficulty of knowing the sensibility of the new service market.

Particularly significant are the results of a research carried out in Texas, USA, about farmers willing to pay the local Weather Service information: between the mean values of information related to two different areas there is a difference greater than 500% in the money the users are willing to pay for the same service.

Consequently the practical applicable method, at least in the beginning phase, is to determine the fees based on information productions costs. In order not to discourage the demand, not yet consolidated because of the innovative character of multiuse monitoring service, it is necessary to reduce users expenses.

The Weather Service that charges for its information must be able to manage all the data in a short time and constitute an official reference of information quality and standardization.

In this sense, the better situation is when in a specific territory there is only one organisation that manages the monitoring network together with users relations.

Table 8.4. Charged products distribution for every users type (%)

Privates	27.5
Professional associations	27.5
Universities (inner activities)	22.5
Associations	7.5
State offices	5
Universities	2.5
Public organizations	2.5
Regional local organizations	2.5
Insurance	2.5

Besides the maximum economical efficiency it implies operative benefits like:

- Only one data base reachable by the users and therefore maximum use of the monitoring systems;
- Only one interlocutor toward the outside (Ministry, other Administrations);
- Only one interlocutor toward the inside (all the eventual users know to whom to make the requests);
- Multiuse better realisation, with the new monitoring systems;
- The possibility of making technologies maximum use.

Veneto Region experience on charged services activation had shown the validity of the idea, specially regarding the attention given to users, and had shown the need of rules to define a specific experimental period.

Chapter 9

The Importance and the Problems of Cartography – An Example: The Cartography of Natural Constraints on a Territory of 74 km^2 in Belgium (the Sprimont Territory)

A. Pissart · D. Closson

9.1
The Authors of the Research

A team of scientists from the Department of Physical Geography at the University of Liège have worked with A. Pissart on the present research; Damien Closson carried out the Arc-Info cartography and the studies on the protection of water catchments, on the sewage possibilities and on the fertility of soils; Camille Ek studied the karstic dangers; Michel Erpicum and Georges Mabille the climatological issues, while François Petit was concerned with the flood hazards.

9.2
Presentation of the Studied Area

The district of Sprimont located south of Liège in Belgium covers 74.28 km^2 and is a part of a natural region called "Condroz". The substratum is formed by Palaeozoic hard rocks folded and giving a typical Appalachian morphology. The ridges, levelled by an Oligocene peneplain, correspond with Devonian sandstones. The depressions are on Carboniferous limestones. The ridges are elongated along a south-west/north-east direction which corresponds to the orientation of the geological structure. Two important rivers (Ourthe and Amblève) are surimposed in the Palaeozoic structure and flow through the morphology without any connection with the geology. The bottoms of the valleys of these rivers are around 100 m a.s.l.; the highest summit is close to 315 m. The gentle slopes feature a discontinuous cover of eolian loesses which are usually reworked by slope processes.

This district is situated approximately at a 20 km distance from the centre of Liege and is part of the suburban belt which exists around this city. The landscape shows deciduous forests and grasslands in which there are villages in expansion due to the opening, a few years ago, of a new highway crossing the whole district.

9.3
The Cartography

For the administration in charge of planning in Wallonia, we carried out a cartographic research under the title of "Research of a methodology to identify which physical constraints are important for planning good land use". The aim was to recognize the physical elements which may play a role in land use, and to present maps collecting all these elements.

During this research we considered all the physical factors which may be interesting for a planner. They are first the hazards connected with floods, landslides, falls of rocks, and karstic dangers; secondly the protection of water catchments, the problems of sewage; third the exposure to the sun, the fertility of the soils and the locations where buildings may spoil the landscape.

This research is a typical physical study which considers not only floods and landslides hazards, but also all the physical parameters which concern the land use problems. It is an extension of the problem considered here from the sole point of view of floods and landslides hazards.

Because we do not know any landslide in the Sprimont district, we will not speak about this hazard. We will consider mainly the cartography of the flood hazards and later, very briefly, the other questions. About the flood cartography, we will explain the regulations which are in force in other countries (mainly in France) where the problem of cartography is officially the first priority and for which rules were enacted by the government.

9.4
The Flood Hazards and the Cartography of Prone Areas

We will consider two different questions. First the cartography of hazards in the alluvial plains of the largest rivers of our district, and secondly the cartography of the storm hazards in small basins (catchments of a few square kilometres or less), hazards which exist when heavy rains occur, a risk which is called "ruissellement pluvial urbain" by the French Ministry for Environmental Problems.

9.4.1
The Cartography of the Flood Extension of Large Rivers

In France, the cartography of the areas which experience floods has been imposed by the administration since 1982. The realization is very slow, as at the end of 1994, the maps of only 500 amongst the 2 000 districts of first priority had been drawn. The aim of this cartography is to have a tool to decide where 1) to prohibit the erection of any new building because the level of water may be high, 2) to limit, by adequate regulations for construction, the vulnerability of new buildings which are located under a lower water level, 3) to preserve the zones in which the water may have accumulated during the floods.

This cartography is difficult: the administration requests that a centennial flood be taken into consideration. Also for this flood, the zones where the level of the water will be higher than 1 m, between 0 and 1 m, and out of the flood, should be defined on the map. The problems of such a cartography are addressed in a booklet published by the French administration under the title of "La cartographie des plans d'exposition aux risques" (Garry 1988). These problems are numerous:

a   The first one is to have a very good cartographic base not only in planimetry but also in altimetry. Three quarters of the work is done if good maps are available. In France, on the IGN map, the accuracy is very often 1 m for the points with a marked elevation value and 2–3 m for the contour lines. The French administration writes that such accuracy is not high enough for mapping the floods.
b   The second difficulty consists in collecting information on the preceding floods. If a flood occurs in the period studied, it is possible to take aerial pictures and to level the high water marks in the field.

c   Another problem lies in the estimation of a centennial flood; it is necessary to have discharge observations during many years to calculate the discharge which may occur every 100 years. With this value it is possible to estimate the level of the centennial flood everywhere, if very accurate maps exist. Such highly accurate altimetric maps are very expensive. In France, one "departement" (administrative subdivision) which has a budget of 1.4 million franc for the whole cartography received proposals just to prepare the base topographic map for 0.75–1.5 million franc!

We have prepared the cartography of this hazard in the Sprimont district. Two rivers, the Ourthe and the Amblève River, are flowing and join on the district of Sprimont. They have basins of 2 666 km² for the Ourthe River and 1 052 km² for the Amblève River at their confluent. The mean discharge of the Ourthe 20 km down this confluence, is 61 m³/s but with peaks which reach of 500 m³/s for a decennial flood, and 750 m³/s for a centennial maximum. As for many rivers, the floods are more and more costly mainly because there are more buildings on the alluvial plain. Two kinds of projects are under study to limit the hazards: a) the construction of dams in the upper part of the basin to control the highest discharges, b) local works to protect inhabited areas.

During our study, we were lucky enough to have high floods of these main rivers. We collected many observations during the floods: a) aerial pictures taken from a small plane for a level of the river with a recurrence of eight years and half for the Ourthe River and for six years for the Amblève River (a flight was impossible when the flood was at its peak), b) information about the high-water marks collected in the field and about the maximum water levels in the houses (this information gives the level of water for a recurrence of 70 years which was the maximum for this flood), c) because we had the best topographic and altimetric maps of the alluvial plain, it was possible to calculate the level of the rivers for the centennial discharge (750 m³/s) and to map the depth of the water everywhere on the alluvial plain.

9.4.2
The Cartography of Storm Flow Hazards

Let us consider now another flood hazard: the hazard related to short but very heavy rains. An example from France is the catastrophic event which resulted from a precipitation of 300 mm in 7 h which fell on Nîmes in October 1988 (Desbordes and Noyelle 1994). The occurrence of such heavy rainfall is rare in Belgium (it sometimes happens during summer thunderstorms) where they have a much lower frequency than in other climatic environments and especially under the Mediterranean climate. The comparison of the curves of the rain characteristics (intensity/length/frequency) in Uccle (near Brusells) (Laurant 1976) with the curve of the decennial rain in the Mediterranean parts of France (Desbordes and Noyelle 1994), clearly shows that Belgium is under an another climatic environment where the hazards are not so great. However, catastrophic events may occur in Belgium, like in Dison in 1957 where we experienced a fall of 146 mm of water in 45 min. The curves of the recurrence of rains clearly show that the greatest differences in the occurrence of heavy rains in Belgium and in Mediterranean countries correspond to events longer than 1 hour.

A catastrophic flood occurs when the runoff coefficient is high and this factor is mainly controlled by the intensity of the rain. A graph given by the French administration for this kind of hazard (Desbordes and Noyelle 1994) shows that the runoff coefficient is high when the intensity is over 100 mm/h. On the intensity/length/frequency graph of Spa (which is the closest meteorological station to Sprimont) a rain with an intensity of 100 mm during 30 min occurs every 100 years (Laurant 1973). In France, the protection against floods has been calculated since around 1940 for events with 10-year recurrence. However, because storm flows may be really catastrophic events when they occur in cities, the proposal has been made to calculate the hazard for longer recurrence periods: 50 or 100 years. This is the reason why we consider here for Sprimont a recurrence of 100 years which, as noted above, is 100 mm per hour during 30 min.

For such a rain, the maximum discharge will occur from a basin with a concentration time of 30 min. It is not easy to calculate the surface from where the water is collected in 30 min; it depends on the values of the slopes, on the cover of the vegetation and on the length of the channels. The equations which are proposed in the "Handbook of hydrology" (Maidment 1992) give concentration times which range from 16–48 min for a basin which we took as a sample in the Sprimont district. This basin was three kilometres long. Subsequently, we will take the value of 3 km as a crude estimate in the district of Sprimont as the length of basins with a concentration time of 30 min.

It is almost impossible to have a rough estimate of the discharge which is likely to result from a rain of 100 mm/h during 30 min on a natural basin because the runoff coefficient depends not only on the permeability of the soil, but also on the previous rains and degree of humidity of the soils. However, the primary objective of the cartography we are discussing is to locate the basins in which problems of storm flows may occur. For this purpose, it is not necessary to have values of discharges but to appreciate where the most dangerous places are located. In our study we have suggested calculating an index of danger for different basins. This index is based on the measurement of the areas with different values of slope and different land uses. It is easy to make the calculations of the different surfaces with a geographical information system. On the other hand it is not necessary to calculate such an index for a large number of places because on a slope map showing the limits of the basins, the few places where a danger is possible immediately become visible.

9.5
Map of the Slopes and Derived Maps

If you want to use a GIS to answer some questions related to storm floods, we need a map of the slopes. It is possible to have a very useful document for this use by digitalizing all the information of the topographical map and exploiting it with a geographical information system like Arc-Info. For this purpose, we do not need the same accuracy as for the cartography of the alluvial plains. With the map the GIS will calculate not only the mean, but also the percentages of slopes of different inclinations. We need these values to calculate the index representative of the storm flow hazards (see Table 9.1).

The map of the slopes gives us, on the other hand, the places where the slopes are steeper than 58%. This slope value corresponds approximately to the equilibrium angle for natural materials and therefore, it is the limit for the fall of rocks. Such a hazard needs to be considered by planners.

Table 9.1. Proposed table for the calculation of a storm flow likelihood index (% of surface with different values of slope * value for land use)

Values for slopes[a]		Values for land uses[b]	
0 – 1%	1	Woodland	0.3
1 – 2%	2	Pasture	0.35
2 – 5%	5	Cultivated	0.4
5 – 10%	10	Residential	0.75
10 – 20%	20		
> 10%	50		

[a] From Desbordes and Noyelle 1994;
[b] From Dunne and Leopold 1978.

Since it is possible to superimpose the map of slopes on the map of the thickness of loose materials and on the map of degree of humidity of the ground, the GIS may help to recognise the places where landslides may occur. However, in the district of Sprimont, we do not know of any great thickness of loose material on slopes and the risk is very limited. No present day or fossil landslide were recognised in the district.

The map of the slopes can be of other interest to the planner: it allows the calculation of the amount of solar energy received during the different seasons (Fig. 9.1), an indication which is of great interest to save the heating energy and for the comfort of the people.

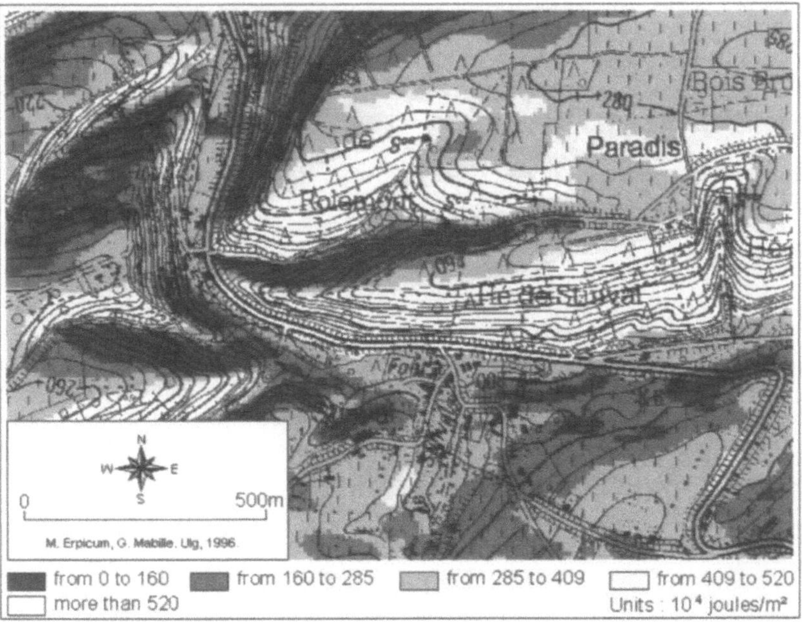

Fig. 9.1. Part of the map giving the amount of heat received on the ground during the day of the winter solstice

9.6
Map with Protection Zones of Water Catchments

Easy to draw, this map only presents the limits of the application of administrative regulations which are imposed to protect the water catchments. This map is very important, as it is forbidden to let the water from sewages sink into the soil of these protected areas, even if the water is rehabilitated by a purification system.

9.7
Map of the Karstic Dangers

The cartography of the karstic phenomena indicates the dangerous spots for the occurrence of caving-in of the ground. It is mainly the proximity of previous karstic holes which are considered as dangerous (Fig. 9.2). With the dangers of floods and of rock falls, they are the most important hazards in the district of Sprimont.

9.8
Map of the Sewage Possibilities

This map gives a picture of the most constraining problem for the country today. In a few years, all the sewage waters will have to be purified in accordance with the regula-

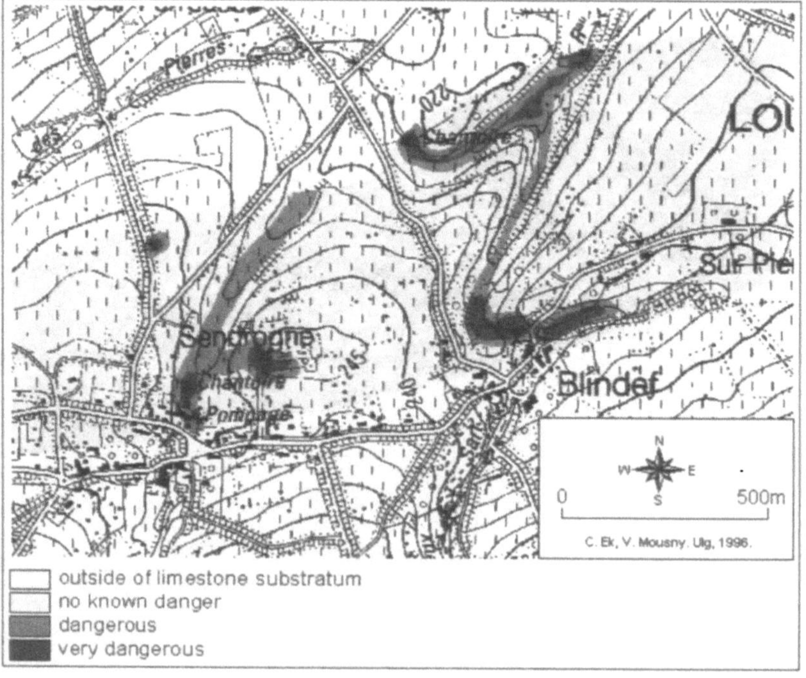

Fig. 9.2. Part of the map showing the karstic dangers

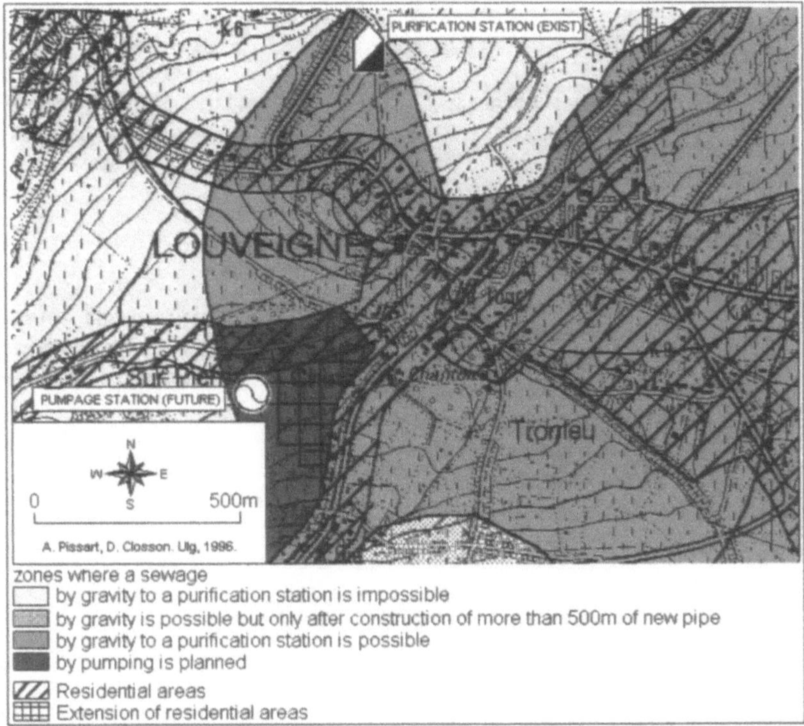

Fig. 9.3. Part of a map of Sprimont showing the sewage possibilities related to the locations of planned purification stations

tions of the European Union. The cost will probably be higher than 100 milliard Belgian francs (3 milliard US dollars), a sum which is very important for Wallonia. To pay for the purification of liquid waste, the price of the water will increase tenfold in the next few years.

The location of houses in the topography is the main constraint to connect the sewers together and with the purification stations. It is necessary for the planners to have a general view of this problem to limit, as much as possible, the individual purification systems on which people do not usually keep a close eye and which do not work correctly most of the time. We drew a map showing where it is possible to flow by gravity the sewers to the purification stations and where it is impossible to do so in accordance with the locations of the stations (Fig. 9.3).

9.9
Maps of the Fertility of Soils

On the basis of the pedological maps of Belgium, a map of the fertility of the soils was presented. Crossed with the slope map (mechanical agriculture is difficult on slopes steeper than 8%), these maps indicate the best soils for different uses.

9.10
Map of the Summits

Buildings on the highest parts of the landscape may destroy the sights of the landscape. For the planners, it is necessary to know the places where they are concerned with this problem.

9.11
A Synthetic Map with All the Physical Constraints

The final map shows all the different constraints put together. It is a useful tool on which it is possible to see immediately where the physical constraints are and to recognize the kind of constraint which is at stake. The caption of this map is presented below:

1. *Areas with natural risks.*
   a   Floods with a probability of 10-year intervals; floods with a probability of 25-years intervals;
   b   Dangers of caving-in related to karstic processes; axes where the solution processes are the most effective; karstic depressions where floods may occur;
   c   Slopes where rock falls may occur.
2. *Areas with constraints related to planning.*
   d   Zones of protection of water catchments;
   e   Sensitive crests for landscape preservation.
3. *Areas where buildings should not be located.*
   f   Soils of great fertility;
   g   Slopes steeper than 15%;
   h   Slopes receiving low solar energy.
4. *Areas with geotechnical constraints related to sewage.*
   i   Areas where the sewage by gravity to a purification station is not possible.

References

Desbordes M, Noyelle J (1994) (Ministère de l'Environnement, Direction de la prévention des Pollutions et des Risques), Ruissellement pluvial urbain. Guide de prévention. Eléments de méthode. La documentation française, 29–31, Quai Voltaire, 75340 Paris Cedex 07, 85 pp

Dunne T, Leopold LB (1978) Water in environmental planning, W.H.Freeman and Company, San Francisco, 818 pp

Garry M (1988) (Ministère de l'Equipement, du Logement, de l'Aménagement et des Transports), La cartographie des plans d'exposition au risque inondation. La documentation française, 29–31, Quai Voltaire, 75340 Paris Cedex 07, 115 pp

Laurant A (1973) La récurrence des intensités maximums de précipitations dans la région de Spa-Eupen. Courbes d'intensité-durée-fréquence. Annales des travaux publics de Belgique, N° 5, pp 319–330

Laurant A (1976) Nouvelles recherches sur les intensités maximum des précipitations à Uccle. Courbes d'intensité-durée-fréquence. Annales des travaux publics de Belgique, N° 4, pp 1–9

Maidment DR (ed) (1992) Handbook of hydrology. Mc Graw-Hill Inc, 1400 pp

From Meteorological Modelling to Flood Forecasting and the Management of Emergencies

E. Todini

10.1
Introduction

In recent years, thanks among other things to increasing computational and measurement capabilities, there has been a growing awareness in various areas of applied research that it is possible to predict emergency situations with a reasonable degree of approximation, especially in the fields of meteorology and hydrology. With the arrival of new measuring systems (Meteorological Radar and Meteosat Satellites), and especially with the new possibilities offered by Limited Area Models (LAM), it has finally become apparent that it is possible to create complex forecasting systems on a series of different time scales. Nevertheless, there is the feeling that the Public Administration is still reluctant to set up an organisational structure for emergency management based on the early but uncertain forecasting provided by hydrological models or by combined meteorological-hydrological models, and for the time being it is confining itself simply to the monitoring of the events under way. It was therefore felt necessary to clarify a number of basic aspects of meteorological and hydrological measuring and forecasting, in the light, among other things, of the experience gained in the development of a series of projects funded by the Commission of the European Communities, as the precondition for a correct approach to the forecasting and management of emergencies in a series of warning and alert stages with differing degrees of space-time resolution depending on the forecasting advance time.

10.2
Precipitation Data Acquisition Systems

At present, there are essentially three basic systems for providing precipitation measurements, which can be used for real-time flood forecasting.

The most commonly and widely used rain sensors for developing operational real-time flood forecasting systems are the conventional ground based telemetering raingauges, generally linked to a central station by means of telephone lines or by radio links (VHF or UHF) or, less frequently, by means of Meteor-burst equipment or via satellite through Data Collection Platforms (DCPs). There are several reasons in favour of the conventional equipment based upon raingauges. Firstly, National Services have a long tradition in using raingauges, which means that long historical records are generally available for calibrating the rainfall-runoff models; secondly, in real-time flood forecasting there is also need of other ground based hydro-meteorological measurements, such as for instance water levels in rivers and air temperatures close to the soil, which sensors may be integrated into the overall data acquisition system so that the cost of the additional rain sensors becomes marginal. Finally, in developing countries, training of local personnel and maintenance result more technically and economically accessible with ground based equipment rather than with other sources, such as weather radars or satellites. The density of raingauge networks depend on several factors (WMO 1981) and must be determined specifically for a single case depending upon the orography and the spatial correlation of observations. Techniques such as Factor Analysis or Kriging are generally used to provide an economically viable but sufficiently dense measurement network. The spatial description of the rain field based upon raingauges may not be accurate at very small scales (i.e. 100×100 m^2), but tends to be sufficiently accurate, for flood forecasting purposes, on larger scales of the order of $10\,000 \times 10\,000$ m^2.

Another precipitation measurement system is the weather radar system, which importance has grown in the last decade, particularly after the introduction of the dual polarisation systems and the Doppler radars. There are now over 100 countries operating more than 600 weather radars and development programmes have been established in several countries as well (Rosas Dias 1994). The European Union sponsored COST72 (COST Project 72 1985) and COST73 (Collier 1990) for establishing a weather radar network in the participating countries. In the USA NEXRAD (NEXRAD 1984) is a programme for establishing a network of 175 S-band Doppler weather radars and in the UK the FRONTIERS programme combines radar and METEOSAT images to produce very short precipitation forecasts (Browning 1979; Browning and Collier 1982). There are two major benefits in using radars: the first one is a finer spatial description of the precipitation field and the second one lies in the possibility of observing approaching storms sometimes before arriving over the catchment of interest. A major disadvantage lies in the need for recalibration of parameters used for converting reflectivity to rain, which generally also requires the installation of a conventional ground based raingauge network.

The third potentially useful measurement system is based upon the analysis of clouds shown by the geo-stationary satellite images (Milford and Dugdale 1989). This approach has been successfully used for the development of the Nile Flood Early Warning System (Grijsen, Snoeker, Vermeulen, Mohamed El Amin Moh Nur and Yasir Abbas Mohamed 1992). Unfortunately, the methodology developed by the Dept. of Meteorology of the University of Reading is adequate for long time intervals and very large catchments (the Blue Nile catchment area is on the order of $500\,000$ km^2) in the Tropical region, while there have been no convincing applications of the technique for smaller catchments in the sub-tropical or in the temperate zones.

When planning an operational flood forecasting system, there is the need of choosing among these precipitation measurement systems which have extremely different characteristics from the point of view of the information content. This requires an analysis of the size and of the nature of the problem which may start by giving an answer to the following question: "When is the improved spatial description of precipitation provided by radar an essential requirement?". The answer: "For very small size catchments, and in particular for small mountain or urban catchments". For these catchments in fact, given the very small sampling time required for runoff forecasting (5'-15') one radar image may be viewed as a snapshot of the spatial distribution of rain while, in order to obtain the same result an extremely dense raingauge network would be required. On the other hand, when the flood forecast is needed for small (>100–200 km^2) or medium size (1 000–20 000 km²) catchments the measurements provided by the conventional raingauge networks tend to be more than adequate.

With a view to providing a reference framework for the measuring systems to be adopted, it should however be stressed that reliance should be placed not on just a single precipitation data acquisition system but on several systems of different kinds: remote raingauges, Meteorological Radar, Meteosat, etc., each of them designed in such a way as to carry out the measurements independently of the other systems. In this way, in addition to the possibility thus offered to exploit the complementary nature of the information, there would also be the added advantage of turning to account the redundancy of the measurements in the case of intense meteorological events, when the probability of malfunctions in the single system solution is greater, and with it the attendant grave risk of losing the information.

10.3
The Predictability of Hydro-Meteorological Events

Today the predictability of hydrological-meteorological events, according to the meteorologists themselves, is "good"; having said this, however, it should be made clear in which situations and contexts such forecasts may be regarded as "good".

Mesoscale models (with mesh sizes of 100 × 100 km²) were basically conceived as descriptive models of climate and weather. As these models evolved, the "precipitation" variable was added to them, which is not however one of those quantities which model designers call "state variable", i.e. a variable which is vital for the description of the physical state of the system. It is rather, in meteorological models, a derived variable or a by-product, a function of the humidity content, the energy content, the condensation point, etc. obtained by means of more or less empirical relations. This should not be taken to mean that precipitation forecasting is basically wrong; rather, the point being made is that a great deal of research work still needs to be done with regard to this variable. It should also be borne in mind that orography, which affects the precipitation quantity, is described in the mesoscale models by a mean value on 100 × 100-km² meshes (i.e. 10 000 km² equal to 7 meshes encompassing the whole of the River Po drainage basin, including the Alps and Apennines) with the result that its effects are marginally felt. A further advance in the representation of rainfall fields was achieved with the advent of the so-called LAM's or Limited Area Models, with mesh sizes as small as the design limit of representation for hydrostatic models of 10 × 10 km², which on the one hand allow the orography to be introduced in greater detail and on the other allow a finer discretisation of the forecast precipitation quantities, and therefore a more realistic spatial variability.

At the present time, atmospheric forecasting models must be viewed as valid qualitative-quantitative rainfall forecasting tools at 24, 48 and 72 h (insofar as absolute precision is not required at these forecasting horizons, but rather an order of magnitude) for events of great intensity and when these phenomena occur on a considerable scale and size (like for example the flood event of November 1994); nevertheless, they cannot yet be regarded as providers of quantitative rainfall forecasts in the short term (6–12 h) to be used directly for flood forecasting purposes as an exciting force on hydrological models, since the quantitative forecasting of precipitation, on the time and space scales commensurate with the dynamic of the hydrological phenomena, has not yet achieved that degree of precision necessary to avoid on the one hand the non-forecasting of exceptional small-scale situations and, on the other, the issuance of unwarranted alarms.

10.4
The Predictability of Flood Events

As far as the predictability of flood events is concerned, it is necessary first of all to debunk a common misconception. Whilst it is true that there are hundreds of hydrological rainfall-runoff models, it has however been abundantly demonstrated that, for the purposes of flood forecasting, the most important effect is the dynamic of the overall soil filling and depletion mechanisms and the resulting variation in the size of the saturated basin area, as a result of which the rainfall that falls in the basin has a direct effect on the flood discharges which is not attenuated by the soil's absorption capacity. These concepts have spawned models which are very widely used today and have been extensively described in the literature, such as the TOPMODEL (Beven and Kirkby 1979) and the ARNO model (Todini 1996a).

Using the ARNO model, the author has developed various real-time operational flood forecasting systems, essentially based on remote measurement data collection systems. Table 10.1 shows the calibration results, in terms of explained variance, of a number of models which are currently in operation on basins of different size in various parts of the world. It should be borne in mind that in the operational phase, an error model is added to the calibrated models in order to further reduce the forecasting uncertainty.

An explained variance of the order of 94–98% shows that the hydrological forecasting problem has in itself actually been resolved: the existing problems are frequently upstream and downstream of the hydrological model. In point of fact, as far as the various Italian rivers are concerned, there is a lack of reliable and up-to-date rating curves and the geometrical descriptions are frequently approximate and/or rarely controlled; at the same time, no Italian authority responsible for flood management has thus far seemed to show any real interest, during flood events, in an "uncertain" discharge forecast for an 8–12 hour horizon, preferring instead to limit themselves to a blind adherence to the provisions of the above-mentioned Royal Decree N° 2 669 of 1939, which in practice means sending an Hydraulic Engineering Official to observe and report on the development of the events and the passing of the warning stages. But of even greater concern is the fact that so far it has not been possible to convince the competent authorities to carry out any trials with the systems which have been developed and are fully functional, in order to assess their real forecasting capability.

Table 10.1. Size of basins and performance of the ARNO model expressed in terms of explained variance

Basin	Area (km^2)	Explained variance
Fuchun at Lan Xi	18 236	0.96
Tiber at Corbara	6 100	0.98
Arno at Nave di Rosano	4 179	0.98
Danube at Berg	4 037	0.94
Reno at Casalecchio	1 051	0.96

10.5
Factors Contributing to the Management of Flood Events

It has been pointed out in the American literature that it is necessary to have real-time flood forecasts with an advance warning time of 6–8 h in order for these forecasts to be of any use and to allow measures to be taken to reduce the danger level or the damage connected with the flooding. All this is true and a consequence of the availability of tried and tested plans and organisational procedures. In Italy, where the only blueprint for action is still the Royal Decree N° 2 669 of 1939, which does not take the slightest account of a forecasting phase but requires instead that action be taken on the basis of the "observed" passing of the preset warning stages, the possibility of implementing defensive measures in advance of the events depends strictly on the size of the catchment areas affected by the phenomenon. In other words, while, on the main stream of the Po and on the basis of the travel times, it is easy to allow even 24 h of advance warning based on the level measurements in the upstream sections, in the minor basins measuring as little as a few thousands of square kilometres, it is necessary to make flood forecasts on the basis of rainfall observations, and in smaller basins measuring as little as a few hundred square kilometres it is of vital importance to use rainfall forecasts in addition to the latest recorded measurements.

It follows that when a real-time flood forecasting system has to be developed for a hydrographic basin, it is necessary to provide for a detailed and congruent structure that can be integrated in the problem according to its scale. As a rough guide, and in view also of the fact that there are not at this time any definite procedures for the local management of emergency measures in the case of flood events, it is necessary to set an advance time to which a quantitative forecast of the flood discharge must refer. On the basis of the many trials and experiences that have been conducted, it is reasonable to set this *forecasting advance time at 12 h* so as to enable an organisational early warning phase to be mounted before the actual operational emergency is triggered.

It should be stressed that, in order to organise a flood service at various warning levels, the sequence of forecasts must generally be structured in the following way, with the precision of the forecasts steadily increasing as the forecast advance time diminishes:

- *Mesoscale qualitative-quantitative rainfall forecasts at 48–72 h* with a degree of precision that is modest at the local level but which is revealing in terms of how the overall event will unfold and develop.
- *Qualitative-quantitative limited area rainfall forecasts with an advance time of 24–48 h.*
- *Quantitative flood forecasts with an advance time of up to 12 h*, and where necessary with reduced precision, that is more than anything an indication of trend, up to 24 h.

In order to predict the flood discharge quantities (case C, not to be confused with the rainfall forecasts of cases A and B) and to allow a quantitative forecast up to 12 h in advance, the basins must be subdivided into three broad classes:

- Large basins (>10 000 km²), for which the flood forecast with up to 12 h advance warning in one section can basically be made on the basis of the water levels (or discharge quantities) recorded in one or more upstream sections. In reality this situation is found in Italy only for the lowland reach of the Po, where the travel time of the flood waves is approximately 72 h from Becca to Pontelagoscuro.

- Intermediate basins (1 000–10 000 km²), for which the flood forecast with up to 12 h advance warning can be made on the basis of the precipitation measurements alone; in other words, the future rainfall has a negligible effect on the discharges which will occur in the next 12 h as a result of the absorption times, the release of meteoric water by the soil and the displacement times of the runoff on the slopes and through the drainage network.
- Small basins (<1 000 km²), for which the flood forecast with up to 12 h advance warning can be made only when a precipitation forecast is available, such as one that might for example be generated by a Limited Area Model.

As can be seen from these observations, the problem of the short-term predictability of a flood event (up to 12 h) can already be resolved for medium-large basins provided one has a real-time telemetering network in place. However, the forecasting problem remains unresolved for a large variety of minor basins which would require quantitative future rainfall forecasts of a quality far superior to that provided by the present models. It should also be noted that all of this is well outside the field of what is known as "nowcasting", which is very much in vogue among meteorologists and which could be of some use in the intermediate basins category.

Within the framework of the AFORISM project (A Comprehensive Forecasting System for Flood Risk Mitigation and Control), financed by the European Union (Todini 1996b), the possibility was first put forward of a hook-up between limited area meteorological models and real-time flood forecasting models; in this context, attention was drawn to the need to subject the meteorological forecasts to post-processing in order to adapt them to the real-time flood forecasting. More recent developments in research have shown that it is possible, using a Kalman Filter, to combine the limited area precipitation forecasts with the recorded measurements on the ground in order to obtain a realistic meteorological forecast to be used effectively as an input to the flood models.

As an example of systems implemented for large basins, various water level real-time monitoring stations have for some years now been installed on the main stream of the Po from Becca to Pontelagoscuro and the data needed to make the forecasts is therefore, available:

- Cross sections describing the river;
- Rating curves;
- Flood levels in various sections of the river and its tributaries.

For this section an automatic flood early warning system has been designed and in operation for over two years. The system, installed with the Po Water Authority, is based on the EFFORTS package (European Flood Forecasting Operational Real Time System), which allows flood forecasts to be made at all the discharge measuring stations between Becca and Pontelagoscuro (Fig. 10.1) and to make hydraulic evaluations in all the intermediate reaches (Fig. 10.2) (Todini 1992). The hydraulic flood routing is performed on the PAB programme (Todini and Bossi 1986) which has been shown to be

Fig. 10.1. The Po River real-time flood forecasting system based on EFFORTS

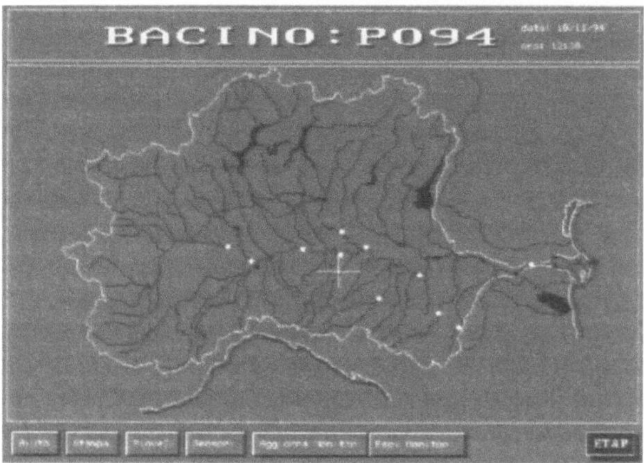

Fig. 10.2. The flood forecast and the cross-section at Pontelagoscuro (10.11.1994)

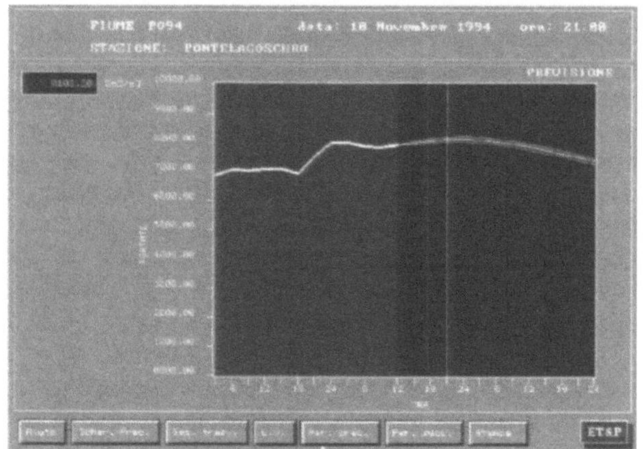

Fig. 10.3. Real-time flood forecasts up to 36 h in advance at Pontelagoscuro during the November 1994 flood in the Po River

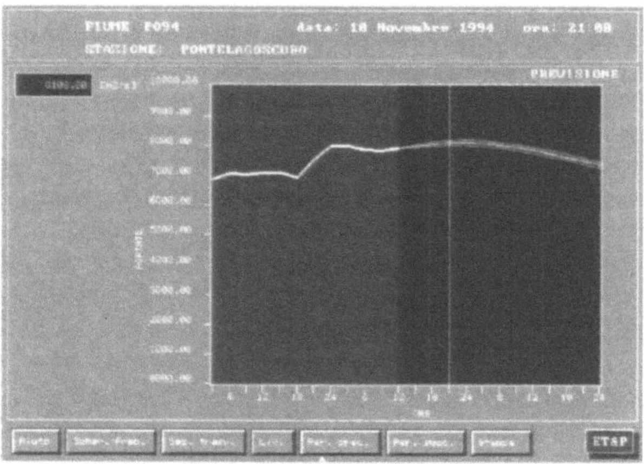

unconditionally stable – a vital condition for a system which is required to operate in automatic mode in real time – and which has furnished excellent results on the Po and on other water courses. Downstream of the hydraulic model is a Kalman Filter which allows the model to be continuously adjusted to the recorded measurements. The flood event in November 1994 was correctly forecast both in terms of its size and in terms of time at Pontelagoscuro (Fig. 10.3), even though it was operating under impaired conditions as a result of the non-reception of the data from Becca, the upstream station and the most important site for the purposes of the forecasting. The system has since been enhanced with the possibility of transferring the data and the forecasts to another computer for follow-up analyses and for disseminating both through a local network and via remote modem link the information and displays available at the Po Water Authority's computation centre: at the present time the two peripheral centres of Rovigo and Mantua are linked up.

Examples of real-time flood advance warning systems for intermediate basins including both the rainfall-runoff models and the flood propagation models are provided

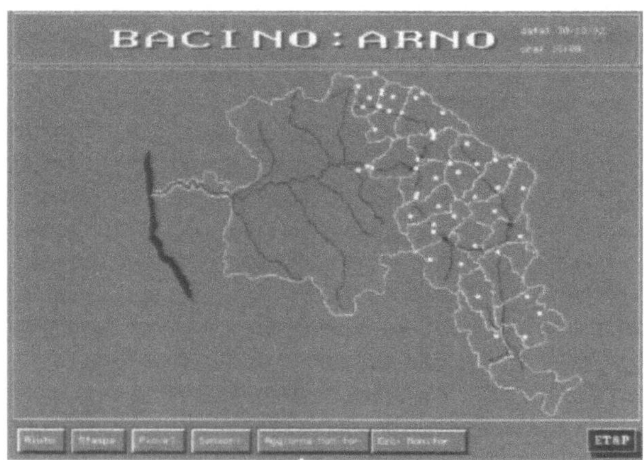

Fig. 10.4. The Arno River real-time flood forecasting system based on EFFORTS

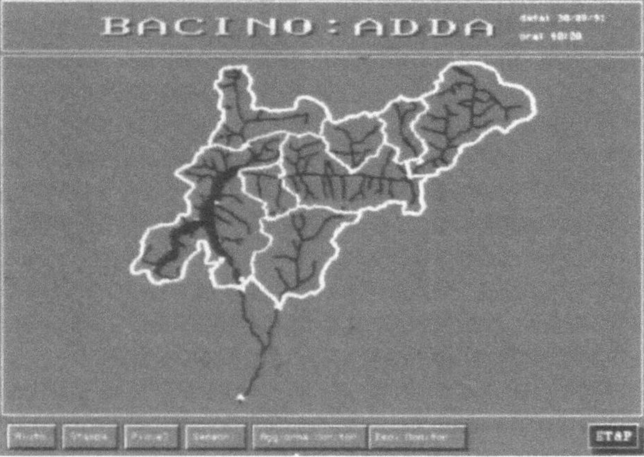

Fig. 10.5. The Adda River real-time flood forecasting system based on EFFORTS·

by the systems designed, installed and in operation on the Arno at the Tuscany Public Works Office (Fig. 10.4), on the Adda at the Adda Consortium (Fig. 10.5) for the control of Lake Como, on the Tiber at ENEL for control of the Corbara reservoir and currently being extended as far as Rome for the Special Office for the Tiber and Agro Romano (Fig. 10.6), and lastly on the Danube, installed at the Flood Forecasting Centre at Baden Württemberg (Fig. 10.7).

With regard to the forecasting problem for minor basins, a study has recently been undertaken on the integration of the rainfall forecasting data performed on a LAM with the rainfall-runoff models for the forecasting of flood events in the minor sub-basins of the Danube. As a consequence of this study, carried out for the Flood Forecasting Centre at Baden Württemberg, it was concluded that it is still not possible to make direct use of the meteorological forecasts, even though they are of the limited area type, as input for the rainfall-runoff models. Therefore, provision has been made within the EFFORTS package for a computational system based on Kalman filters which allows three successive forecasts provided by the German Meteorological Service (DWD) to

Fig. 10.6. The Tiber River real-time flood forecasting system based on EFFORTS

Fig. 10.7. The Danube River real-time flood forecasting system based on EFFORTS

be used on their limited area model (13.5 × 13.5 km²) in order to extrapolate, up to a time horizon of 12 h, the precipitation measured on the ground by the remote raingauge network. In this way it was possible to make a realistic quantitative flood forecast for a horizon of 12 h in advance for the minor sub-basins as well.

The following figures illustrate what has been described above. Figure 10.9 shows the discharge quantities observed at the outlet of the mountain sub-basin measuring approximately 290 km² as indicated in Fig. 10.8, and the forecasts made at 4, 8 and 12 h in advance on the assumption that there is no more rainfall following the last measurement time: the pronounced underestimation of the discharge values provides verification that if the forecast is to be extended to short time horizons, then the future rainfall must be known. Figure 10.10 shows the flood forecast performed on the basis of the rainfall predicted by the DWD without any processing: in this case it is seen that the discharges are markedly overestimated to such an extent that they cannot realistically be used for an operational flood forecasting system designed for the management of emergency situations. By contrast, Fig. 10.11 shows that, after processing the data, the rainfall forecast, and consequently the flood forecast, generates values that are only slightly different from the observed data and can therefore, be used at the operational level.

To return to the central question of the factors involved in the forecasting and management of flood events, it was seen that there are three forecasting levels, comprising firstly two meteorological forecasts (A and B) and then one flood forecast (C). In op-

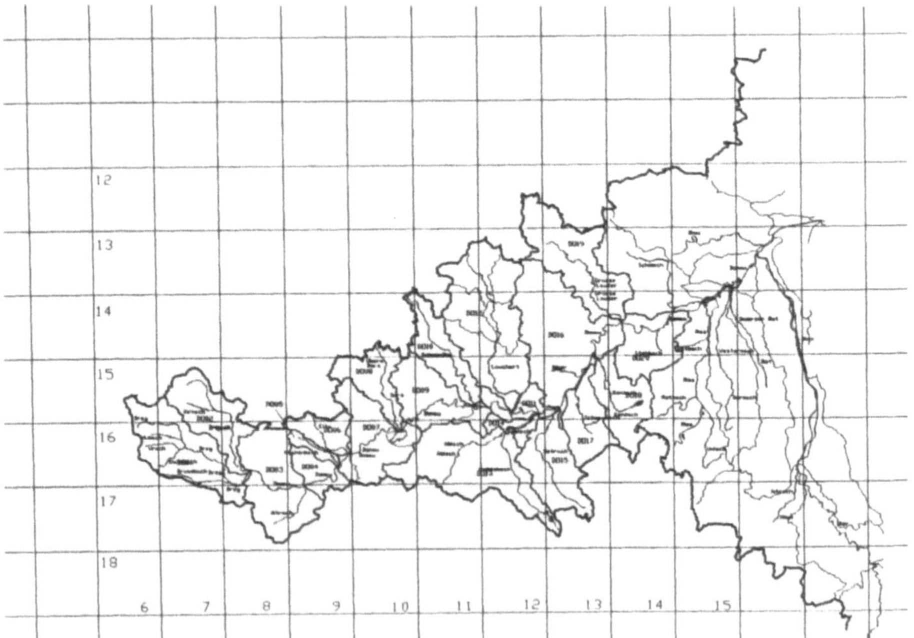

Fig. 10.8. The catchment and the sub-catchments in the Danube real-time flood forecasts model and the DWD Limited Area Model grid points

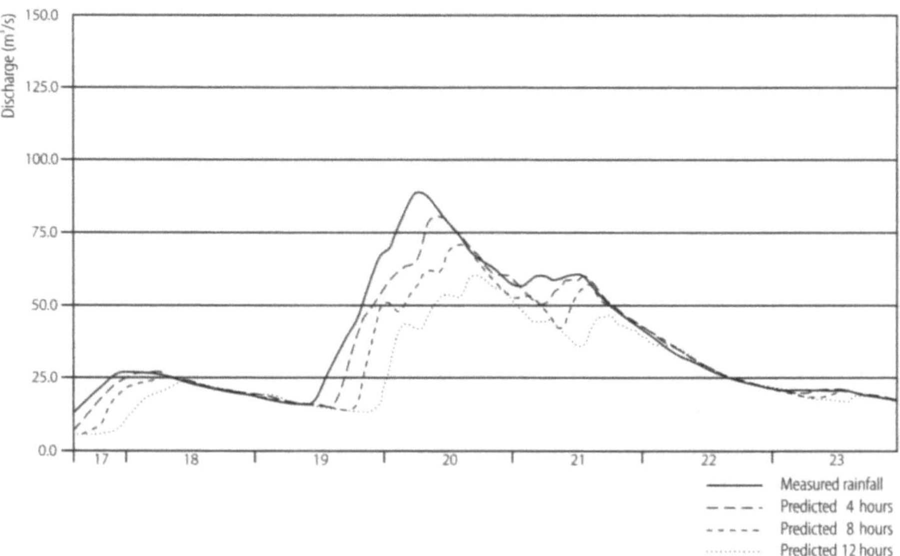

Fig. 10.9. 4-8-12 h in advance real-time flood forecasts on the Danube at St. Donaueschingen setting future rain equal zero

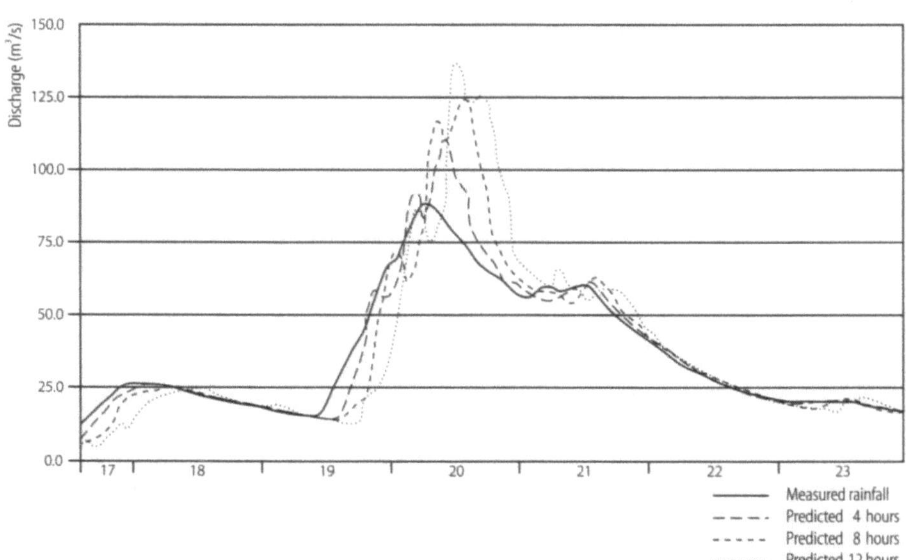

Fig. 10.10. 4-8-12 h in advance real-time flood forecasts on the Danube at St. Donaueschingen using DWD precipitation forecasts

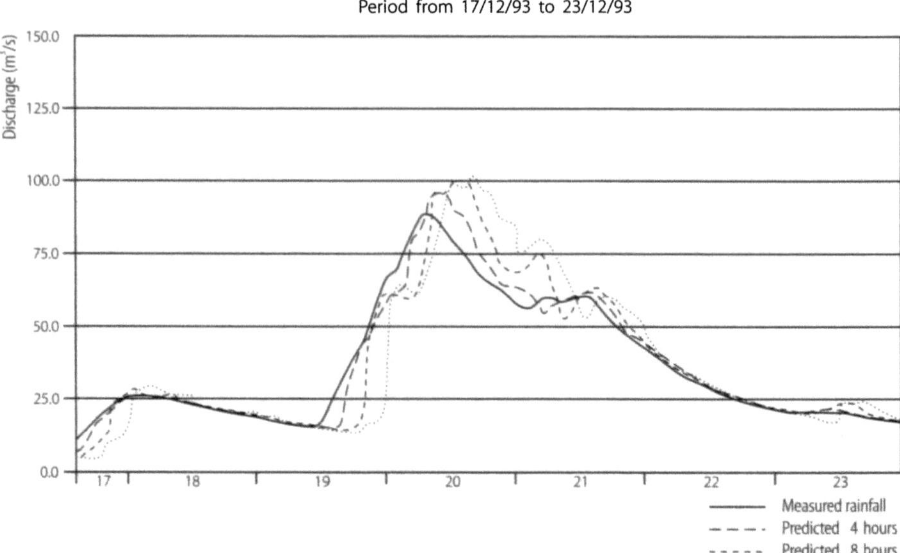

Fig. 10.11. 4-8-12 h in advance real-time flood on the Danube at St. Donaueschingen using Kalman filter corrected DWD precipitation forecasts

erational terms, it therefore follows that a flood monitoring, forecasting and management system is required which provides for different degrees of detail and which must therefore, be structured around several modules: from the monitoring of the data and the meteorological forecasting to the automatic flood forecasting and advance warning of risk situations, and finally to a detailed analysis of the courses of action and decision-making support, as has been highlighted by the AFORISM project.

Downstream of the meteorological forecasting, the various levels at which it is necessary to organise the system comprising the monitoring, early warning, forecasting and decision-making support to the management of flood event emergencies, must adhere to the following scheme:

i.   Real-time data acquisition;
ii. Real-time analysis, verification and, if necessary, reconstruction of the missing data;
iii. Automatic performance of the rainfall forecasts based on the latest measurements and on any forecasts that may have been generated by the Limited Area Models;
iv. Automatic performance of flood forecasts, based on the observed measurements and the rainfall forecasts, using continuous forecasting models, in which the saturation condition of the soil is constantly updated;
v.   Automatic performance of flood forecasts and routing along the river reaches using hydraulic models which provide for the identification of the key critical areas;
vi. Implementation of possible intervention simulations and verification of their effects; in this last phase of analysis which corresponds to a declared state of emergency, all

of the activities included in points ii, iii, iv and v above, which represented the early warning phase and which necessarily had to be implemented in automatic form, can be reviewed and modified in assisted semi-automatic form, in order to verify the effects of possible alternative courses of action.

Note that point (ii) above, which is frequently overlooked, is by contrast of paramount importance. In real-time, it is in fact essential that there are no breaks in the flow of input data to the forecasting model and the reconstruction of missing data, where necessary, must be performed using reliable procedures and in fully automated fashion: it is of course unthinkable that an operator should sit at the terminal around the clock and reconstruct the missing data using more or less manual procedures.

10.6
Concluding Remarks

As has been described, today it is possible to create systems which furnish, in operational terms and in automatic form, flood forecasts characterised by a high degree of precision for a time horizon of 12 h on medium and large basins (>1 000 km²), while research efforts aimed at achieving the same degree of accuracy and advance warning in forecasts on small basins are now at an advanced stage of development. However, one frequently feels that the Public Administration has still not come to terms with and does not fully appreciate the very real possibilities offered by the flood forecasts performed, albeit with the uncertainty inherent in every forecast, with an advance prediction time sufficient to organise emergency action and assistance, and that it persists in limiting itself merely to the monitoring of the events in progress.

In this connection, it should be reiterated once again that it is possible to make flood forecasts using limited area models and to achieve meteorological forecasting advance warning times of 72 h, and that these forecasts are found to be reliable when relating to basins of considerable size (the Po basin in the November 1994 event is a good example). Flood forecasting can today be performed with a substantial margin of reliability, for time horizons of 12 h, in both large and intermediate-sized basins, while experimentation is under way aimed at using the precipitation forecasts generated by LAM's directly as input in flood forecasting models in the context of a new project financed by the European Union, under which Bologna University and the Emilia Romagna Regional Meteorological Service are working in collaboration with the Irish and Swedish Meteorological Services and with Dublin University. As has been said, the forecasts of the Limited Area Models are unfortunately still not able to guarantee the spatial disaggregation and the dependability necessary for short-term flood forecasting (8–12 h), so as to permit the forecasting and operational management of flood event emergencies in minor basins and small sub-basins. However, even in the case of minor basins, an effective solution has already been found which entails the use of the LAM forecasts for the extrapolation of the rainfall data recorded on the ground in real-time. Bearing in mind the fact that today many Italian basins are already equipped with remote raingauge networks and real-time flood forecasting models (Po, Arno, Tiber, Adda, Oglio, Toce, Agno Guà, and Reno), many of which have been operational for over a year, it is easy to see that the use of limited area meteorological forecasting, even in the case of minor basins, can be implemented rapidly and with relatively simple procedures.

Downstream of the meteorological and flood forecasting, however, the question of the planning and coordination of emergency situations is still unresolved: in fact it would be essential to have the on-line availability of a system of definite and unified procedures allowing all the parties responsible for the management of flood events to be aware of the situation, of what the other parties have done up until that point, and of the possible operational options that are available to them. It is in this direction that attention must now be focused, and most notably in supplanting Royal Decree N° 2 669 of 1939 which, in not taking into account a forecasting phase, not only has not permitted the use of forecasting models in the past, but in some way has even inhibited the interest of the competent authorities, which are still more concerned with the monitoring of emergency situations than with the operational forecasting of these same events.

References

Beven KJ, Kirkby MJ (1979) A physically-based variable contributing area model of basin hydrology, Hydrological Sciences Bull 24(1):43–69

Browning KA (1979) The FRONTIERS plan: A strategy for using radar and satellite imagery for short range precipitation forecasting. Met Mag 108:161–184

Browning KA, Collier CG (1982) An integrated radar-satellite nowcasting system in the UK. In: Browning KA (ed) Nowcasting. Academic Press, London, pp 47–61

Collier CG (1990) COST 73: The development of a weather radar network in Western Europe. In: Collier CG, Chapuis M (eds) Weather radar networking seminar on COST Project 73, Kluwer Academic Publishers, Netherlands EUR 12414 EN–FR

COST Project 72 (1985) European Commission Report EUR 10353 EN–FR

Grijsen JG, Snoeker XC, Vermeulen CJM, Mohamed El Amin Moh Nur, Yasir Abbas Mohamed (1992) An information system for flood early warning. In: Saul AJ (ed) Floods and flood management. Kluwer Academic Publishers, pp 263–289

Milford JR, Dugdale G (1989) Estimation of rainfall using geostationary satellite data. Applications of remote sensing in agricultural sciences. University of Nottingham, Butterworth London

NEXRAD (1984) Next generation weather radar. Programmatic environmental impact statement R400-PE201, NEXRAD Joint System Program Office, US Dep of Commerce

Rosas Dias Manuel P (1994) Radar measurements of precipitation for hydrological purposes. In: Almeda-Teixeira ME, Fantechi R, Moore R, Silva VM (eds) Advances in radar hydrology. European Commission Report EUR 14334 EN

Todini E (1992) Verso la realizzazione di un sistema di supporto operativo di piena per il Po. Proceedings of workshop on "Modelli Matematici per il Bacino del Fiume Po", Quad. Ist. Ric. Acque, N° 95, pp 3.1–3.28

Todini E (1996a) The ARNO rainfall-runoff model. J Hydr 175:1–4

Todini E (1996b) AFORISM – A comprehensive forecasting system for flood risk mitigation and control. Final Report of Contract EC-EPOC-CT90-0023

Todini E, Bossi A (1986) PAB (Parabolic and Backwater) – An unconditionally stable flood routing scheme particulary suited for real time forecasting and control. J Hydr Res 24(5):405–424

WMO (1981) Guide to hydrological practices, WMO publications, N° 168, 4th edn, Geneva

Chapter 11

Real-Time Flood Forecasting Systems: Perspectives and Prospects

R.J. Moore

11.1
Introduction

Advances in flood forecasting concern much more than innovation in model formulation. This paper aims to provide a perspective on developments in the UK over the last twenty years, initially by reviewing models in current operational use but progressing to discuss updating methods, the use of weather radar and lastly the introduction of integrated flood forecasting systems. Updating methods allow real-time measurements, for example of river level received via telemetry, to be used to improve model performance. Weather radar allows point measurements of rainfall, from an often sparse raingauge network, to be complemented by spatially continuous measurements of the rainfall field and thereby, provide improved input to flood forecasting models. Most recently, flood forecasting systems have been developed capable of coordinating the construction of forecasts at many points across a possibly complex region and being generic in their use of models and their configuration to any set of river networks without expensive recoding. These developments are examined with an acknowledged emphasis on procedures developed at the Institute of Hydrology and particularly those methods which are incorporated in its River Flow Forecasting System or RFFS. However, these methods are reviewed with reference to techniques developed elsewhere and which are in use operationally, thereby providing an overview of the present state-of-the-art in the UK. Prospects for future improvement are considered, focusing on the potential value of digital terrain models to formulate a new generation of distributed model appropriate for operational use in combination with radar rainfall and satellite-derived thematic data, for example on land-use.

11.2
Hydrological Models for Flood Forecasting

11.2.1
Introduction

Improvements in flood forecasting derive classically from improved model formulations, These are reviewed here under the headings of rainfall-runoff models, channel flow routing models and snowmelt models. Rather than review these in a comprehensive way emphasis is put on describing the model component forms, rather than the detail of particular "brand name" models. Those models available as part of the RFFS are used as the primary models for purposes of illustration.

11.2.2
Rainfall-Runoff Models

11.2.2.1
Conceptual Models and Model Components

In the UK there are three main conceptual rainfall-runoff models used for operational real-time flood forecasting. In the Severn-Trent Region extensive use is made of a development of the Institute of Hydrology Conceptual Model or IHCM originated by Dickinson and Douglas in 1972 (Bailey and Dobson 1981). This model employs a soil moisture storage component to effect a separation between fast response runoff to the channel system and downward percolation, the latter forming an input to a nonlinear storage representation of groundwater. A linear channel is used to delay and spread out the total flow and a further nonlinear storage provides for additional channel storage attenuation.

In the Thames basin a conceptual model has been tailored to accommodate the variety of hydrological responses seen in this mixed urban and rural region with significant aquifers and artificial influences. The structure of the Thames Conceptual Model, or TCM (Greenfield 1984), is based on subdivision of a basin into different response zones representing, for example, runoff from aquifer, clay, riparian and paved areas and sewage effluent sources. Within a given zone the same vertical conceptualisation of water movement is used, the different characteristic responses from the zonal areas being achieved through an appropriate choice of parameter set, some negating the effect of a particular component used in the vertical conceptualisation. The zonal flows are combined, routed if required through a discrete kinematic wave channel flow routing model (optional), to form the total basin runoff. Within a given zone, water movement in the soil is controlled by the classical Penman storage configuration (Penman 1949) in which a near-surface storage, of depth equal to the rooting depth of the associated vegetation (the root constant depth), drains only when full into a lower storage of notional infinite depth (Fig. 11.1). Evaporation occurs at the Penman potential rate (E), whilst the upper store contains water and at a lower rate (E_a), when only water from the lower store is available. The Penman stores are replenished by rainfall, but a fraction ϕ (typically 0.15) is bypassed to contribute directly as percolation to a lower "unsaturated storage". Percolation occurs from the Penman stores only when the total soil moisture deficit has been made up. The total percolation forms the input to the unsaturated storage which behaves as a linear reservoir. Its outflow, termed "recharge", forms an input to a quadratic storage representing storage of water below the phreatic surface in an aquifer. Withdrawals are allowed from this storage to allow pumped groundwater abstractions to be represented. A more generalised form of the model, as well as incorporating updating facilities, has been developed by the Institute of Hydrology: this extended model has been termed, generically, the Penman Store Model or PSM and incorporated within the RFFS Model Calibration Facility.

The third model is called the Probability-Distributed Moisture or PDM model (Moore 1986). This model has been applied throughout the Yorkshire region, to the Lincoln area of Eastern England, to the White Cart Water near Glasgow and in Hong Kong and constitutes the rainfall-runoff model provided as a standard within the River Flow Forecasting System discussed in more detail later. The PDM model provides a

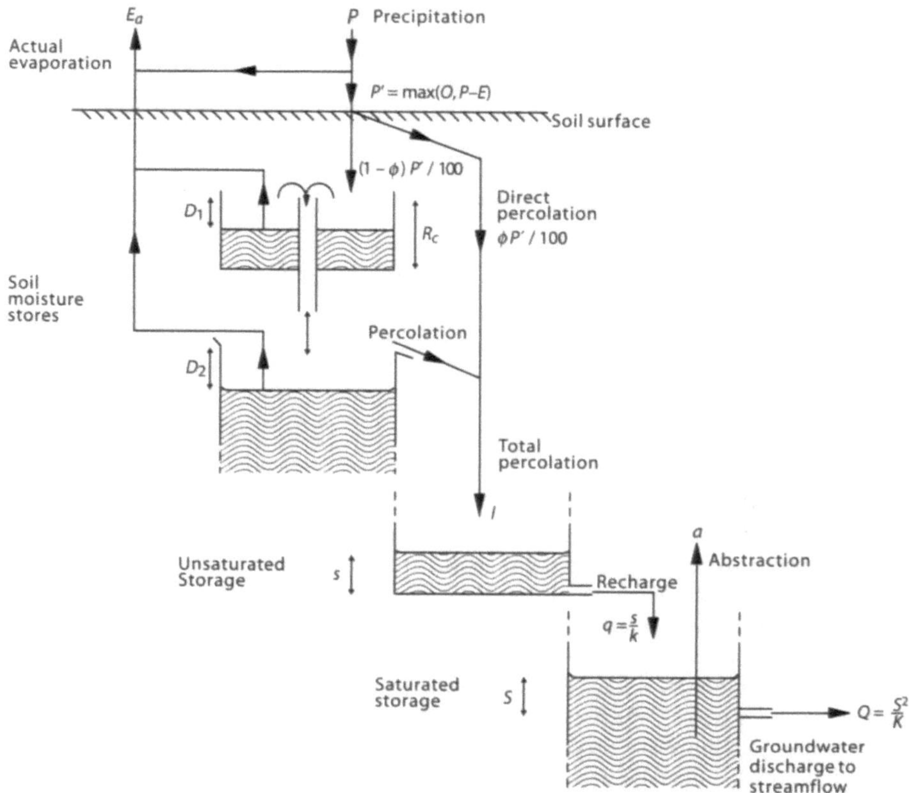

Fig. 11.1. The Thames Conceptual Model single zone representation

general technique for forecasting flow from rainfall at the basin scale. The model is formulated for implementation at a short time interval, (typically 15 min) and allows real-time measurements of flow to be incorporated in order to improve forecasts of flow. In addition, the model is capable of operating at longer time intervals (a day or more) and over a range of flows so that it is potentially useful for drought management, as well as, for flood forecasting and short-term river management. Central to the development of the PDM has been a view that the complex hydrological response of river basins is best represented by models which represent the components of run-off production and translation in a conceptual manner. This is not to decry the utility of simple black box models for certain applications but derives from experience gained on a number of river basins that conceptual models, when properly constructed, are able to better reproduce the nonlinear behaviour of the rainfall-runoff process. The model parameterisation has been developed in a fairly general way to allow representation of a broad range of hydrological responses. At the same time care has been taken to adopt a model parameterisation with as few parameters as possible and which is also amenable to automatic optimisation of the model parameters. A continuous time formulation ensures that the model may be readily operated for a range of time intervals

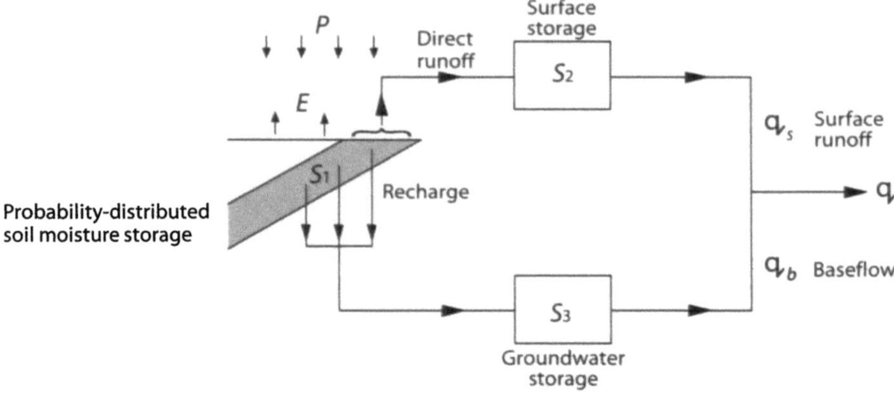

Fig. 11.2. The PDM model

from, say 5 min to a day or longer. The model incorporates within its internal structure a set of updating procedures which utilise recent measurements of flow available from telemetry to improve the accuracy of model forecasts; These procedures are based on an empirical state updating methodology which attributes the model error to different components of the model and adjusts the water content of each component accordingly.

The configuration of storage elements used in the PDM resembles that originally suggested by Dooge (1973) where rainfall is partitioned into storm response and groundwater runoff components. Figure 11.2 depicts the form of the model schematically. A probability-distributed soil moisture storage (Moore 1985) is used to partition rainfall into direct runoff and subsurface runoff and also performs a routing function for the drainage of soil water. Direct runoff is routed through a surface storage prior to contributing to basin runoff. The remaining rainfall enters soil storage where it is depleted by evaporation at a rate dependent on the available soil moisture storage and the potential evaporation rate. Further depletion of the soil water storage by recharge to groundwater occurs, and this water contributes to basin runoff after being routed by the groundwater storage. Consequently, there are two primary routing elements in the model, one for direct runoff and which essentially performs a channel storage routing function, and the other for recharge and which represents the release of water from groundwater storage. Although differing in the nature of their response, these two routing elements may be considered as common model components and similar mathematical functions may be used to represent their behaviour. It is seen that the PDM consists of a configuration of model components for soil moisture accounting and flow routing.

A feature of the above three conceptual models is that they are made up of rather similar components, differing essentially in the detail of their configuration. This feature is common to other conceptual models developed in other countries, most "brand name" models being essentially variations on a common set of components, Indeed, the PDM was in part developed to exploit the generic nature of model components commonly employed in conceptual rainfall-runoff models. With this in mind the remaining discussion on rainfall-runoff models will outline the model components available in the PDM software in more detail, whilst making reference to their application in other model formulations. The use of the probability-distributed storage concept used in the PDM model will be introduced in the next sub-section which is followed by an outline of non-linear storage models. A discussion of the role of Transfer Function or TF models, which are not commonly regarded as conceptual model components, concludes the section.

11.2.2.2
Probability-Distributed Storage Capacity Component

Consider that runoff production at any point within a river basin may be conceptualised as a single storage, or tank, of capacity c', representing the absorption capacity of the soil column at that point. The storage takes up water from rainfall (P), and loses water by evaporation (E), until either the storage fills and spills, generating direct runoff (q), or empties and ceases to lose water by evaporation. Figure 11.3a depicts such a storage, whose behaviour may be expressed mathematically by

$$q = \begin{cases} P - E - (c'-S_0) & P > c'+E \\ 0 & P \le c'+E \end{cases} , \tag{11.1}$$

where S_0, is the initial depth of water in storage, and where P, E and q represent the depth of rainfall, evaporation and the resulting direct runoff over the interval being considered. Now consider that runoff prediction at every point within a river basin may be similarly described, each point differing from one another only with regard to the storage capacity. The storage capacity at any point (c) may then be considered as a random variate with probability density function, $f(c)$, so that the proportion of the river basin with depths in the range (c, $c + \mathrm{d}c$) will be $f(c)\mathrm{d}c$.

The water balance for a river basin assumed to have storage capacities distributed in this way may be constructed as follows. First imagine that stores of all possible different depths are arranged in order of depth and with their open tops arranged at the same height: this results in a wedge-shaped diagram as depicted in Fig. 11.3b. If the basin is initially dry so that all stores are empty and rain falls at a net rate P for a unit duration, then stores will fill to a depth P unless they are of lesser depth than P when they will fill and spill. During the interval the shallowest stores will start generating direct runoff and at the end of the interval stores of depth P will just begin to produce runoff, so that the hachured triangular area denotes the depth produced from stores of different depth over the unit interval. Since, in general, there are more stores of one depth than another the actual runoff produced over the basin must be obtained by weighting the depth produced by a store of a given depth by its frequency of occurrence, as expressed by $f(c)$. Now, at the end of the interval, stores of depth less than P are generating runoff: let this critical capacity below which all stores are full at some time t be denoted by $C^* \equiv C^*(t)$ ($C^* = P$ in the present example). The proportion of the basin containing stores of capacity less than or equal to C^* is

$$\mathrm{prob}(c \le C^*) = F(C^*) = \int_0^{C^*} f(c)\mathrm{d}c . \tag{11.2}$$

The function $F(.)$ is the distribution function of store capacity and is related to the density function, $f(c)$, through the relation $f(c) = \mathrm{d}F(c) / \mathrm{d}c$. This proportion is also the

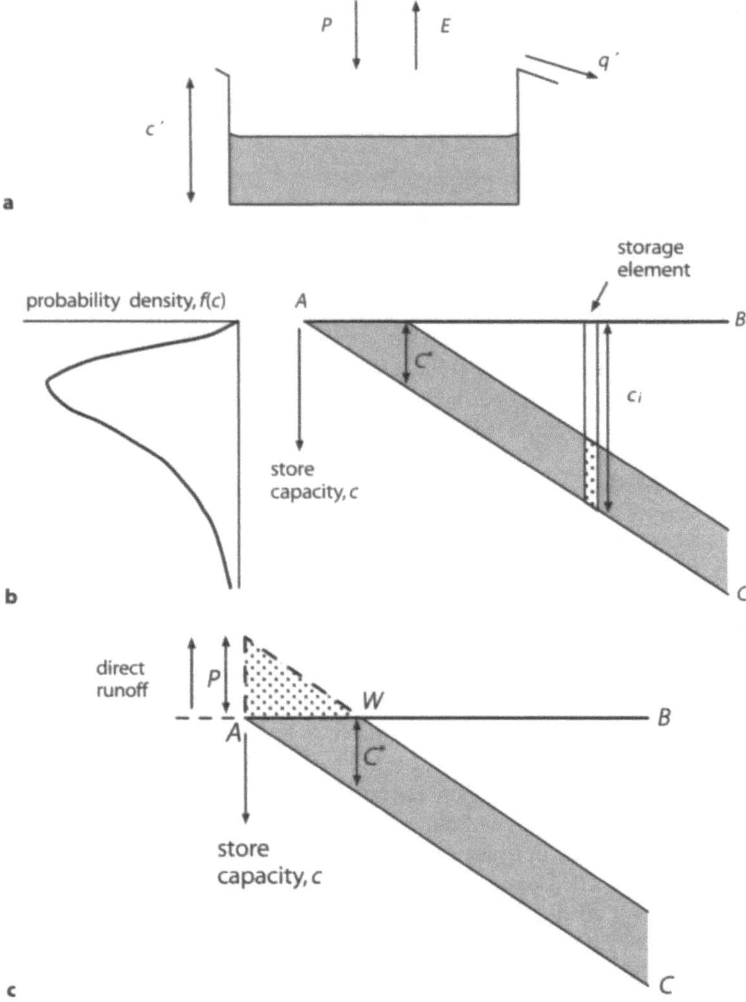

Fig. 11.3. Definition diagrams for the probability-distributed interacting storage capacity component. **a** Point representation of runoff production by a single store; **b** Basin representation by storage elements of different depth and their associated probability density function; **c** Direct runoff production from a population of stores

proportion of the basin generating runoff, so that the contributing area at time t for a basin of area A is

$$A_c(t) = F(C^*(t))A \ .$$

$$(11.3)$$

The instantaneous direct runoff rate per unit area from the basin is the product of the net rainfall rate, $\pi(t)$, and the proportion of the basin generating runoff, $F(C^*(t))$; that is

$$q(t) = \pi(t)F(C^*(t)) \ . \tag{11.4}$$

During the i'th wet interval, $(t, t + \Delta t)$, suppose rainfall and potential evaporation occur at constant rates P_i, and E_i, so that net rainfall $\pi_i = P_i - E_i$. Then the critical capacity, $C^*(\tau)$, will increase over the interval according to

$$C^*(\tau) = C^*(t) + \pi_i(\tau - t) \qquad t \leq \tau \leq t + \Delta t \ , \tag{11.5}$$

the contributing area will expand according to Fig. 11.3, and the volume of basin direct runoff per unit area produced over this interval will be

$$V(t + \Delta t) = \int_t^{t+\Delta t} q(\tau)d\tau = \int_{C^*(t)}^{C^*(t+\Delta t)} F(c)dc \ . \tag{11.6}$$

During dry periods potential evaporation will deplete the water content of each storage. It will be assumed during such depletion periods that water moves between storages of different depths so as to equalise the depth of stored water at different points within the basin. Thus, at any time all stores will have a water content, C^*, irrespective of their capacity, unless this is less than C^* when they will be full: the water level profile across stores of different depths will therefore, always be of the simple form shown in Fig. 11.3c. The assumption which allows redistribution of water between storages of different size during depletion periods is particularly important for real-time applications of the model where the possibility of updating the store contents is envisaged. Moore (1985) shows how this assumption, when not invoked, leads to a more complex water accounting procedure which is less amenable to real-time empirical state adjustment schemes. Particularly important is that a unique relationship exists between the water in storage over the basin as a whole, $S(t)$, and the critical capacity, $C^*(t)$, and in turn to the instantaneous rate of basin runoff production, $Q(t)$. Specifically, and referring to Fig. 11.3c, it is clear that the total water in storage over the basin is

$$
\begin{aligned}
S(t) &= \int_0^{C^*(t)} cf(c)dc + C^*(t)\int_{C^*(t)}^{\infty} f(c)dc \\
&= \int_0^{C^*(t)} (1 - F(c))dc
\end{aligned}
\tag{11.7}
$$

For a given value of storage, $S(t)$, this can be used to obtain $C^*(t)$ which allows the volume of direct runoff, $V(t + \Delta t)$, to be calculated using Eq. 11.6 together with 11.7.

The dependence of evaporation loss on soil moisture content is introduced by assuming the following simple function between the ratio of actual to potential evaporation, E_i' / E_i, and soil moisture deficit, $S_{max} - S(t)$:

$$\frac{E_i'}{E_i} = 1 - \left\{ \frac{(S_{max} - S(t))}{S_{max}} \right\}^{b_e} \ ; \tag{11.8}$$

either a linear ($b_e = 1$ so $E_i' = (S(t)/S_{max})E_i$) or quadratic form ($b_e = 2$) is usually assumed.

Here, $S_{\max}$ is the total available storage, and is given by

$$S_{\max} = \int_0^\infty cf(c)dc = \int_0^\infty (1 - F(c))dc = \bar{c} \ , \tag{11.9}$$

where $\bar{c}$ is the mean storage capacity over the basin.

Further loss as recharge to groundwater may be introduced by assuming that the rate of loss, γ_i, depends linearly on basin soil moisture content i.e.

$$\gamma_i = k_b \ S(t) \ , \tag{11.10}$$

where k_b is a groundwater recession constant with units of inverse time. Alternative formulations are available which allow recharge to depend on both soil and groundwater storage. With both losses to evaporation and recharge, the net rainfall (π_i) may be defined in general as

$$\pi_i = P_i - E_i' - k_b \ S(t) \ . \tag{11.11}$$

During a period when no runoff generation occurs then, for this general case, soil moisture storage accounting simply involves the calculation

$$S(\tau) = S(t) + \pi_i(\tau - t) \ , \qquad t \le \tau \le t + \Delta t, \quad 0 \le S(\tau) \le S_{\max} \ . \tag{11.12}$$

When runoff generation does occur then the volume of runoff produced, $V(t + \Delta t)$, is obtained using Eq. 11.6, and then continuity gives the replenished storage as

$$S(t + \Delta t) = \begin{cases} S(t) + \pi_i\Delta t - V(t + \Delta t) & S(t + \Delta t) \le S_{\max} \\ S_{\max} & \text{otherwise} \end{cases} \ . \tag{11.13}$$

If basin storage is fully replenished within the interval $(t, t + \Delta t)$ then $V(t + \Delta t)$ should be computed from continuity as

$$V(t + \Delta t) = \pi_i\Delta t - (S_{\max} - S(t)) \ . \tag{11.14}$$

The above completes the procedure for soil moisture accounting and determining the value of runoff production according to a probability-distributed storage capacity model. Figure 11.4 provides a graphical representation of this procedure for a wet interval $(t, t + \Delta t)$ during which soil moisture storage is added to by an amount $\Delta S(t + \Delta t) = \pi_i + \Delta t - V(t + \Delta t)$, and a volume of direct runoff, $V(t + \Delta t)$, is generated.

A specific application of the procedure can be developed for a given choice of probability density function. Analytical solutions of the integrals in the probability-distributed storage capacity model component (specifically Eqs 11.6 and 11.7) are presented in Institute of Hydrology (1992) for a range of possible distribution types. After a number of trials on alternative distributions, a Pareto distribution of storage capacity is now most widely used in practice and will be used to illustrate application of the method. The distribution function and probability distribution function for this distribution are

$$F(c) = 1 - (1 - c / c_{\max})^b \ , \qquad 0 \le c \le c_{\max} \ , \tag{11.15}$$

Fig. 11.4. The storage capacity distribution function used to calculate basin moisture storage, critical capacity, and direct runoff according to the probability-distributed interacting storage capacity model

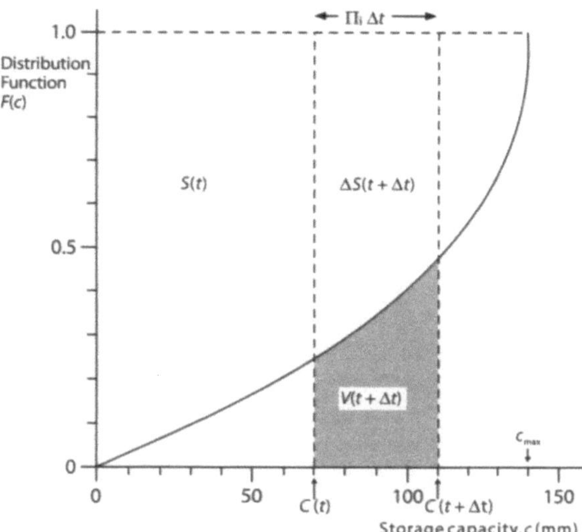

$$f(c) = \frac{dF(c)}{dc} = \frac{b}{c_{\max}}\left(1 - \frac{c}{c_{\max}}\right)^{b-1} , \qquad 0 \le c \le c_{\max} , \tag{11.16}$$

where parameter $c_{\max}$ is the maximum storage capacity in the basin, and parameter b controls the degree of spatial variability of storage capacity over the basin. These functions are illustrated in Fig. 11.5. Note that the rectangular distribution is obtained as a special case when $b = 1$, and $b = 0$ implies a constant storage capacity over the entire basin. The following relations apply for Pareto distributed storage capacities:

$$S_{\max} = c_{\max} / (b + 1) , \tag{11.17a}$$

$$S(t) = S_{\max}\left\{1 - (1 - C^*(t)/c_{\max})^{b+1}\right\} , \tag{11.17b}$$

$$C^*(t) = c_{\max}\left\{1 - (1 - S(t)/S_{\max})^{1/(b+1)}\right\} , \tag{11.17c}$$

$$V(t + \Delta t) = \pi_i \Delta t - S_{\max}\left\{(1 - C^*(t)/c_{\max})^{b+1} - (1 - C^*(t + \Delta t)/c_{\max})^{b+1})\right\} . \tag{11.17d}$$

The relationship between rainfall and runoff implied by the above expressions, for given conditions of soil moisture, is presented in Fig. 11.6. A related, if not similar, procedure forms the basis of the Xinanjiang model developed by Ren Jun Zhao and co-workers in China (Zhao and Zhuang 1963; Zhao et al. 1980) and most recently popularised and extended by Todini and co-workers in the form of the Arno model in Italy (Todini, personal communication). Indeed, Moore (1985) traces back the origins of such probability-distributed principles in hydrology to the pioneering contribution of Bagrov in 1950, working in what was then the USSR.

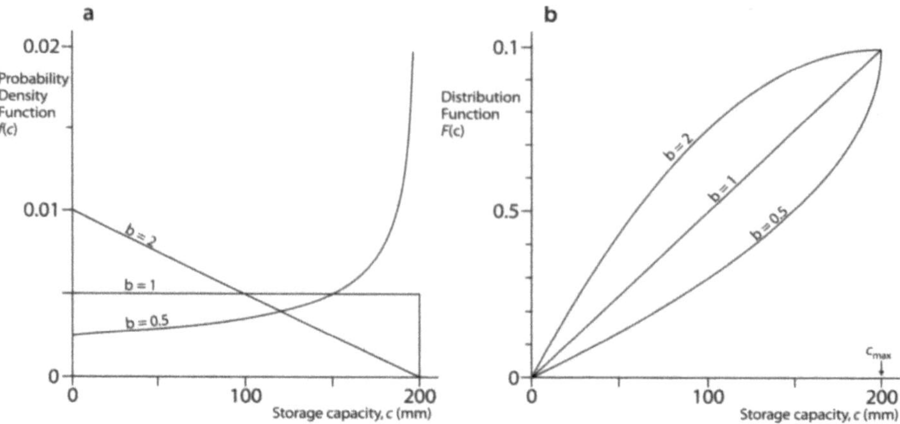

Fig. 11.5. The Pareto distribution of storage capacity. **a** Probability density function; **b** Distribution function

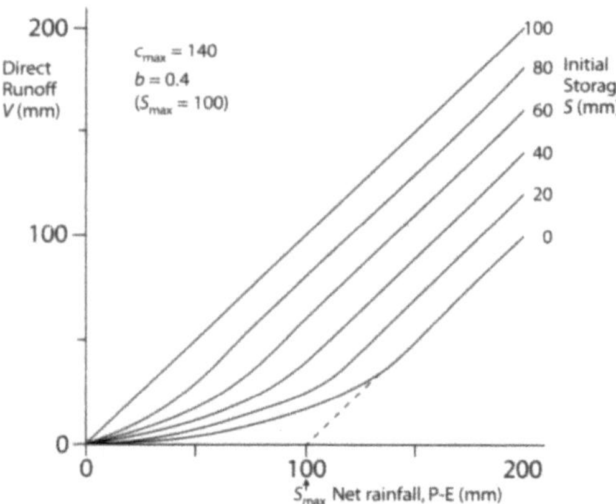

Fig. 11.6. Rainfall-runoff relationship for the probability-distributed interacting storage capacity model

11.2.2.3
Nonlinear Storage Components

Storage elements may be used to reproduce the effect of channel storage on direct runoff and of groundwater storage on recharge as water moves through the basin to the basin outlet. These storage elements may be defined mathematically through some empirical relation between storage and discharge. One simple relationism that discharge from storage (q) depends nonlinearly on the volume of water in storage (S), such that

$$q = k\, S^m\ , \qquad k > 0, \quad m > 0\ , \tag{11.18}$$

where k and m are parameters. Combining this with the equation for continuity

$$\frac{dS}{dt} = u - q \ , \tag{11.19}$$

where u is the input to the storage element (direct runoff or groundwater recharge in this instance), gives

$$\frac{dq}{dt} = a(u - q)q^b \ , \qquad a > 0, \quad -\infty < b < 1 \ , \tag{11.20}$$

where $a = m\,k^{1/m}$, $b = (m - 1)\,/\,m$. This nonlinear storage model is sometimes referred to as the Horton-Izzard model (Dooge 1973; Moore 1983).

For the recession case when the input u is zero, Eq. 11.20 may be solved to give

$$q(t + \Delta t) = \begin{cases} (q(t)^{-b} + ab\Delta t)^{-1/b} & b \neq 0 \ , \\ \exp(-k\Delta t)q(t) & b = 0 \ . \end{cases} \tag{11.21}$$

Exact and approximate solutions for a number of different values of m for the case when the input u is not zero, but assumed constant over the measurement interval, can be derived. Specifically for use in representing groundwater storage an exponent m equal to 3 has been found most satisfactory. An approximate solution in this case may be obtained by using a method due to Smith (1977) which gives the following recursive equation for storage:

$$S(t + \Delta t) = S(t) - \frac{k}{3S(t)^2}\Big\{\exp(-3kS(t)^2\Delta t) - 1\Big\}(u - kS(t)^2) \ . \tag{11.22}$$

Discharge may then be obtained simply using the nonlinear relation

$$q(t + \Delta t) = kS(t + \Delta t)^3 \ . \tag{11.23}$$

Further details of this approximation for a cubic storage function are given in Appendix 2 of Institute of Hydrology (1992). It is this form of element that is most commonly used to represent the slow storage component of the PDM model, although other choices are available.

Two other special cases are, or have been used; in the UK for operational flood forecasting. A choice of $m = 2$, giving the quadratic storage function model was proposed for use in design by Mandeville (1975) as part of the UK Flood Study (NERC 1975). As the Isolated Event Model, or IEM, it is in use for flood forecasting in London and the Thames basin (Haggett et al. 1991) and in Haddington, Scotland (Brunsden and Sargent 1982). In this form it serves as a fast response routing function in combination with a loss function. In its classic form this loss function depends exponentially on soil moisture deficit as calculated using a Penman-type soil moisture accounting procedure; however, in the Scottish implementation of the IEM "initial flow prior to the storm" is used as an index of antecedent conditions in preference to soil moisture deficit. The same quadratic storage function is employed in the Thames Conceptual Model, or TCM, where it functions as a lower zone routing operator (Fig. 11.1). This model is also used for forecasting in the Thames basin.

The second special case, the exponential storage model, is obtained when $b = 1$ in the Horton-Izzard model of Eq. 11.20. This implies the logarithmic storage relation

$$\log q = \gamma + aS \quad \text{or} \quad q = \exp(\gamma + aS) \ , \tag{11.24}$$

where a is the same parameter as appears in Eq. 11.41, and γ is an intercept parameter. It is this form that has been most associated with the work of Lambert (1972) and which was widely used in the River Dee, North Wales as a rainfall-runoff model for real-time flood forecasting (Central Water Planning Unit 1977). In fact, piecewise storage functions were often used with both linear and exponential storage forms.

The analytical recursive relations employed in the implementation of these two special cases are given below:

- *Quadratic storage case:*

$$q(t + \Delta t) = u\left(\frac{z-1}{1+z}\right)^2 \ , \tag{11.25}$$

where

$$z = \exp(a\Delta t u^{1/2})\left[\frac{1 + (q(t)/u)^{1/2}}{1 - (q(t)/u)^{1/2}}\right] \ , \tag{11.26}$$

or alternatively

$$q(t + \Delta t) = u\left[\frac{(q(t)/u)^{1/2} + \tanh\left\{(uk)^{1/2}\Delta t\right\}}{1 + (q(t)/u)^{1/2}\tanh\left\{(uk)^{1/2}\Delta t\right\}}\right]^2 \ . \tag{11.27}$$

- *Exponential storage case:*

$$\begin{aligned}
q(t + \Delta t) &= \frac{q(t)}{(q(t)/u) + (1 - q(t)/u)\exp(-a\Delta t u)} \\
&= \frac{q(t)}{\exp(-a\Delta t u) + (q(t)/u)(1 - \exp(-a\Delta t u))} \ .
\end{aligned} \tag{11.28}$$

Again, these solutions along with other special cases are available as options in the PDM software.

In the Thames Conceptual Model allowance is made for groundwater abstraction from the quadratic storage and the possibility of negative values for the input u. Then the following expression holds:

$$q(t + \Delta t) = u\tan^2\left\{\tan^{-1}(g(t)/(-u))^{1/2} - (-uk)^{1/2}\Delta t\right\} \tag{11.29}$$

and flow will cease after a time

$$\Delta t' = (1/(-uk))^{1/2}\tan^{-1}(q(t)/(-u))^{1/2} \ , \tag{11.30}$$

when a volume deficit begins to build up which at the end of the interval is equal to $u(\Delta t - \Delta t')$. Inflow to the quadratic storage from the linear unsaturated soil store is calculated as the mean outflow over the interval from a linear storage; that is

$$\overline{q}(t + \Delta t) = (k\Delta t)^{-1}(1 - \exp(-k\Delta t))q(t) + \left\{1 - (k\Delta t)^{-1}(1 - \exp(-k\Delta t))\right\}u \cdot \qquad (11.31)$$

11.2.2.4
Transfer Function Components

Transfer function, or TF, models as popularised by Box and Jenkins (1970) are well known to hydrologists but their relevance to real-time flood forecasting is perhaps less well understood. A direct application of this class of model to rainfall-runoff model-ling is arguably inappropriate because of its simple linear form where an output $q_t \equiv q(t)$ is related to an input $u_t \equiv u(t)$ by a discrete time difference equation.

$$q_t = -\delta_1 q_{t-1} - \delta_2 q_{t-2} - \dots - \delta_r q_{t-r} + \omega_0 u_{t-r} + \omega_1 u_{t-r-1} \dots + \omega_{s-1} u_{t-r-s+1} \, , \qquad (11.32)$$

with dependence on r past outputs and s present and past inputs delay by τ time units.

If the input-output pair is rainfall-runoff then the nonlinearity known to exist by hydrologists is clearly not represented. This problem has been addressed by using a nonlinear loss function to transform rainfall to "direct runoff" or "effective rainfall" and using this as the input variable u_t. Functionally, the transfer function serves as a simple linear routing function. Alternatively, a parallel system of two transfer function models can be envisaged together with a partitioning rule which directs rainfall to the two functions which operate as slow and fast translation pathways. A variety of non-linear loss functions and parallel TF model functions were investigated in the UK for use in flood forecasting (Moore 1980, 1982). Most recently, an improved estimation scheme for this class of parallel TF model has been developed by Jakeman et al. (1990) which overcomes some of the problems encountered in this earlier work.

Other workers have sought to circumvent the shortcomings of the linear transfer func-tion by allowing the parameters to be time-variant and tracking the variation using a re-cursive estimation scheme. For example, Cluckie and Owens (1987) employ a TF model in such a way that a single gain parameter, controlling the proportion of rainfall that becomes runoff, is recursively estimated. Transfer function models also form the basis of the Nith flood forecasting system in Scotland, developed at the University of Lancaster in associa-tion with the Solway River Purification Board (Lees et al. 1993). They are used to relate upstream level to downstream level and smoothed effective rainfall (defined by a nonlin-ear operation involving the product of rainfall and the previous river level) to river level. The model steady state gains were found to be time variant and are tracked using recur-sive least squares, assuming a random walk process for the parameter variability, in a similar way to the approach adopted by Owens and Cluckie in Northwest England. A drawback of this recursive approach is that the variation is merely "tracked" and not "anticipated". Our understanding of hydrological science, for example, tells as that antecedent wet-ness can influence the gain or runoff proportion and that soil moisture accounting model components can be used to anticipate this effect. This leads one to recognise that the role of the transfer function is primarily that of a linear routing operation and can be incorporated, as such, into a conceptual model as merely one component form.

Indeed it is useful to proceed further and to seek equivalencies between the so-called "black box" form of the TF model and simple hydrological conceptualisations. Consider, for example, that routing of direct runoff to the basin outlet may be represented by a cascade of two linear reservoirs, so

- *Momentum:*

$$q_i = k_i S_i \ , \qquad i = 1, 2 \ ; \tag{11.33}$$

- *Continuity:*

$$\frac{dS_1}{dt} = u - q_1 \ , \qquad \frac{dS_2}{dt} = q_1 - q_2 \ . \tag{11.34}$$

Differentiation and substitution yields the following second order differential equation in terms of discharge from the second storage, q_2, and the input to the first storage, u:

$$\frac{d^2 q_2}{dt^2} + (k_1 + k_2)\frac{dq_2}{dt} + k_1 k_2 q_2 = k_1 k_2 u \ . \tag{11.35}$$

This may be factorised to yield the equation

$$(1 + k_1^* D)(1 + k_2^* D)\, q_2(t) = u(t) \ , \tag{11.36}$$

where D is the differential operator, $D^m q_2(t) = d^m q_2(t) / dt^m$, and $k_1^* = k_1^{-1}$, $k_2^* = k_2^{-1}$ are the storage time constants with dimension time. Solving over an interval, $(t - \Delta t, t)$, assuming that the input over this interval is constant and equal to u_t gives the discrete time difference equation

$$q_t = -\delta_1 q_{t-1} - \delta_2 q_{t-2} + \omega_0 u_t + \omega_1 u_{t-1} \ , \tag{11.37}$$

where the suffix 2 has been dropped since it is clear that it relates to output from the second reservoir, and the time suffix t is introduced to emphasise that $\{q_t\}$ refers to flow at discrete time intervals, ..., t, $t + 1$, $t + 2$, ... It may be shown (O'Connor 1982) that the parameters of this difference equation relate to those of the differential equation as follows:

$$\delta_1 = -(\delta_1^* + \delta_2^*) \ , \qquad \delta_2 = \delta_1^* \delta_2^* \ ,$$

where

$$\delta_1^* = \exp(-k_1 \Delta t) \ , \qquad \delta_2^* = \exp(-k_2 \Delta t) \ ,$$

$$\omega_0 = \frac{k_1^*(\exp(-k_1 \Delta t) - 1) - k_2^*(\exp(-k_2 \Delta t))}{k_2^* - k_1^*} \ , \qquad k_1 \neq k_2 \ ,$$

$$\omega_1 = \frac{k_2^*(\exp(-k_2\Delta t)-1)\exp(-k_1\Delta t) - k_1^*(\exp(-k_1\Delta t)-1)\exp(-k_2\Delta t)}{k_2^* - k_1^*} \ , \qquad k_1 \neq k_2 \ ,$$

$$\omega_o = 1 - (1 + k_1\Delta t)\exp(-k_1\Delta t) \ , \qquad k_1 = k_2 \ ,$$

$$\omega_1 = (\exp(-k_1\Delta t) - 1 + k_1\Delta t)\exp(-k_1\Delta t) \ , \qquad k_1 = k_2 \ . \tag{11.38}$$

Note that the difference equation has four parameters but the differential equation has only two: we therefore see that the link between the original continuous time formulation and its discrete time equivalent allows for a more parsimonious parameterisation of the TF model. This is in part accounted for by the fact that the relation ensures that the two models preserve continuity whereas, the general form the TF model would not. The continuity preserving nature of this TF model when constrained to be coincident with the continuous time model at discrete time intervals means that the model's steady state gain is unity and therefore $\omega_o = 1 + \delta_1 + \delta_2 - \omega_2$.

Note that by linking the TF model with the continuous time model means that discrete time transfer function models may be specified for any sampling interval Δt. This means that the model may be operated at different time intervals appropriate to the level of hydrological activity. However, care must be exercised here since the assumption of constant input over the sampling interval will affect the flow response if a time interval coarser than the input sampling interval is used during periods of changing input. Of course, no problem arises in using a coarser time interval for the transfer function model if the input is zero.

The TF model of the above constrained type is available as a component option in the PDM software and is commonly invoked to represent routing of direct runoff along a fast response pathway, such as an open channel. Cluckie and Han (personal communication) have used stability theory associated with TF models to develop a Physically Realisable Transfer Function or PRTF model whose impulse response is constrained to be both stable and non-oscillatory. This is achieved by constraining the δ_i parameters such that $\delta_i = C_r^{r-i}(-\beta)^{-i}$ where C is a combinatorial and β a parameter which can be related to the time-to-peak of the impulse response. The PRTF model is in use as a rainfall-runoff model in parts of the Wessex and Anglian regions of the UK for operational flood forecasting.

11.2.3
Channel Flow Routing Models

The kinematic wave model, in its basic form, is usually inadequate for forecasting flows over the full range experienced in natural river channels. However, it does provide an important point of departure from which to develop more realistic, but still simple, representations of channel flow for application in real-time for flood warning and control, reservoir flow augmentation and intake management. Initial developments using a simple one parameter model have been reported by Moore and Jones (1978) and an extension to a four parameter nonlinear form by Jones and Moore (1980). The model has since been extended to incorporate additional functionality following experience gained in its use for modelling the River Wyre in North Lancashire and, most notably the entire River Ouse network above its tidal limits in Yorkshire. Documentation of the Computer program KW, which encompasses this model formulation and its developments, is available as RFFS Technical Note N° 10 (Institute of Hydrology 1992c). Features of the model include:

i. Discharge-dependent wave speed, with a variety of functional forms to choose from;
ii. A simple representation of overflows into washlands using a threshold storage function which may be constrained to preserve continuity of water or used to accommodate gains or losses to the system to compensate for, for example, faulty ratings at high flows or losses to mine workings;
iii. Incorporation of a stage-discharge relation as part of the overall model calibration Task in order to allow model calibration to proceed using level-only records.

The basic model formulation begins with the 1-D kinematic wave equation in partial differential equation form where channel flow (q) and lateral inflow per unit length of river (u) are related through

$$\frac{\partial q}{\partial t} + c\frac{\partial q}{\partial x} = cu \; , \tag{11.39}$$

where c is the kinematic wave speed. Consider time (t) and space (x) to be divided into discrete intervals Δt and Δx such that k and n denote positions in discrete time and space. Invoking forward difference approximations to the derivatives in Eq. 11.39 we have.

$$q_k^n = (1 - \theta)q_{k-1}^n + \theta(q_{k-1}^{n-1} + u_k^n) \; , \tag{11.40}$$

where the dimensionless wave speed parameter $\theta = c\,\Delta t\,/\,\Delta x$ and $0 < \theta < 1$. This is a recursive formulation which expresses flow out of the n'th reach at time t, q_k^n, as a linear weighted combination of the flow out of the reach at the previous time and inflows to the reach from upstream (at the previous time) and as the total lateral inflow along the reach (at the same time). If the stretch of river to be modelled has length L, and is subdivided into N reaches of equal length, so $\Delta x = L\,/\,N$, then a condition for stability is that $c < L\,/\,(N\,\Delta t)$.

An alternative derivation of Eq. 11.40 can be sought from a simple hydrological storage approach. The n'th reach can be viewed as acting as a linear reservoir with its outflow related linearly to the storage of water in the reach such that

$$q_k^n = \kappa\, S_k^n \; , \tag{11.41}$$

where κ is a time constant with units of inverse time. If S_k^n is the storage in the reach just before flows are transferred at time k then continuity gives

$$S_k^n = S_{k-1}^n + \Delta t(q_{k-1}^{n-1} - q_{k-1}^n + u_k^n) \tag{11.42}$$

and the equivalence to Eq. 11.40 follows, given $\theta = \kappa\Delta t$.

A major weakness of the basic formulation is that a single parameter θ controls both the speed and attenuation of flows for a given choice of space and time step and when constant a fixed wave speed is imposed. A simple generalisation is to assume that the wave speed parameter varies with time, so θ is replaced by θ_k^n in Eq. 11.40 allowing the parameter to vary at each time step and for each reach. In practice, a length of river is chosen for which it is a reasonable approximation to allow the speed to be constant from reach to reach (i.e. speed is constant for all n). A relationship is then invoked between the wave speed (θ_k) and the modelled flow out of the furthest downstream reach at the previous time step, q_k^N. The choice of functional form draws on hydraulic theory and empirical evidence that wave speed increases within the channel section with increasing discharge (e.g. Price 1977). As the discharge approaches bankfull and spillage onto washland areas occurs then a reduction in overall wave speed occurs with further increase in discharge. This trend later reverses as the flood plain begins to function as an open channel and speeds increase again with discharge. A parametrically efficient representation of this behaviour is provided by a simple cubic function, for which we have

$$\theta_k = a + bq_{k-1}^N + c(q_{k-1}^N)^2 + d(q_{k-1}^N)^3 \; , \tag{11.43}$$

where a, b, c and d are parameters. A second functional form available for use is the "exponential" model:

$$\theta_k = a + q_{k-1}^N(b + c\exp(-dq_{k-1}^N)) \; , \tag{11.44}$$

defined by the four parameters a, b, c and d. As an alternative to these parametric forms the option to define the speed-discharge function as a set of piecewise linear segments is provided in the KW software. In certain circumstances survey data providing cross-sectional information at various points along a reach of river may be available. RFFS Technical Note N° 2 showed how hydraulic theory together with such survey data may be used to calculate the speed-discharge relation without recourse to fitting parametric functions (Institute of Hydrology 1989). Pre-calculated relations of this kind can be incorporated in the KW program by invoking the "look-up table" option, which as its name suggests allows any relation available in tabular form to be used. Two parameters are available to "distort" the basic tabulated relation to achieve a better fit. It should be remembered that whilst this option is the most physically-based there is little guarantee that it will provide the best model performance. In general, the piecewise linear, cubic or constant forms are preferred for real-time use.

The speed-discharge relations represent in broad terms the change in speed and attenuation of a flood wave when it propagates through a compound channel made up of in-bank and flood plain elements. The representation applies to the whole stretch of river modelled. In practice, natural river channels often exhibit a tendency to overflow at restricted localities – where the flood banks are lower or at confluence points affected by backwater – and water may be stored temporarily in preferential areas on the flood plain. Water on the flood plain may be stored to eventually contribute to the downstream hydrograph or may be lost to the channel system by evaporation or lost, for example, by filling mine workings. Threshold storage functions have been developed as a means of representing such localised behaviour which effects both the attenuation and mass balance of propagating channel flows.

A threshold storage function may be invoked either at the junction of two model reaches or to the lateral inflow to a reach. In its simplest form the function operates as a simple transformation of flow such that below a certain flow threshold value, typically related to the bankfull discharge, no change occurs; above the threshold, a storage mechanism operates. The implied relation between transformed flow (q) and storage (S) for the simplest threshold storage functions is defined as

$$q = \begin{cases} S & S \le q_o \\ q_o + b(S - q_o) \end{cases} , \qquad \begin{aligned} S \le q_o \\ S > q_o \end{aligned} , \tag{11.45}$$

where the two parameters, q_o and b, are the flow threshold value and a scaling factor respectively. Two generalisations of this are available which can accommodate more general scaling and an additional threshold level.

In the above, the storage S is regarded as equal to the inflow to the threshold storage and the formulation implies that some water is lost when flows exceed the flow thresholds q_o and q_1. The inflow might be the model flow out of the n'th reach, q_k^n, or the lateral inflow to the n'th reach, u_k^n. A conserving form of these functions can be invoked by considering S to be the sum of the inflow to the storage and the water retained by the threshold storage at the previous time step. This forces flows held back in storage to progressively return to the channel system.

It is not an uncommon occurrence that river level data exist for a site on a river for which there is no corresponding rating relation. This may arise simply through the absence of a routine current metering programme or because hydraulic conditions, notably backwater effects and unstable sections, make developing a unique rating problematic. In real-time applications it is usually worthwhile developing a rating, however approximate, in order to maximise the use of real-time telemetered level data. An attractive solution to this problem is to incorporate the rating relation as an integral component of the channel flow model formulation and to estimate its parameters along with those of the KW model. The rating is formulated in the usual way as a set of power law relations between flow (Q) and level (H) over different level ranges with appropriate datum adjustments. Thus, the m sets of relations defining a rating can be expressed as

$$Q = \kappa_i (H + \alpha_i)^{p_i} , \qquad l_i \le H \le u_i \qquad (i = 1, 2, \ldots, m) , \tag{11.46}$$

where the coefficient κ_i the datum correction α_i and the exponent p_i are parameters and l_i and u_i define the lower and upper limits of the i'th relation. In practice, alternative parameterisations of this rating are provided in the KW program. Incorporation of the rating within the overall KW model formulation allows the model mass balance to have considerable freedom and considerable care must be exercised in the use of this facility.

The basic formulation of the kinematic wave model allows for lateral inflows to each reach through the term u_k^n. In many cases the lateral inflows will be ungauged and will have to be estimated from available data. The main approach adopted is to transfer flows from a neighbouring and/or a hydrologically similar basin; this may be the basin which forms the upstream boundary to the modelled stretch. A simple scaling of the flows by a "weight" which accounts for the areas and standard annual average rainfalls of the two basins provides a physically-based means of data transfer. Formally, the weight is defined for basin j for transfer of flows from basin g as $w_j = A_j P_j / (A_g P_g)$ where A and P, with their respective basin subscripts, denote basin area and Standard Average Annual Rainfall. In practice, these physical weights may be used as starting values for subsequent optimisation or optimisation may be relied upon entirely to establish the weight values.

An alternative hydrological channel flow routing model allowing for variable wave speed and out-of-bank flows, called the DODO model, is in use in the Severn-Trent region (Douglas and Dobson 1987). Reach outflow is regarded as a linear weighted sum of lagged inflow and storage in the reach; out-of-bank flows are similarly modelled in parallel to the in-bank flows but with different weights. Wave speed is assumed to vary as a power function of the reach inflow and used to define a variable time lag for the inflow. Special account is taken of the drainage of static floodplain storage back into the main channel, for example to mimic submerged flapped outfalls.

Modelling of a tidal river with the RFFS is accomplished using a model algorithm called HYDRO, based on the United States National Weather Service's DWOPER/NET-WORK program (Fread 1985), and which employs a four-point implicit solution of the Saint Venant equations. It has been adapted for use within the RFFS for real-time use and includes extensions to represent "static washlands", multi-branched networks and automatic rules for gate operation, as well as, improvements to achieve faster model execution and more flexible specification of model configuration data. Its application to the tidal Ouse in Yorkshire incorporates rules for operating a tidal barrier; most recently HYDRO has been applied to the tidal Shenzhen River in Hong Kong.

11.2.4
Snowmelt Models

An analysis of currently available snowmelt models reveals that they are made up of essentially four components (Harding and Moore 1988):

i. An input transformation, correcting for the representativeness of climatic inputs, especially precipitation;
ii. A surface melt component, often using a simple excess temperature mechanism as a substitute for a full energy budget controlled melt formulation;
iii. A snowpack storage mechanism, controlling how surface melt is retained within the pack; and
iv. A drainage term, defining the release of water from the pack, and often formulated as an integral part of (iii).

The PACK model incorporated into the RFFS is based on these four components and employs representations particularly suited to UK conditions and for real-time application. The model is based on a subdivision of a snowpack into "dry" and "wet" stores, a temperature excess melt equation controlling the passage of water between them. An areal depletion curve is invoked to allow shallow packs to only cover a fraction of the basin area. Figure 11.7 presents the general schematic form of the model. The model is available for calibration to snow survey data within the RFFS Model Calibration Facilities (Institute of Hydrology 1992c) and point and basin forms of the model are incorporated as Model Algorithms within the operational RFFS.

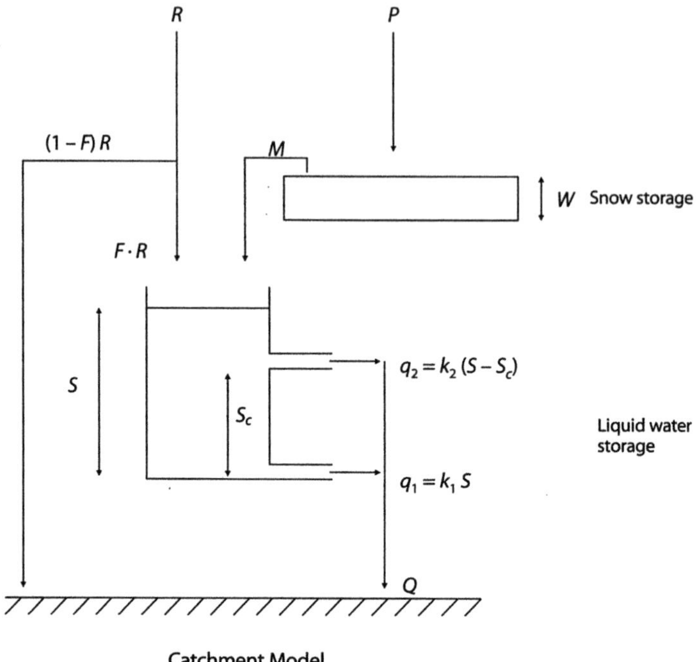

Fig. 11.7. The PACK snowmelt model

The model formulation may be summarised as follows.

Input Transformation. Any precipitation measurements are first corrected for representativeness using a factor c for example to correct for gauge loss or altitude and aspect effects. A temperature threshold (T_s) is used to discriminate precipitation, p (after correction for representativeness) into rainfall (R) and snowfall, P. Thus we have $R = p$ and $P = 0$ when the standard air temperature $T \geq T_s$ and $R = 0$ and $P = p$ otherwise. It will be assumed at this stage that measurements of precipitation in the form of snow are available. A typical value for T_s under UK conditions is 1° C.

Melt Equation. A simple temperature excess representation of the rate of melt (M) is used of the form:

$$M = \begin{cases} f(T - T_{\mathrm{m}}) & T > T_m \\ 0 & \text{otherwise} \end{cases}, \tag{11.47}$$

where T_{m} is a critical temperature above which melt occurs and f is a melt factor in units of mm/d/° C. The critical temperature T_{m} is usually taken to be 0° C. An extension to incorporate wind speed may be important for UK conditions (Harding and Moore 1988) but the added level of complexity has not been incorporated in the current implementation of the model.

Snowpack Storage and Drainage. Water accounting within the snowpack is accomplished by introducing the concept of "dry" and "wet" stores. New snow falling on the pack contributes to the dry store and water continuity is defined by the equation

$$\frac{dW}{dt} = P - M \; , \tag{11.48}$$

where W is the water equivalent of the snow in the dry store which is added to by snowfall (P) and is depleted by melt (M).

A second wet store receives water as rainfall (R), as melt from the dry store (M) and loses water via losses to drainage (Q), which can subsequently form the input to a catchment scale rainfall-runoff model. Continuity in this case gives

$$\frac{dS}{dt} = R + M - Q \; , \tag{11.49}$$

where S is the water equivalent of the snow in the wet pack store.

Release of water from the wet pack as drainage occurs at a slow rate proportional to the wet pack storage. When the total water equivalent of the pack exceeds a critical limit, referred to as the Critical Water Capacity (S_c) then release of water begins to occur at a faster rate. This mechanism can be conceptualised by viewing the wet store as a tank with two orifices, one located at the base of the store and releasing water slowly and a second one at a height, which varies dynamically as a function of the total water equivalent of the pack. The total drainage from the store can be expressed mathematically as

$$Q = q_1 + q_2 = \begin{cases} k_1 S \\ k_1 S + k_2(S - S_c) \end{cases} , \qquad \begin{cases} S \leq S_c \\ S > S_c \end{cases} , \tag{11.50}$$

where k_1 and k_2 are storage time constants with units of inverse time. The dynamically varying level of the upper orifice S_c is defined through the relation

$$S_c = S_c^*(S + W) \; , \tag{11.51}$$

where S_c^* is the maximum liquid water content (expressed as a proportion of the pack water content), a fixed parameter of the model. This formulation is similar to that used in the Japanese Tank Model (Sugawara et al. 1984). As a special, simplified case the lower orifice may be removed and the upper orifice replaced by an open tank by setting k_1 to 0 and k_2 to 1.

Solution of the equations of continuity in conjunction with the dynamics as represented above is achieved using a simple discrete time formulation. Representing the dry and wet store contents at the discrete time point t by W_t and S_t then the set of water accounting equations for use between the discrete time points $t-1$ and t of duration Δt are:

$$W_t = W_{t-1} + (P_t - M_t)\Delta t \; ,$$

$$S_t' = S_{t-1} + (R_t + M_t)\Delta t \; ,$$

$$S_c = S_c^*(S_t' + W_t) \ ,$$

$$Q_t = \begin{cases} k_1 S_t' & S_t' \leq S_c \\ k_1 S_t' + k_2(S_t' - S_c) & S_t' > S_c \end{cases} , \qquad (11.52)$$

$$S_t = S_t' - Q_t \Delta t \ .$$

A further refinement is introduced to inhibit drainage during cold periods. A cold period is defined to be below a temperature of T_c taken to be $0°C$ in the present implementation.

Areal Depletion Curve. A further extension of the model to incorporate the phenomenon that shallow snowpacks may occupy only a fraction F of the basin may be important in some instances. The fraction of snow cover may be allowed to vary as a function of the total water equivalent of the pack $\theta = S + W$ ranging from zero when $\theta = 0$ to unity when θ exceeds a critical value θ_c and complete snow cover occurs. The functional form for $F \equiv F(\theta)$ adopted here derives from

$$\theta = (\theta_c + 1)^{F(\theta)} - 1 \ , \qquad (11.53)$$

suggested by Laramie and Schaake (1972). In terms of $F(q)$ we have

$$F(\theta) = \frac{\log(\theta + 1)}{\log(\theta_c + 1)} \ . \qquad (11.54)$$

Given the current water content of the pack θ the fraction of the basin covered by snow is readily calculated from the above. In the event of a fresh snowfall, $\Delta\theta$, it is assumed temporally that the fraction reverts to 1 until a fraction $(1 - \alpha)\Delta\theta$ has melted. A linear reversion to the original point on the areal depletion curve occurs in melting the remainder of the new snow, $\alpha\Delta\theta$. Normally the proportion α, is set to 0.25. Any rain falling on the fraction devoid of snow is available immediately for input to the rainfall-runoff model for the basin.

The PACK model is available within the River Flow Forecasting System in calibration form for parameter estimation using snow survey data and as two model algorithms, for point and basin-scale use, for operational snowmelt flood forecasting.

11.3
Real-Time Updating Procedures

11.3.1
Introduction

The use of current or recent measurements, for example of river level received via telemetry, to improve model performance is called "updating". Conventionally a model is a mathematical operation which transforms a set of inputs (e.g. rainfall and potential evaporation) to a set of outputs (e.g. soil moisture deficit and basin runoff) without reference to the measured outputs, except for the purposes of model initialisation. The potential clearly exists to improve the model forecasts using measurements of the outputs in some way. Three forms of updating are in common use:

i. *State updating*, for example where corrections are made to the water content of internal storages to achieve closer correspondence between measured and predicted flow;
ii. *Error prediction*, where the tendency for model errors to persist is exploited in order to predict future errors;
iii. *Parameter updating*, where the model parameters are adjusted to achieve better agreement between observed and model predicted values.

A fourth possibility exists to adjust the input but this will not be discussed further here. In many respects, state and parameter updating are similar and are often based on some form of Kalman filter scheme (Jazwinski 1970; Gelb 1974; Moore and Weiss 1980). However, the use of parameter updating for flood forecasting appears less attractive because parameter variability is often a reflection of an inadequate model structure, which the updating scheme can merely track. For example, it is usually preferable to model the dependence of runoff production on soil moisture than to track the time variation of a runoff coefficient (or gain parameter) using a parameter updating scheme. There has also been some discussion of this problem in the context of TF models in Section 11.2.2.4. For these reasons attention will focus here on state updating and error prediction methods.

11.3.2
State Updating

The term "state" is used to describe a variable of a model which mediates between inputs to the model and the model output (Szollosi-Nagy 1976). In the case of the PDM rainfall-runoff model the main input is rainfall and basin flow is the model output. Typical state variables are the water contents of the surface and groundwater stores, S_2 and S_3 and of the probability-distributed soil storage S_1 (using the notation of Fig. 11.2). The flow rates out of the conceptual stores can also be regarded as state variables: examples are q_s the flow out of the surface storage, and q_b the flow out of the groundwater storage.

When an error, $\varepsilon = Q - q = Q - (q_s + q_b)$, occurs between the model prediction (q) and the observed value of basin runoff (Q), it would seem sensible to "attribute the blame" to mis-specification of the state variables and attempt to "correct" the state values to achieve concordance between observed and model predicted flow. Mis-specification may, for example, have arisen through errors in rainfall measurement which, as a result of the model water accounting procedure, are manifested through the values of the store water contents, or equivalently the flow rates out of the stores. A formal approach to "state correction" is provided by the Kalman filter algorithm (Jazwinski 1970; Gelb 1974; Moore and Weiss 1980). This provides an optimal adjustment scheme for incorporating observations, through a set of linear operations, for linear dynamic systems subject to random variations which may not necessarily be Gaussian in form.

For nonlinear dynamic models, such as the PDM, an extended form of Kalman filter based on a linearisation approximation is required which is no longer optimal in the adjustment it provides. The implication of this is that simpler, intuitive adjustment schemes can be devised which potentially provide better adjustments than the more complex and formal extensions of the Kalman filter which accommodate nonlinear dynamics through approximations. We will call such schemes which make physically sensible adjustments "empirical state adjustment schemes". A simple example is the apportioning of the error (ε) between the surface and groundwater stores of the PDM in proportion to their contribution to the total flow. Mathematically this may be expressed as

$$q_b^* = q_b + \alpha g_b \varepsilon \;,$$ (11.55a)

$$q_s^* = q_s + (1 - \alpha) g_s \varepsilon \;,$$ (11.55b)

where

$$\alpha = q_b / (q_s + q_b)$$ (11.56)

and the superscript * indicates the value after adjustment. The "gain" coefficients, g_b and g_s when equal to unity yield the result that $q_b^* + q_s^*$ equals the observed flow (Q), thus achieving exact correction of the model flow to equal the observed value. Values of the coefficients other than unity allow for different adjustments to be made, and g_b and g_s can be regarded as model parameters whose values are established through optimisation to achieve the "best" fit between state-adjusted forecasts and observed flows. A generalisation of the above is to define α to be

$$\alpha = \frac{q_b}{\beta_1 q_s + \beta_2 q_b}$$ (11.57)

and to choose the incidental parameters β_1 and β_2 to weight the apportionment towards or away from one of the flow components; in practice β_1 and β_2 are assigned values of 10 and 0.1 to apportion more of the error adjustment to the surface store. Note that the adjustment is carried out at every time step and the time subscripts have been omitted for notational simplicity. The scheme with α defined by Eq. 11.56 is referred to as the proportional adjustment scheme and that defined by Eq. 11.57 is the super-proportional adjustment scheme. Replacing α and $(1 - \alpha)$ in Eq. 11.55 by unity yields the simplest non-proportional adjustment scheme.

Other state variables within the PDM can also be adjusted. With the surface store characterised by the cascade of two linear reservoirs, shown to be discretely coincident to a TF model in Section 11.2.2.4, the outflows from the two reservoirs can be identified as q_{s1} and q_s. Then q_{s1} can be adjusted according to the rule

$$q_{s1}^* = q_{s1} + (1 - \alpha) g_{s1} \varepsilon \;\cdot$$ (11.55c)

An adjustment may also be made to the direct runoff, $u_s \equiv V$, entering the surface store; this takes the form

$$u_s^* = u_s + (1 - \alpha)g_u \varepsilon \ .$$ (11.55d)

Finally, an adjustment to the probability-distributed soil moisture, $S \equiv S_1$, may be made, either of the proportional form

$$S^* = S + \alpha g \varepsilon$$ (11.55e)

or the direct form of gain with α equal to unity.

It should be noted that all the above forms of adjustment utilise the same basic form of adjustment employed by the Kalman filter in which an updated state estimate is formed from the sum of the current state value and the model error multiplied by a gain coefficient. However, instead of defining the gain statistically, as the ratio of the uncertainty in the observation to that of the current state value, it is first related to a physical apportionment rule multiplied by a gain factor. This gain factor acts as a relaxation coefficient which is estimated through an off-line optimisation using past flood event data. An essentially similar form of empirical state updating has been used by Moore et al. (1989) to update the Thames Conceptual Model: adjustments to the storages controlling the output from the zonal components of flow are made in proportion to their contribution to the total flow.

Another example of empirical state updating is provided by its use in the PACK snowmelt model outlined in Section 11.2.4. Correction of the snowpack to accord with snow survey measurements can be done at times when measurements of both the pack water equivalent and density are available: typically at 09.00 on days of lying snow. Because these measurements relate to a snow survey measurement point and not to the basin for which snowmelt is to be forecast, the correction process is not straightforward. A correction is first calculated for a "point snowmelt model" and this is subsequently transferred as a correction to the basin scale snowmelt model or models for which the point model is regarded as representative. Thus, updating a snowmelt forecasting model involves running two models in parallel, first a point model at the snow survey site using climate data (precipitation, temperature) in the vicinity and second a basin model whose water balance accounting employs climate data representative of the basin and which is updated by transfer of state-correction information from the point model.

The state-correction information computed within the point snowmelt model at the survey site comprises two quantities. The first is the snow correction factor f, computed as the ratio of the measured water equivalent of the pack θ_m to the modelled value θ; that is

$$f = \theta_m / \theta \ .$$ (11.58)

The second is the proportion of dry snow in the pack expressed as

$$\beta = \frac{\rho_m - 1}{\rho_w - 1} \ ,$$ (11.59)

where ρ_m, ρ_w are the measured snow pack density and the density of dry snow, assumed equal to 0.1.

The two state variables of the point snowmelt model, W the water equivalent of the dry pack and S the water equivalent of the wet pack, are then updated as follows:

$$W^\dagger = \beta\theta_m \ , \tag{11.60a}$$

$$S^\dagger = \theta_m - W^\dagger = (1 - \beta)\theta_m \ , \tag{11.60b}$$

where the superscript dagger is used to denote the updated quantity. Note that this correction ensures that the water equivalent and density of the modelled and measured packs agree, and also establishes the partition between wet and dry pack water storage in the modelled pack.

Transfer of the state correction information, f and β, to the basin scale model used for snowmelt forecasting is straightforward. First the water equivalent of the model pack, $\theta = W + S$ (no change in notation will be introduced as it is clear that here we are referring to the basin scale model), is factored using the point snow correction factor f, such that

$$\theta^\dagger = f\theta \ . \tag{11.61}$$

$$W^\dagger = \beta\theta^\dagger \ , \tag{11.62a}$$

$$S^\dagger = \theta^\dagger - W^\dagger = (1 - \beta)\theta^\dagger \ . \tag{11.62b}$$

The correction is thus, in proportion to that applied in the point model, relative to the water equivalent of point and basin packs, and also establishes a partition between wet and dry packs by maintaining the same density.

11.3.3
Error Predictors

State updating techniques have been developed based on correcting the water content of conceptual storage elements in the belief that the main cause of the discrepancy between observed and modelled runoff will arise from errors in estimating basin average rainfall, which in turn accumulate as errors in water storage content. Rather than attribute the cause directly and devise empirical adjustment procedures we can analyse the structure of the errors and develop predictors of future errors based on this structure which can then be used to obtain improved flow forecasts. A feature of errors from a conceptual rainfall-runoff model is that there is a tendency for errors to persist so that, sequences of positive errors (underestimation) or negative errors (overestimation) are common. This dependence structure in the error sequence may be exploited by developing error predictors which incorporate this structure and allow future errors to be predicted. Error prediction is now a well established technique for forecast updating in real-time (Box and Jenkins 1970; Moore 1982). Error prediction is available as an alternative to empirical state updating in the PDM software. Predictions of the error are added to the deterministic model prediction to obtain the updated model forecast of flows. In contrast to the state adjustment scheme, which internally adjusts values within the PDM model, the error prediction scheme is wholly external to the deterministic PDM model operation. The importance of this is that error prediction may be used in combination with any model, be it of TF, conceptual or "physics-based" form, and for representing rainfall-runoff or channel flow processes. Indeed, error prediction is used as the standard updating method in the KW and hydrodynamic models used to represent non-tidal and tidal channel flow in the RFFS.

Consider that $q_{t+\ell}$ is the forecast of the observed flow $Q_{t+\ell}$ at some time $t+\ell$, made using, for example, a conceptual rainfall-runoff model. Since $q_{t+\ell}$ will have essentially been obtained by transformation of rainfall into flow through some model conceptualisation of the catchment, it will not have used previous observed values of flow, except perhaps for the purposes of model initialisation. It will consequently be referred to as a simulation-mode forecast to distinguish it from a real-time, updated forecast which incorporates information from observed flows.

The error, $\eta_{t+\ell}$ associated with this simulation-mode forecast is defined through the relation

$$Q_{t+\ell} = q_{t+\ell} + \eta_{t+\ell} \ . \tag{11.63}$$

If the simulation-mode error $\eta_{t+\ell}$ may be predicted using an error predictor which exploits the dependence structure of these errors, then an improved forecast may be obtained.

Let $\eta_{t+\ell|t}$ denote a prediction of the simulation-mode error, $\eta_{t+\ell}$ made ℓ steps ahead from a forecast origin at time t using an error predictor. (The suffix notation $t+\ell|t$ should be read as a forecast at time $t+\ell$ given information up to time t.) Then a real-time forecast $q_{t+\ell|t}$ made ℓ time units ahead from a forecast origin at time t may be expressed as follows:

$$q_{t+\ell|t} = q_{t+\ell} + \eta_{t+\ell|t} \ . \tag{11.64}$$

The real-time forecast error is

$$a_{t+\ell|t} = Q_{t+\ell} - q_{t+\ell|t} \ , \tag{11.65}$$

which, depending on the performance of the error predictor, should be smaller than the simulation-mode forecast error

$$\eta_{t+\ell} = Q_{t+\ell} - q_{t+\ell} \ . \tag{11.66}$$

Turning now to an appropriate form of error predictor it is clear that a structure which incorporates dependence on past simulation-mode errors is required. Thus, the autoregressive (AR) model

$$\eta_t = -\phi_1\eta_{t-1} - \phi_2\eta_{t-2} - \ldots - \phi_z\eta_{t-z} + a_t \tag{11.67}$$

is an obvious candidate, where a_t is the residual error (uncorrelated), and $\{\phi_i\}$ are parameters. However, a more parsimonious form of model is of the autoregressive-moving average (ARMA) form

$$\eta_t = -\phi_1\eta_{t-1} - \phi_2\eta_{t-2} - \ldots - \phi_p\eta_{t-p} + \theta_1 a_{t-1} + \theta_2 a_{t-2} + \ldots + \theta_q a_{t-q} + a_t \quad , \tag{11.68}$$

which incorporates dependence of past residual errors, $a_{t-1}, a_{t-2} \ldots$.

In general, the number of parameters $p + q$ associated with the ARMA model will be less than the number z associated with the AR model, in order to achieve as good a level of approximation to the true simulation-mode error structure. The ARMA model may be used to give the following error predictor

$$\begin{aligned}
\eta_{t+\ell|t} = &-\phi_1\eta_{t+\ell-1|t} - \phi_2\eta_{t+\ell-2|t} - \ldots - \phi_p\eta_{t+\ell-p|t} + \theta_1 a_{t+\ell-1|t} + \theta_2 a_{t+\ell-2|t} \\
&+ \ldots + \theta_q a_{t+\ell-q|t} \quad , \qquad \ell = 1, 2, \ldots
\end{aligned} \tag{11.69}$$

where

$$a_{t+\ell-i|t} = \begin{cases} 0 & , \quad \ell - 1 > 0 \\ a_{t+\ell-i} & , \quad \text{otherwise} \end{cases} \tag{11.70}$$

and $a_{t+\ell-i}$ is the one-step ahead prediction error

$$\begin{aligned}
a_{t+\ell-i} \equiv a_{t+\ell-i|t+\ell-i-1} &= \eta_{t+\ell-i} - \eta_{t+\ell-i|t+\ell-i-1} \\
&= Q_{t+\ell-i} - q_{t+\ell-i|t+\ell-i-1}
\end{aligned} \tag{11.71}$$

and

$$\eta_{t+\ell-i|t} = \eta_{t+\ell-i} = Q_{t+\ell-i} - q_{t+\ell-i} \qquad \text{for } \ell - i \le 0 \ . \tag{11.72}$$

The prediction Eq. 11.69 is used recursively to produce the error predictions $\eta_{t+1|t}, \eta_{t+1|t} \ldots, \eta_{t+\ell|t}$ from the available values of $a_t, a_{t-1}, \ldots$ and $\eta_t, \eta_{t-1}, \ldots$.

Using this error predictor methodology, the conceptual model simulation-mode forecasts $q_{t+\ell}$ may be updated using the error prediction $\eta_{t+\ell|t}$ obtained from Eq. 11.69 (and the related Eqs 11.70–11.72), to calculate the required real-time forecast $q_{t+\ell|t}$ according to Eq. 11.64. Note that this real-time forecast incorporates information from the most recent observations of flow through the error predictor, and specifically through calculation of the one-step ahead forecast errors $a_{t+\ell-i}$ according to Eq. 11.71.

Alternative error predictor schemes may be devised by working with other definitions of the basic errors: for example by using proportional errors. One such scheme can be formulated by starting with the logarithmic model

$$\log Q_{t+\ell} = \log q_{t+\ell} + \eta_{t+\ell} \quad , \tag{11.73}$$

so that the simulation-mode error is now defined as

$$\eta_{t+\ell} = \log(Q_{t+\ell} / q_{t+\ell}) \ . \tag{11.74}$$

An error predictor for $\eta_{t+\ell}$, may be formulated in the normal way using Eqs 11.69 and 11.70 with the one-step ahead prediction error given by

$$a_{t+\ell-i} = \eta_{t+\ell-i} - \eta_{t+\ell-i\,|\,t+\ell-i-1} \quad . \tag{11.75}$$

Instead of Eq. 11.64 the real-time forecast, $q_{t+\ell\,|\,t}$ takes the form

$$q_{t+\ell\,|\,t} = q_{t+\ell} \exp(\eta_{t+\ell\,|\,t}) \quad . \tag{11.76}$$

Whilst error prediction provides a general technique which is easy to apply, its performance in providing improved forecasts will depend on the degree of persistence in the model errors. Unfortunately in the vicinity of the rising limb and peak of the flood hydrograph this persistence is least and errors show a tendency to oscillate rapidly and most widely; dependence is at its strongest for errors on the falling limb, where improved forecast performance matters least. In addition, timing errors in the model forecast may lead to erroneous error prediction corrections being made, a problem which is also shared by the technique of state adjustment. The general applicability and popularity of error prediction as an updating tool commends its use here as an "off-the-shelf" technique. Empirical state adjustment schemes should also be considered as viable alternatives to the use of error prediction.

The Transfer Function Noise (TFN) Modelling Package provided within the RFFS is used to identify the form of ARMA error predictor and to estimate its parameter values. Also a means exists within both the PDM and KW software to estimate the ARMA error predictor parameters for an assumed model structure. Often a third order autoregressive, with dependence on three past model errors, provides an appropriate choice for UK conditions and a 15 minute model/data time interval.

11.4
Spatial Variability and Weather Radar

11.4.1
Introduction

Progress in real-time flow forecasting has so far been discussed in terms of model formulation and updating techniques. Further advances, particularly in model formulation, was recognised to be constrained by the use of point estimates of rainfall obtained from a usually sparse raingauge network. The ability of weather radar to provide a continuous estimate of rainfall over space was clearly recognised by hydrologists as a potential path for improvement. The national radar network, operated by the UK Meteorological Office, serviced well the primary function of the daily weather report and forecast: the visual images of the spatial extent and propagation of storms are now a familiar feature of such reporting on television. However, the need for quantitative rainfall estimates to be used as input to flood forecasting models was not so well met. This led the Institute of Hydrology to develop local methods for radar calibration and forecasting (Moore et al. 1989), particularly in collaboration with the National Rivers Authority's London Flood Warning Centre located at Waltham Cross (Haggett 1986). Here, the potential advantages of a local radar calibration procedure were apparent in that data from some 30 gauges were available via telemetry in addition to the 5 gauges used for calibration at the radar site. The calibration procedure developed was shown, for an area 60 km square containing up to 30 gauges, to provide an improvement in accuracy of 22% relative to the uncalibrated radar (Moore et al. 1993a). The calibration system has been operational at the London Flood Warning Centre since March 1989. The success of this calibration study was followed by a radar rainfall forecasting project (Moore et al. 1993b). This aimed to infer the speed and direction of movement from two time-displaced radar images which would be used, with the current image, as the basis of forecasting. The resulting algorithm was shown to outperform not only a persistence (no-change) forecast but also the national FRONTIERS rainfall forecasts (Moore et al. 1991); however, the latter complements the local method in providing forecasts for longer lead times when the local method, based on a single radar, suffers from exhaustion of the field being advected forward. This system has been in operational use at the London Flood Warning Centre since November 1991.

The above developments created two distinct, but uncoupled, tools whereby weather radar could be used to better support flood warning. The forecasting study also produced algorithms to preprocess the radar data to suppress transient clutter and to remove more permanent anomalies, due to blockages in the radar beam. There was a clear need to integrate these preprocessing, calibration and forecasting procedures within a single integrated HYDdrological RADar system and in this way the HYRAD system was conceived. The kernel to the system comprises the above procedures together with a post-processor to derive catchment average rainfall time series data (utilising digitised catchment boundary data) and an interface to the River Flow Forecasting System. Additional features include an interface to real-time radar reception software, a radar data archiving facility and a Windows 3.1 (or NT) radar display system incorporating hydrologically appropriate overlay information. The latter provides for client-server interaction between PCs and a host VAX or UNIX computer running the kernel, reception and archiving software.

11.4.2
Radar Calibration

The algorithm used for radar calibration is based on fitting a surface to the calibration factors calculated at n raingauge locations. For the i'th raingauge the factor is defined as

$$z_i = \frac{R_g^i + \varepsilon_g}{R_r^i + \varepsilon_r} \, , \qquad (11.77)$$

where R_g^i and R_r^i are the gauge and coincident radar measurement of rainfall for a 15 minute interval; ε_g and ε_r are regarded as incidental parameters which ensure that z_i is defined for zero radar values. The type of surface used is an extended form of the multiquadric surface proposed by Hardy (1971). If $x_i = (u_i, v_i)$ denotes the grid reference of the i'th gauge then a surface may be formed as the weighted sum of n distance measures (basis functions) centred on each gauge; thus

$$s(\underline{x}) = \sum_{j=1}^{n} a_j g(\underline{x} - \underline{x}_j) + a_0 \, , \qquad (11.78)$$

where $(a_j, j = 0, 1, 2, \ldots, n)$ are the parameters of the surface. A simple Euclidean distance measure is chosen such that

$$g(\underline{x}) = \|\underline{x}\| = \sqrt{(u^2 + v^2)} \; , \tag{11.79}$$

so that the surface is constructed from n right-sided cones, each centred on a raingauge location, Estimation of the weights a_j, using the calculated z_i values is described in Moore et al. (1993a, 1989). The procedure incorporates a constraint for flatness at large distances from the raingauge network and an implicit allowance for errors through allowing the surface not to pass exactly through the calculated calibration factor values. An illustration of the application of the algorithm to a single 15 minute time-frame is shown in Fig. 11.8. A variant of the method is available for use in HYRAD to derive a rainfall field estimate from raingauge network data, either as a complement to the calibrated radar field or to replace it when radar data are unavailable. In this case, an exponential form of the Euclidean distance is used to define $g(.)$.

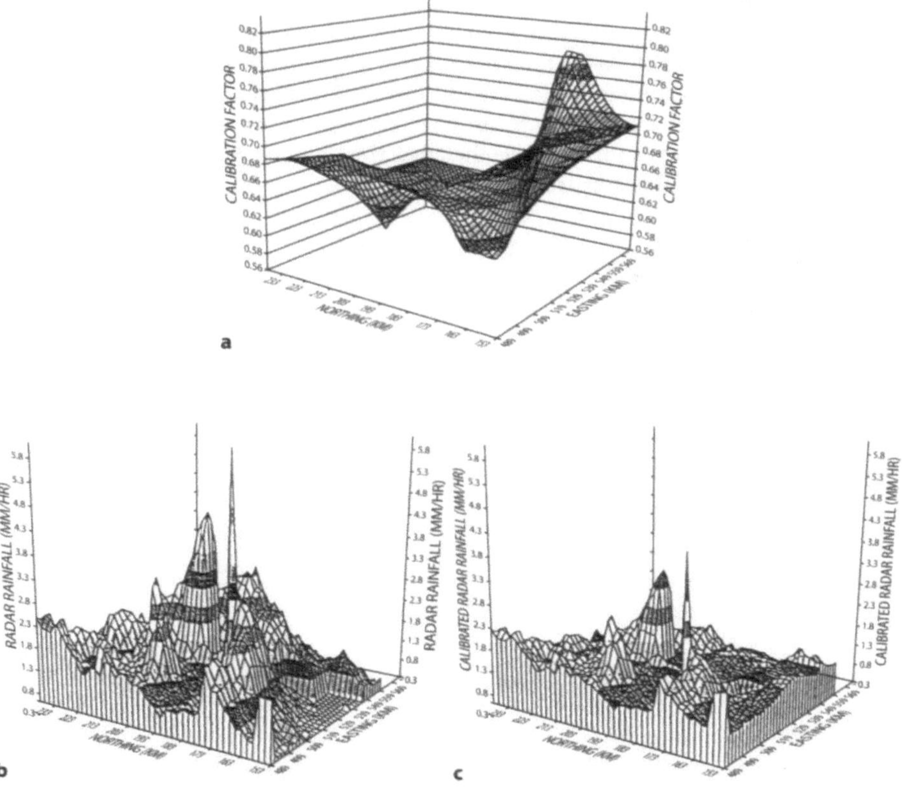

Fig. 11.8. Illustration of the local rainfall calibration algorithm for one 15 minute time-frame (London Weather Radar, 16.30 14th March 1989). **a** Calibration factor surface; Rainfall field prior to calibration (**b**) and after calibration(**c**)

11.4.3
Radar Rainfall Forecasting

The basis of the radar rainfall forecasting algorithm can be summarised as follows. Let c_u denote the easterly component of velocity of the rainfall field and consider the translation of the field from time t to time $t + \tau$. Then the position $u(t)$ along a west-east axis at time t will move to position

$$u(t + \tau) = u(t) + c_u \tau \tag{11.80}$$

if a simple advection model is assumed to hold; a similar expression holds for the northerly component giving the position on the south-north axis, $v(t + \tau)$, resulting from a movement with velocity c_v. An optimal choice for the velocities c_u and c_v is arrived at by displacing a radar image 15 min before, using an assumed trial velocity, and calculating the discrepancy with the current image. A root mean square log error criterion is used to assess the degree of correspondence. The choice of trial velocity pairs follows a shrinking grid search procedure which initially defines a coarse, but extensive grid of values. This is reduced in extent and coarseness over three steps, utilising only velocities which result in exact displacements of one grid over the other, thus avoiding the need for interpolation. At the final step a direct interpolation in the error field, based on a 4-point interpolation formula, is used to arrive at the velocity to be used for forecasting. The forecasts are constructed by advecting the current radar image using the inferred velocity vector to first derive the advection forecast $\hat{R}_{ij}$ for radar grid square (i, j). The final forecast is constructed as

$$\tilde{R}_{ij} = \begin{cases} \overline{R} + a(\hat{R}_{ij} + \overline{R}) \ , & \hat{R}_{ij} > 0 \\ 0 & \hat{R}{ij} = 0 \end{cases} \tag{11.81}$$

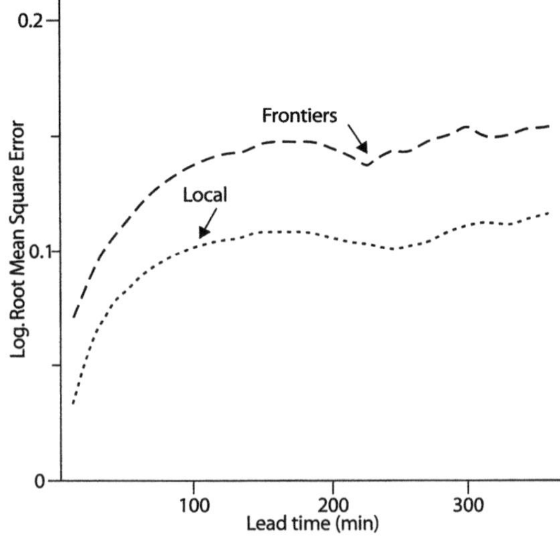

Fig. 11.9. Rainfall forecast performance for a range of lead times obtained using the Local Rainfall Forecast and Frontiers methods; 15 minute rainfall totals for 14 storm events over London

where $\bar{R}$ is the field average value and $a = f^\tau$ is a shrinkage factor, reducing with increasing lead time τ, with f a constant value. Thus, the forecast is based on an advection forecast which is progressively shrunk towards the field average value with increasing lead time. Figure 11.9 summarises how the accuracy of the method diminishes with increasing lead time but continues to outperform the Frontiers forecasts in terms of the root mean square log error criterion.

11.5
Integrated Flood Forecasting Systems

11.5.1
The RFFS

An initial alert of possible flooding can be based on measurements of heavy rainfall from a raingauge network and from weather radars. It can also derive from rainfall forecasts derived from weather radars, from synoptic analyses and from mesoscale models. More precise information, for use as a basis of flood warning to specific localities, require hydrological and hydrodynamic models of river basins capable of forecasting the rising flood hydrograph at possibly many locations within a river basin, or across a region comprising of several river networks. The complexity of the forecasting and flood warning dissemination task demands that automated systems are developed to coordinate the activity and provide simple support to decision making under conditions of uncertainty. The RFFS, provides a generic River Flow Forecasting System capable of coordinating the task of flood forecast construction for both simple and complex river networks (Cottingham and Bird 1991; Moore and Jones 1991; Moore et al. 1992; Moore 1993). It incorporates an interface to the HYRAD system, discussed in the previous section, as well as, providing facilities for telemetry data acquisition, forecast display, decision support and dissemination of flood warning messages by fax.

The Information Control Algorithm, or ICA, has been developed to invoke and manage the coordination of the acquisition of data needed for forecast construction and the order-of-execution of models working from the headwaters to the sea. Together with model calibration facilities, the ICA forms the kernel to the RFFS. It is serviced by the RFFS shell which functions to receive data from telemetry, fax, telex and manual entry for subsequent access by the ICA (Fig. 11.10). The shell also receives forecasts constructed by the ICA kernel for display, general reporting and to support the issuing of flood warnings. The ICA tackles a possibly complex overall forecasting problem through division into a number of simpler sub-problems, or "Model Components". The results of one or more model components are fed as input series into a subsequent Model Component. These input series will in practice be made up of observed values in the past, model infilled values in the past where observations are missing and model forecast values in the future. The infilled or forecast values will usually be constructed using preceding Model Components. At the extremities of the information network, beyond the river network, are special model components which can provide missing values for their input series through the use of backup profiles. Rainfall series completion is a typical example. Thus, a particular forecasting problem may be viewed as a node and link network in which each node is associated with a forecast requirement

Fig. 11.10. The RFFS shell and
ICA kernel

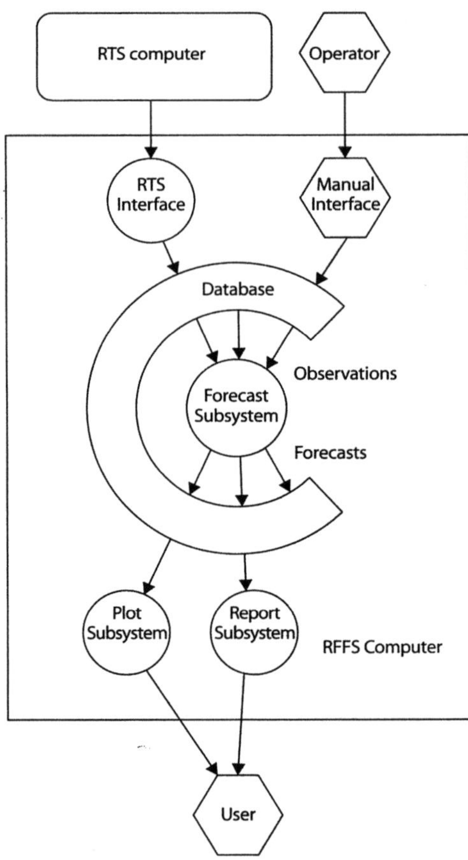

(e.g. to forecast flow at site x) and each link to a Model Component which contains a set of Model Algorithms used to construct the forecasts.

A particular configuration of forecast points within a river system is described within the ICA by a set of description files. These files take two main forms:

i. A *Model Component file* which defines the form of model structure, through the specification of Model Algorithms to be employed, and the data inputs to be used to make forecasts for a particular location or set of locations;
ii. A *Forecast Requirement file* which defines for each forecast point the Model Component to be used to construct the forecast for that point, the type of forecast (e.g. river level, flow, snowmelt) and the connectivity with other model components.

A Model Component is typically made up of a number of Model Algorithms, for example for snowmelt modelling, rainfall-runoff modelling and real-time updating. Model Algorithms comprise forecasting procedures which can be used to create forecasts for several sites in the region, a particular procedure typically only differing from site to site through the model parameters used. The model algorithms to be used are

Fig. 11.11. A model component and its associated model algorithms

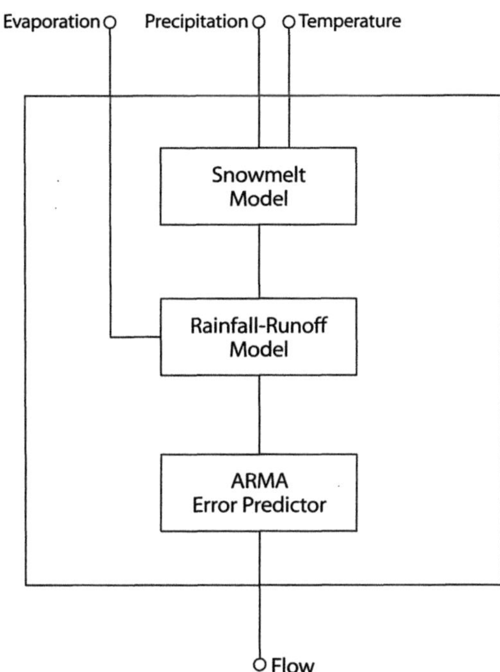

defined within the Model Component tile description, together with the parameter values appropriate to the site(s) for which the component is to provide a forecast. Figure 11.11 illustrates a typical Model Component and its associated Model Algorithms and Fig. 11.12 illustrates the connectivity between model components. This connectivity allows the ICA to represent river systems with complex dendritic structures including bifurcations.

The specifications for a Model Component can be replaced with those for a modified version without disturbing the specification of the rest of the Model Component network and, essentially, without having even to consider more than the local part of the rest of the network. This provides for flexibility and ease of modification of an existing configuration.

The Model Algorithms are formulated within a generic subroutine structure which allows new algorithms to be coded and accessed by the ICA without recoding of the ICA itself. The generic structure is sufficiently general to allow algorithms of varying complexity to be coded. For example, algorithms can be as simple as calculating catchment average rainfall as a weighted average of raingauge data or may be as complex as ones which incorporate control rules for river gate settings as part of a hydraulic model of a tidal river. The generic features of the algorithm interface are made specific to a particular model algorithm through a third type of description file. This Model Algorithm file defines an identifier, the name and number of the parameters, states, types of input/output and other features which are specific to a particular algorithm. The values given to a particular model algorithm parameter set in its application to a specific site is given in the Model Component file.

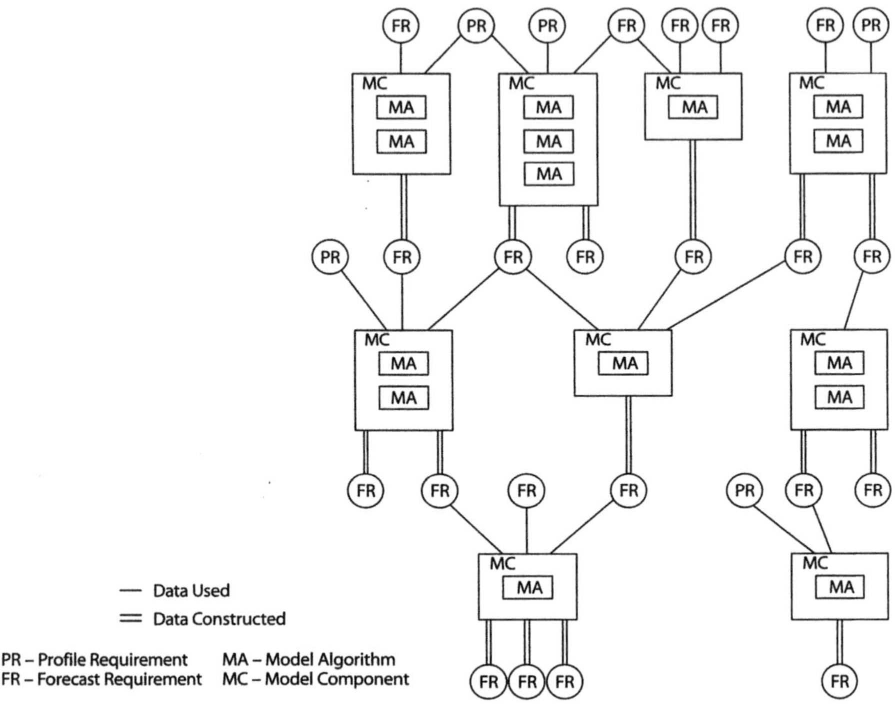

Fig. 11.12. Connectivity between model components

It is worth emphasising the importance of Model Component construction in the effective configuration of the ICA to a forecasting problem. One option is to equate Model Component and Model Algorithm and to have many Model Components. A more effective construction is to identify model algorithms which commonly occur together and use these as Model Component building blocks. This preserves flexibility whilst simplifying configuration and understanding of the resulting system. It is clearly easier to build a house from large bricks than small ones, provided the bricks are not too large to handle or too varied.

Having constructed a set of Model Component and Forecast Requirement description files to define a structure for the particular forecasting problem, the ICA initially employs these to construct a file used to order the sequence of Model Component executions. This "order-of-execution" list need only be constructed once for a given forecast network configuration. Operational running of the ICA deploys this list to get the data it requires to make the forecast run and then to execute each Model Component.

The ICA works down the tree network of the river system, in the order dictated by the list, so that forecasts of flow or level are used as input to the next Model Component downstream. At run time, the lead time of the forecasts can be changed, as well as, various settings controlling the input used by the Model Components.

The ICA allows the user to dynamically define "subnetworks" within the overall model network configuration. These can be defined, for example, to only execute the non-tidal part of the model, or to execute a selected set of rapidly-responding catchments requiring a flash-flood warning, updated at frequent intervals. The subnetwork concept could be used for each Area of a Region, to separate particular river networks, to separate each hydrometric area or to correspond to chosen "Catchment Areas".

On completion of a forecast run, the "states" of the models required to initialise a subsequent run are stored; the time selected for storing the states is usually 30 min before the present time to allow for delays in receiving telemetry data. The states will be typically the water contents of conceptual stores within snowmelt and rainfall-runoff models or the river levels and flows of channel flow routing models. A subsequent run at a later forecast time origin will start forecasting forwards from the time of storing the states from a previous run. The set of model states for an algorithm, for a particular time point, contains a sufficient summary of all data previous to that time point to allow the algorithm to continue executing as if it had been run over a warm-up period up to that time point.

Operationally in non-flood conditions the system might be run automatically once a day, say at 7:00 a.m. following routine data gathering by the Regional Telemetry System. This means that the model states are available to provide good initial conditions from which to run the model for a flood event occurring later the same day, thus avoiding the need for a long "warm-up" period for model initialisation. During flood events the system might be run frequently under the control of the operator. For small, fast responding systems a fully automated mode of operation might be preferred, as is the case for an implementation of the RFFS in Hong Kong.

The model algorithms available for use within the ICA fulfil a range of functions. Some serve as simple utilities to set flows to a constant value, for example to represent a fixed compensation release from a reservoir, or to merge data from different sources according to a priority hierarchy to ensure that a data series required for forecasting is complete. The more conventional forms of model algorithm perform some specific hydrological function such as rainfall-runoff modelling, channel flow routing, snowmelt modelling, hydraulic modelling of the tidal river and error prediction. Specifically, the PDM, KW, PACK, HYDRO and ARMA models described in Section 11.2 are available as model algorithms.

The ICA has been designed so that the generation of flow forecasts can be resilient to data loss. This is accomplished for a point "internal" to the network by ensuring that the model component which constructs forecasts for the point will also fill in missing values in the past data. For "external" points, typically rainfall and other forms of climate data, model algorithms are used to merge data time series from a variety of sources. In the event of no data being available, provision is made to supply a backup profile. A hierarchy of priority of data source can be imposed in the case of data being available for a given time from more than one source. For example, in the case of rainfall, the priority for a given catchment rainfall might be radar data, raingauge data from n raingauges and then any combination of less than n (allowing for raingauge system malfunction), and a backup rainfall profile. For future times when no observation data are available the priority might be a Local Radar Rainfall Forecast from HYRAD, a FRONTIERS forecast and a synoptic forecast (provided automatically to the modelling computer by a computer/telex facility from a Weather Centre) and finally a backup rainfall profile. The rainfall profiles can be selected to be seasonally dependent and categorised into light, moderate and heavy with the option of invoking a selection at run time. They may also be subdivided into different synoptic regions over a region.

Another important use for profiles is to provide boundary conditions for the tidal hydraulic model. Astronomical tidal predictions would be made available in profile form and augmented with tide residual forecasts supplied by the Storm Tide Warning Service. Other uses for profiles are for potential evaporation and temperature to support rainfall-runoff and snowmelt modelling.

The first implementation of the RFFS was to provide forecasts at some 150 locations within the 13 500 km² region of Yorkshire, England. This system has operated on a trial basis since February 1992 and by April 1993 experience gained had been sufficient to allow forecasts upstream of York to be used for routine flood warning. Demonstration of the RFFS's configurability and transportability has been demonstrated through its use as the basis of the White Cart Water Flood Forecasting System (Anderson et al. 1992), providing flood warning to the Southern part of Glasgow, Scotland. This demonstrated how the kernel software is largely machine independent (transfer from a VAX to a Data General computer) and suitable for both large, regional warning schemes and small catchment situations (6 forecast points supported by telemetry of 6 river gauging and 4 raingauge sites). Most recently the system has been transported to a UNIX workstation to forecast at sites in a 70 km² basin in the New Territories, Hong Kong. Another UNIX application is proposed for the 27 000 km² Anglian Region of England where nearly 600 locations have been identified as needing forecasts, the majority for flood warning but also for managing water transfers and licensing at lower flows (Moore et al. 1993c). The aim in this case is to provide a unified windows-based environment for water management during floods, droughts and pollution incidents. Extension to water quality is clearly possible given the generic nature of the ICA.

11.5.2
Other Systems in the UK

Whilst the RFFS is arguably the most advanced integrated flood forecasting system in operational use in the UK it has a number of precursors. An early example was the Dee Flood Forecasting System implemented in 1974 which also incorporated weather radar (Central Water Planning Unit 1977). A more recent example was the Haddington flood warning system in Scotland which was the first low-cost microprocessor controlled system in the UK (Brunsden and Sargent 1982; Sargent 1984). These earlier systems were developed as case-specific applications with the mathematical modelling software pre-configured to the specific river network of interest. The development of the Severn-Trent Flood Forecasting System (Bailey and Dobson 1981; Dobson 1993) was a significant landmark in being the first region-wide system in operation in the UK and having a software design which was highly modular. However, the RFFS brought further features including data-defined model configuration, a generic model algorithm interface and improved ways of providing resilience to missing data. The CASCADE flood forecasting system in use by NRA Thames Region is arguably the most advanced in its deployment of radar (Haggett et al. 1993). It incorporates the local calibration and forecasting systems previously discussed along with FRONTIERS rainfall forecasts and a range of processing, display and warning dissemination tools. Whilst it can make model forecasts for specific catchments, developments have not yet been made to coordinate the construction of forecasts down the river network. The Wessex Flood Forecasting System (Aucott et al. 1993) is based on the WRIP (Wessex Radar Information Project) system and the STORM system for radar data acquisition and display, both developed at the University of Salford. The system makes exclusive use of transfer function models and features a user-friendly X-windows graphical user interface. A river network form of the system is under development. Systems in use in other NRA regions, for example North-west (Smith et al. 1991) and Anglian (Moore et al. 1993c), are in the process of change and will not be reviewed here. Anderson et al. (1992) provides a recent review of flood warning schemes on mainland Scotland which include the White Cart Water RFFS and the systems for Haddington and the Nith previously mentioned.

11.6
Towards More Distributed Models?

The more widespread availability of spatial rainfall data from weather radar together with digital terrain models or DTMs has revived interest in simple distributed rainfall-runoff models for operational flood forecasting. The author, working in collaboration with Alice Gilyen-Hofer of VITUKI Hungary in 1986 under a British Council bursary, formulated a prototype grid-model based on the radar grid and isochrones derived from contour maps in the absence of a DTM. This model has recently been further developed and tested with the support of a DTM, improved radar data and satellite-derived data on land-use (Moore and Bell 1993; Moore et al. 1992). The model uses the length of travel paths of water derived from a digital terrain model (DTM), along with assumed hillslope and channel velocities, to derive times of travel to the basin outlet from any point within a catchment. This allows a catchment to be sub-divided into isochronal strips and associated with a time of travel to the basin outlet. A simple water balance is performed on each radar grid square using the square's radar rainfall as input to generate direct runoff. The grid is conceptualised as a store whose capacity is a function of the average slope in the square, as estimated from the DTM; slow drainage from the base of the store is also allowed. Associating isochrone strips and coincident radar grid square fractions then allows the output from the grid stores to be delayed and summed to form the total catchment response. Initially this convolution operation was performed as a discrete double summation over the isochrones and grid squares. In the current formulation the convolution approach to translation has been replaced by a discrete space/time kinematic wave representation. Translation through each isochrone strip is represented by a simple kinematic wave reach model and thus, a catchment is represented as a cascade of kinematic reach models. The new approach has significant computationally advantages, as well as, providing a simple way of introducing diffusion into the isochrone formulation, which in its original convolution form operates a simple advection delay. Specifically flow out of the n'th reach at time k is represented by

$$q_k^n = (1 - \theta)q_{k-1}^n + \theta(q_{k-1}^{n-1} + r_k^n) \; , \tag{11.82}$$

where r_k^n is the direct runoff generated from the j'th isochrone strip in the interval $(k-1, k)$ and θ is a dimensionless wave speed parameter in the range 0–1. Note that this is the simple form of KW model introduced as Eq. 11.40 but with the lateral inflow u_k^n being defined as the runoff, r_k^n. This is calculated as

$$r_k^n = f \sum_{i=1}^{m} A_{n,i} r_{k,i} \; , \tag{11.83}$$

where $A_{n,i}$ is the area of the i'th radar grid square occupied by the n'th isochrone strip, $r_{k,i}$ is the runoff generated from this square and f is a units conversion factor (in the present case from km² mm/15 min to m³/s).

Two important further developments of the model implementation have been made. The first is to include the estimation of the hillslope and channel velocities, used in the derivation of the isochrones from the DTM, into the overall model parameter optimisation procedure. Second, is the use of Landsat satellite imagery to classify landuse over the catchment and to use this to define impervious areas (with zero storage capacity) over the catchment. The resulting model utilising grid-square radar accounts for 85% of the flow variation compared to only 55% when raingauge data are used, when assessed using 15 minute data over a monthly period for the Mole at Kinnersley Manor catchment covering an area of 145 km² to the south of London. Initial results obtained for hillier catchments of South Wales and Northwest England are not so encouraging and in general depend on the performance of the radars in particular situations, which can be affected by low level enhancement of rainfall, beam blockages, etc. More experience over a range of conditions is required before such grid-models can be recommended for operational use.

11.7
Future Opportunities

The state-of-the-art in operational flood warning suggests that it is difficult to improve upon the simple advection-based rainfall forecasting methods used in HYRAD and the use of lumped conceptual rainfall-runoff models at the catchment scale used in the RFFS. However, in a research context important advances are being made in both rainfall forecasting and catchment scale modelling which offer promise for future improvement of flood warning. Advection-based rainfall forecasts are limited when storms exhibit significant development and decay over short time periods, often in situ, as is often the case with convective storms. Progress is being made to forecast such storms using simplified, but high resolution, representations of the water balance of a column of the atmosphere (Lee and Georgakakos 1991; French and Krajewski 1992). These models integrate information from weather radar, satellite, mesoscale models, radiosonde soundings and climate stations on the ground. Whilst the operational deployment of such integrated rainfall forecasting systems is for the future, use is already being made of mesoscale model products. The Frontiers system in the UK employs the upper air winds as the basis of advection forecasting. Also, the mesoscale model forecasts are subject to synoptic interpretation prior to being sent as rainfall and temperature forecasts for use in the Yorkshire RFFS.

Important advances are being made in distributed modelling for flood forecasting. Previously, more complex formulations, were often not justified because of the difficulty of implementation and the paucity of rainfall information from raingauges, which imposed a limit on model performance. The availability of digital terrain and satellite-derived land use data now makes parameterisation of a distributed model less of an onerous task. Also, the availability of distributed rainfall data from weather radar is beginning to remove the constraints associated with using point rainfall measurements from raingauge networks. Two examples of recent progress in this area are Chander and Fattorelli (1991) and the grid-model reported above. However, much testing in a research environment is needed before these more advanced rainfall and flow forecasting techniques can be reliably used for operational flood warning.

Acknowledgements

Contributions from colleagues at the Institute of Hydrology to the work reported here are gratefully acknowledged, as is the help and support of co-workers in the National Rivers Authority responsible for operational flood warning. Funding support for this work has come from the Commission for the European Communities, the National Rivers Authority and the Ministry of Agriculture, Fisheries and Food.

References

Anderson JL, Burns JC, Curran JC, Fox IA, Goody NP, Inglis T, Sargent RJ (1962) A review of flood warning schemes on mainland Scotland. In: Joint meeting of Scottish branch of the institution of water and environmental management. Scottish Hydrological Group of the Institution of Civil Engineers, Scottish Hydraulics Study Group, Perth, 31 March 1992, 6 pp

Aucott LM, Grigg WL, Han D, Cluckie ID (1993) Developing applications of weather radar in the Wessex Flood Forecasting System, In: Verwoorn H (ed) Hydrological applications of weather radar, Proc. 2nd Int. Symp., Paper J2, 10 pp

Bagrov NA (1950) Analysis of the formation of losses. Meteorology and Hydrology 1

Bailey RA, Dobson C (1981) Forecasting for floods in the River Severn Catchment. J Inst Water Engineers and Scientists 35(2):168–178

Box GEP, Jenkins GM (1970) Time series analysis forecasting and control. Holden-Day

Brunsden GP, Sargent RJ (1982) The Haddington flood warning system. In: Advances in hydrometry (Proc. Exeter Symp.), IAHS Publ. N° 134, pp 257–272

Central Water Planning Unit (1977) Dee weather radar and real-time hydrological forecasting project, Report by the Steering Committee, 172 pp

Chander S, Fattorelli S (1991) Adaptive grid-square based geometrically distributed flood forecasting model. In: Cluckie ID, Collier C (eds) Hydrological applications of weather radar, pp 424–439, Ellis Horwood

Cluckie JD, Owens MD (1987) Real-time rainfall-runoff models and use of weather radar information. In: Collinge VC, Kirby C (eds) Weather radar and flood forecasting, pp 171–190, J. Wiley

Cottingham ML, Bird PB (1991) The operational requirements for a River Flow Forecasting System, In: MAFF Conference of River and Coastal Engineers, Loughborough University, England, 8–10 July 1991, Ministry of Agriculture Fisheries and Food

Dickinson WT, Douglas JR (1972) A conceptual runoff model for the Cam catchment. Report N° 17, Institute of Hydrology

Dobson C (1993) Forecasting flows in the Severn and Trent catchments. Rainfall-runoff modelling, British Hydrological Society National Meeting, Dept. Civ. Eng., Imperial College, London, 23 June 1993, 7 pp

Dooge JCI (1973) Linear theory of hydrologic systems. Technical Bulletin N° 1468, Agriculture Research Service, United States Department of Agriculture, US Government Printing Office, Washington, DC20402, USA, 327 pp

Douglas JR, Dobson C (1987) Real-time flood forecasting in diverse drainage basins. In: Collinge VK, Kirby C (eds) Weather radar and flood forecasting. J. Wiley, pp 53–169

Fread DL (1985) Channel routing. In: Anderson MG, Burt TP (ed) Hydrological forecasting. Chapter 14, J. Wiley, pp 437–504

French MN, Krajewski WF (1992) Quantitative real-time rainfall forecasting using remote sensing. IIHR Limited Distribution Report N° 200, Iowa Institute of Hydraulic Research and Deparment of Civil and Environmental Engineering, University of Iowa, October 1992, 241 pp

Gelb A (ed) (1974) Applied optimal estimation, MIT Press, 374 pp

Greenfield BJ (1984) The Thames water catchment model. Internal Report, Technology and Development Division, Thames Water, UK

Haggett CM (1986) The use of weather radar for flood forecasting in London. Conference of River Engineers, Cranfield 15–17 July, Ministry of Agriculture, Fisheries and Food, 11 pp

Haggett CM, May BC, Crees MA (1993) Advances in operational flood forecasting in London. In: Verwoorn H (ed) Hydrological applications of weather radar. Proc. 2nd Int. Symp., Paper N° 2, 11 pp

Haggett CM, Merrick GF, Richards CJ (1991) Quantitativeuse of radar for operational flood warning in the Thames area. In: Cluckie ID, Collier CG (eds). Hydrological applications of weather radar, pp 590–601, Ellis Horwood

Harding RJ, Moore RJ (1988) Assessment of snowmelt models for use in the Severn Trent flood forecasting system. Contract report to Severn Trent Water Authority, Institute of Hydrology, 41 pp

Hardy RL (1971) Multiquadric equations of topography and other irregular surfaces. J Geophys Res 76(8):1905–1915

Institute of Hydrology (1989) A wave speed-discharge algorithm for compound channels. River Flow Forecasting System: IH Technical Note N° 2, Version 1.0, 3 pp

Institute of Hydrology (1992a) PACK: A pragmatic snowmelt model for real-time use, National Rivers Authority River Flow Forecasting System Developers' Training Course, Version 1.0, March 1992, 10 pp

Institute of Hydrology (1992b) PDM: A generalised rainfall-runoff model for real-time use. National Rivers Authority River Flow Forecasting System Developers' Training Course, Version 1.0, March 1992, 26 pp

Institute of Hydrology (1992c) RFFS Model Calibration Facilities: A user guide. National Rivers Authority River Flow Forecasting System, IH Technical Note N° 10, Version 3.0, March 1992

Jakeman AJ, Littlewood IG, Whitehead PG (1990) Computation of the instantaneous unit hydrograph and identifiable component flows with application to two small upland catchments, J Hydrology 117:275–300

Jazwinski AH (1970) Stochastic processes and filtering theory. Academic Press, 376 pp

Jones DA, Moore RJ (1980) A simple channel flow routing model for real-time use. Hydrological Forecasting, Proc. Oxford Symp., April 1980, IAHS Publ. N° 129:397–408

Lambert AO (1972) Catchment models based on ISO-functions, J Inst Wat Eng 26:413–422

Laramie RL, Schaake JC (1972) Simulation of the continuous snowmelt process. Report N° 143, Ralph M. Parsons Laboratory for Water Resources and Hydrodynamics, Dept. of Civil Engineering, 167 pp

Lee TH, Georgakakos KP (1991) A stochastic-dynamical model for short-term quantiative rainfall prediction. IIHR Report N° 349, Iowa Institute of Hydraulic Research and Deparment of Civil and Environmental Engineering, University of Iowa, June 1991, 247 pp

Lees M, Young PC, Ferguson S (1993,) Flood warning, adaptive. In: Young PC (ed) Concise encyclopedia of environmental systems, Pergamon

Mandeville AN (1975) Nonlinear conceptual catchment modelling of isolated storm events. Ph.D. thesis, University of Lancaster

Moore RJ (1980) Real-time forecasting of flood events using transfer function noise models; Part 2; Contract report to Water Research Centre, 115 pp, Institute of Hydrology

Moore RJ (1982) Transfer functions, noise predictors and the forecasting of flood events in real-time. In: Singh VP (ed) Statistical analysis of rainfall and runoff. Water Resources Publ pp 229–250

Moore RJ (1983) Flood forecasting techniques. WMO/UNDP Regional Training Seminar on Flood Forecasting, Bangkok, Thailand, 37 pp

Moore RJ (1985) The probability-distributed principle and runoff production at point and basin scales. Hydrol Sci J 30(2):273–297

Moore RJ (1986) Advances in real-time flood forecasting practice. In: Symposium on Flood Warning Systems. Winter meeting of the River Engineering Section, Inst. Water Engineers and Scientists, 23 pp

Moore RJ (1989) Radar measurement of precipitation for hydrological application. In: Weather radar and the water industry: Opportunities for the 1990s, BHS Occasional Paper N° 2, Institute of Hydrology pp 24–34

Moore RJ, Bell V (1993) A grid-square rainfall-runoff model for use with weather radar data. In: Advances in Radar Hydrology, Proc. Int. Workshop, Lisbon, Portugal, 11–13 November 1991, Commission for the European Communities, 9 pp

Moore RJ, Jones DA (1978) An adaptive finite-difference approach to real-time channel flow routing. In: Vansteenkiste GC (ed) Modelling, identification and control in environmental systems. North-Holland, Amsterdam, The Netherlands

Moore RJ, Jones DA (1991) A river flow forecasting system for region-wide application. In: MAFF Conference of River and Coastal Engineers, Loughborough University, England, 8–10 July 1991, Ministry of Agriculture Fisheries and Food, 12 pp

Moore RJ, Weiss G (1980) Real-time parameter estimation of a nonlinear catchment model using extended Kalman filters. In: Wood EF (ed) Real-time forecasting/control of water resource systems. Pergamon Press, pp 83–92

Moore RJ, Jones DA, Black KB (1989) Risk assessment and drought management in the Thames basin, Hydrol Sci J 34(6):705–717

Moore RJ, Watson BC, Jones DA, Black KB (1989) London weather radar local calibration study: Final report. Contract report prepared for the National Rivers Authority Thames Region, September 1989, Institute of Hydrology, 85 pp

Moore RJ, Watson BC, Jones DA, Black KB, Haggett C, Crees M, Richards C (1989) Towards an improved system for weather radar calibration and rainfall forecasting using raingauge data from a regional telemetry system. In: New directions for surface water modelling, IAHS Publ. N° 181:13–21

Moore RJ, Jones DA, Bird PB, Cottingham ML (1990) A basin-wide flow forecasting system for real-time flood warning, river control and water management. In: Int. Conf. on River Flood Hydraulics, Wallingford England, 17–20 September 1990, J. Wiley, pp 21–30

Moore RJ, Hotchkiss DS, Jones DA, Black KB (1991) London weather radar local rainfall forecasting study: Final Report. Contract Report to the National Rivers Authority Thames Region, Institute of Hydrology, September 1991, 124 pp

Moore RJ, Bell V, Roberts G, Morris D (1992) Development of distributed flood forecasting models using weather radar and digital terrain data, Interim Report, R&D Project 357/1/T, National Rivers Authority, Institute of Hydrology, 63 pp

Moore RJ, Jones DA, Cottingham M, Tinnion M, Akhondi A (1992) A versatile flow modelling system for real-time use. In: 3rd Int. Conf. on River Flood Hydraulics, Florence, Italy, 24–26 November 1992, preprint, BHR Group Ltd., 7 pp

Moore RJ, Jones DA, Seed D (1993c) Anglian flow modelling system. Systems Analysis, contract report prepared for Servelec Ltd, and the National Rivers Authority Anglian Region, Institute of Hydrology, February 1993, 153 pp

Moore RJ, Hotchkiss DS, Jones DA, Black,KB (1993b) Local rainfall forecasting using weather radar: The London case study, In: Advances in radar hydrology, Proc. Int. Workshop, Lisbon, Portugal, 11–13 November 1991, Commission for the European Communities, 7 pp

Moore RJ, May BC, Jones DA, Black KB (1993a) Local calibration of weather radar over London. In: Advances in radar hydrology, Proc. Int. Workshop, Lisbon, Portugal, 11–13 November 1991, Commission for the European Communities, 10 pp

NERC (1975) Flood Studies Report Volume 1: Hydrological Studies, Natural Environment Research Council, pp 513–531

O'Connor KM (1982) Derivation of discretely coincident forms of continuous linear time-invariant models using tbe transfer function approach. J Hydr 59:1–48

Penman HL (1949) The dependence of transpiration on weather and soil conditions. J Soil Sci 1:74–89

Price RK (1977) Comparison of four numerical methods for flood routing. J Hydraul Div, Amer. Soc. Civ. Engrs, 100, HY7, pp 879–899

Sargent RJ (1984) The Haddington fiood warning scheme. J. Inst of Water Engineers and Scientists 38(2):108–118

Smith JM (1977) Mathematical modelling and digital simulation for engineers and scientists. J. Wiley, 332 pp

Smith WD, Noonan GA, Wrigley A (1991) Flood warning in the National Rivers Authority, North West Region. In: Emergency Planning 1991, Int. Conf., 8–11 September 1991, Lancaster, 23 pp

Sugawara M, Watanabe I, Ozaki E, Katsuyama Y (1984) Tank Model with Snow Component, Research Notes of the National Research Centre for Disaster Prevention N° 65, 293 pp

Szollosi-Nagy A (1976) Introductory remarks on the state space modelling of water resource systems, Int. Inst. for Applied Systems Analysis, RM-76-73, 81 pp

Zhao RJ, Zhuang Y (1963) Regionalisation of the rainfall-runoff relations. Proc. East China College of Hydraulic Engineering (in Chinese)

Zhao RJ, Zhuang Y, Fang LR, Lin XR, Zhang QS (1980) The Xinanjiang model. In: Hydrological forecasting (Proc. Oxford Symp., April 1980). IAHS Publ, No 129, pp 351–356

Chapter 12

Relationships Between Environment and Man in Terms of Landslide Induced Risk

M. Panizza

In order to examine the relationships between environment and man, the first being considered from a geomorphological point of view, two main possibilities can be considered (Panizza 1992) (Fig. 12.1).

1. Geomorphological resources in relation to human activity, where the geomorphological environment is regarded as mainly passive in relation to man (active): in other words, a resource may be altered or destroyed by human activity. The consequences can be defined as *impact*.
2. Geomorphological hazard in relation to area vulnerability, where the geomorphological environment is regarded as mainly active in relation to man (passive): in other words, a hazard may alter or destroy something that exist in relation with man. The consequences can be defined as *risk*.

Geomorphological resources include both raw materials (related to geomorphological processes) and landforms, both of which are useful to man or may become useful depending on economic, social and technological circumstances. Geomorphological

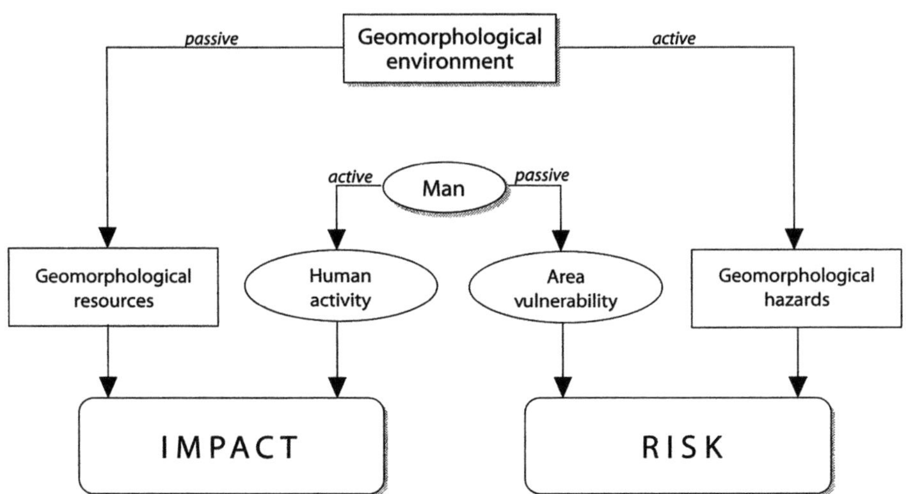

Fig. 12.1. Relationships between geomorphological environment and man

hazards can be defined as the probability that a certain phenomenon of geomorphological instability and of a given magnitude may occur in a certain territory in a given period of time. Human activity is the specific action of man which may be summarised under the headings of hunting, grazing, farming, deforestation, utilisation of natural resources and engineering works. Area vulnerability is the most complex of all that exists in relation to man: population, buildings, infrastructures, economic activity, social organisation and development programmes planned for an area.

In terms of risk mitigation, it can be achieved by reducing either the hazard or the vulnerability.

As regards geomorphological hazard reduction, we can consider two possibilities. Process modification, where an action should be taken regarding the causes of the geomorphological phenomenon: for example, in a process of solifluction, water percolation into the ground could be reduced.

Hazard resistance, where it is possible to prearrange particular passive measures to protect the environment subject to a geomorphological process; for example, against bank erosion, the construction of lateral walls along riversides.

As regards vulnerability reduction, there are two possible choices. Structural measures as passive protection against the risk: for example specific construction norms adopted in areas subject to some form of geomorphological instability. The possibility to predict or forecast a geomorphological hazard: for example a mass movement can be identified by means of monitoring systems or anticipated by simulation models of the geomorphological process.

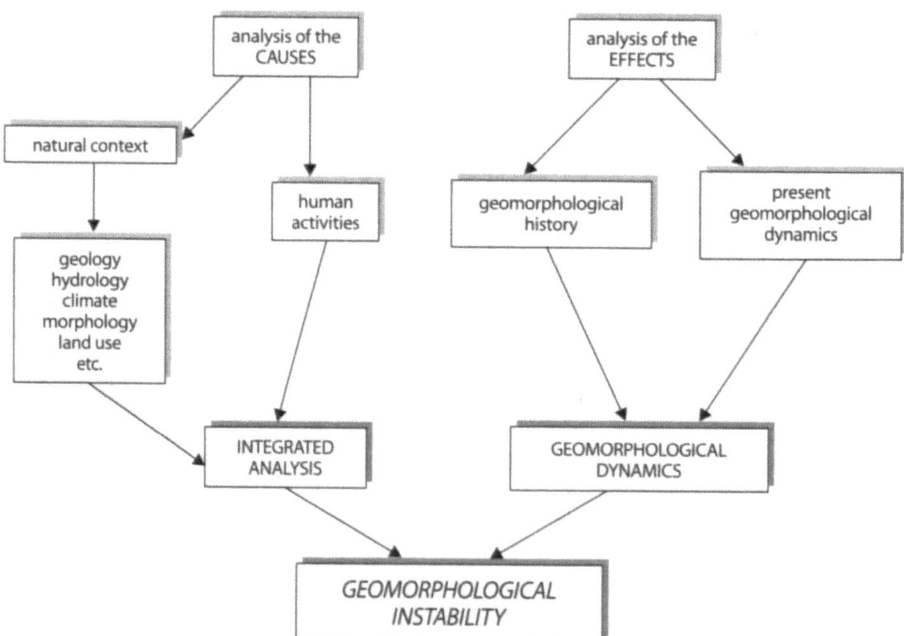

Fig. 12.2. Logical scheme for the determination of geomorphological instability

In order to assess geomorphological hazard, two steps are necessary: first keeping separate, on the one hand, the causes of instability (natural context and human activities) and, on the other hand, the effects of instability (geomorphological history and dynamics); and secondly comparing the two afterwards (Panizza and Piacente 1978) (Fig. 12.2).

Among the causes that favour or prevent hazard, the geological, hydrogeological, topographic, climatic and anthropic could be cited. The analysis of all these parameters will indicate the limits within which geomorphological instability may develop; in other words, it will be possible to define the range of variability, for example, for predictable landslides in a certain area, over a specific time period, both in terms of volume and quantity. The synthesis could be shown on an Integrated Analysis Map, through an overlapping and a comparison of the various basic elements that are the predisposing hazard causes within a given territory.

The analysis of effects is based on a geomorphological investigation which offers an overall picture of the forms and processes of instability, both past and present. The study of past events should lead to a better understanding of the magnitude, frequency and type of phenomena which reflect geomorphological instability; a comparison of present-day processes with past geomorphological events will lead to an estimate of an evolutionary trend for hazard. The effects of geomorphological hazard are essentially traceable back to geomorphological dynamics processes and their deriving forms, distinguished also according to their state of activity. The result of these studies could be shown on a Geomorphological Dynamics Map.

With reference to the Italian territory, the most frequent geomorphological hazard effects are the following:

- Degradational or landslide scarp;
- Erosional river scarp;
- Erosional marine or lacustrine scarp;
- Karst scarp;
- Man-made scarp;
- Trench or crack;
- Deep-seated deformation;
- Rock fall or topple;
- Slide;
- Slump;
- Flow, gelifluction and solifluction;
- Debris flow;
- Deepening riverbed;
- Gully erosion;
- Rill or sheet erosion;
- Karst depression;
- Man-induced degradation.

By analysing this list it is possible to deduce some of them correctly by considering the overlapping of predisposing causes (such as landslides and solifluction); in other situations an evaluation of the hazard effects can be rather difficult and uncertain and a reliable forecast of their occurrence could be made only with difficulty (for example certain processes of man-induced degradation). As a consequence, when a hazardous area has to be recognised and properly assessed by means of the reciprocal influence of the "causes", not all the "effects" listed below will be considered, but mainly those deriving from the various types of slope movements and in particular the landslides. For example, for rock falls mainly the geological, hydrogeological, meteorological and topographic causes will have to be considered.

After the phases of survey and data acquisition regarding the geomorphological effects and of definition and assessment of the predisposing causes, it is necessary to compare them and to elaborate a synthesis Geomorphological Hazard Map; this document will show both the areas actually and potentially unstable, subdivided according to the instability causes.

This procedure, however, implies problems of different kinds. In fact, data collected in a sufficiently objective way (instability processes) are sometimes related *a priori* to a series of parameters that interact in time and space and from one area to the other in various ways. Their real role in causing instability conditions is not easily ascertainable. For this reason, an analysis system concerning the relationships between causes and effects should be provided. Such a system should be "open" and reproducible. For every study area it should, therefore, be possible to identify the type and number of causes playing a role in the various instability processes, by applying a statistical model relating every instability parameter to the unstable forms and processes actually observed in the field.

The conceptual and methodological aspects mentioned above represented the basis for some inter-university European research projects, in which the author participated and is still participating as a coordinator of a research team. In particular, the investigations have been carried out within the following projects.

- "The temporal occurrence of landslides in the European Community" (EPOCH Programme; contract 90 0025; period 1991–1993) (Casale et al. 1994; Soldati 1996).
- "The temporal stability and activity of landslides in Europe with respect to climatic change – TESLEC" (ENVIRONMENT Programme; contract EV5V-CT94: period 1994–1996).
- "New technologies for landslide hazard assessment and management in Europe – NEWTECH" (ENVIRONMENT AND CLIMATE Programme; contract ENV4-CT96-0248; period 1996–1998).

The areas studied within these projects by the Modena team are located in the northeastern Italian Alps: Alpago, Cortina d'Ampezzo, Corvara in Badia.

These areas have been affected, on the one hand, by several mass movements since the retreat of the Würmian glaciers which have been mainly favoured by the large presence of clayey formations (Panizza et al. 1996). On the other hand, the sites considered (especially Cortina d'Ampezzo and Corvara in Badia) have recently undergone a rapid touristic development connected with sport and recreational activities (Panizza 1990). As a consequence, a high density of population comes during certain periods of the year, a continuous spreading of different types of constructions (residences, hotels, ski-pistes etc.) and the development of infrastructures (for summer and winter spots, roads etc.). All this means an increasing vulnerability.

The relationship between landslide hazard and the area vulnerability gives the landslide risk, which results in being quite high in the studied areas.

References

Casale R, Fantechi R, Flageollet JC (eds) (1994) Temporal occurrence and forecasting of landslides in the European Community – Final Report. European Commission, Bruxelles, 957 pp

Panizza M (1990) Geomorfologia applicata al rischio e all'impatto ambientali. Un esempio nelle Dolomiti (Italia). In: Gutiérrez M, Peña JL, Lozano MV (eds) Actas 1 Reunión Nacional de Geomorfología, 17–20 Septiembre 1990, Teruel, pp 1–16

Panizza M (1992) Geomorfologia. Pitagora Editrice, Bologna, 397 pp

Panizza M, Piacente S (1978) Messa a punto concettuale per la realizzazione di una cartografia applicata alla "stabilità del territorio". Geogr. Fis. Dinam. Quat. 1:25–27

Panizza M, Pasuto A, Silvano S, Soldati M (1996) Temporal occurrence and activity of landslides in the area of Cortina d'Ampezzo (Dolomites, Italy). Geomorphology 15(3–4):311–326

Soldati M (ed) (1996) Landslides in the European Union. Geomorphology, Special Issue, 15(3–4):183–364

Part III

Management – Socio-Economic Aspects

Chapter 13

Innovative Approaches to Integrated Floodplain Management

F.N. Correia · M. da Graça Saraiva · F.N. da Silva

13.1
Introduction and Problem Statement

The results presented in this paper correspond largely to the contribution given by the authors to a research project funded by the European Commission under the Environment Programme, titled EUROFlood. The EUROFlood project has been concerned with flood hazard assessment, modelling and management. As such it was involved in determining the nature, extent and severity of flood hazards across the European Union countries, by collecting data on these hazards and modelling their likely future patterns. In addition, however, EUROFlood was also concerned with investigating methods for better management of hazards and populations vulnerable to hazardous events, so as to reduce vulnerability and enhance the use of floodplain areas. A detailed presentation of EUROFlood results can be found in Penning-Rowsell (1996). A full description of the authors' contribution to this research project is presented in Correia et al. (1996).

A framework for the participatory evaluation of flood hazard management policies is addressed in this chapter. This evaluation is based on the GIS-based comparison of different scenarios for urban growth, the hydrologic and hydraulic simulation of flood hazards for these scenarios, and the implementation of several alternative flood control measures. Environmental impacts of flood mitigation measures are also addressed because they are considered an increasingly important component of the decision making process.

The core of this research is to explore the possibility of using Geographic Information Systems (GIS), and complementary multimedia interactive devices, to play a crucial role in a participatory evaluation of floodplain management policies. In fact, dissemination of relevant flood information using GIS, or other computer graphic devices, can help to support individual and institutional decision making processes, either for preventive flood management and for dealing with emergency situations. In both cases public involvement is a major issue for better decision making and policy effectiveness.

Emphasis is being given to the formulation of different scenarios for urban development and its relevance for flood occurrence. Comparison of alternatives for flood mitigation, taking into account the environmental impacts and the public perception is being also considered. Refined hydrologic and hydraulic models allow for the evaluation of the consequences of flood hazards in terms of flood affected areas. Social and economic characterisation of flood affected areas and the use of flood damage models provides us with an evaluation of the consequences for each scenario.

Geographic Information Systems (GIS) have been recognised as a powerful mean to integrate and analyse data from various sources (Correia et al. 1994, 1995). Adequate

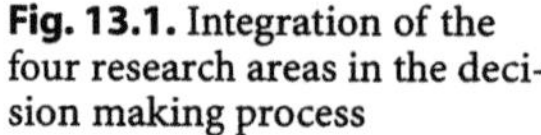

Fig. 13.1. Integration of the four research areas in the decision making process

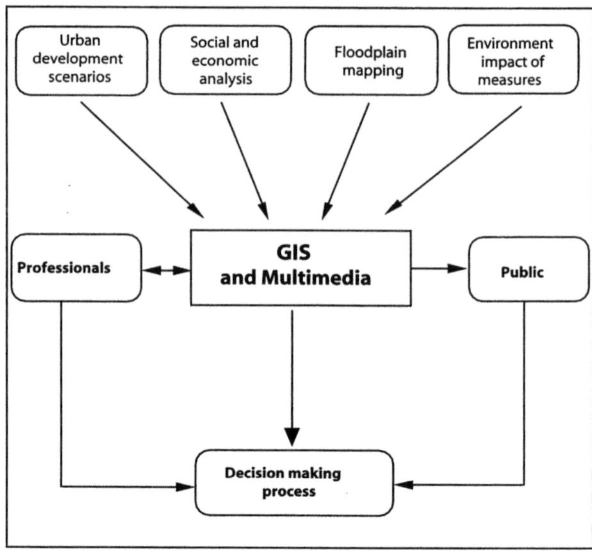

information and prediction capability is vital to improve the decision making processes. Therefore, emphasis is put on improving the possibility of simulating the consequences of the alternative scenarios for urban development and for flood mitigation strategies.

More specifically, the following steps are being pursued in four important fronts of work:

- Social and economic characterisation of the affected people and property in the floodplain;
- Consideration of alternative scenarios for urban development and evaluation of consequences on the flood regime;
- Refinement of hydrologic and hydraulic modelling aiming at a more accurate flood mapping for different scenarios;
- Preliminary evaluation of environmental impacts of flood mitigation measures.

The integration of these fronts of work in the GIS, used as a technical tool by the professionals and also as a tool for public involvement, is presented in Fig. 13.1.

Livramento River basin, in Setúbal, approximately 40 km South of Lisbon was used as a very comprehensive case study. The characteristics of this case study are described in detail by Correia et al. 1994, 1995.

13.2
Background and Methodology

Many activities are necessary for the formulation of floodplain management policies. It is important to have a global picture of these activities in order to understand how they relate to each other and how they contribute for the decision process involved in

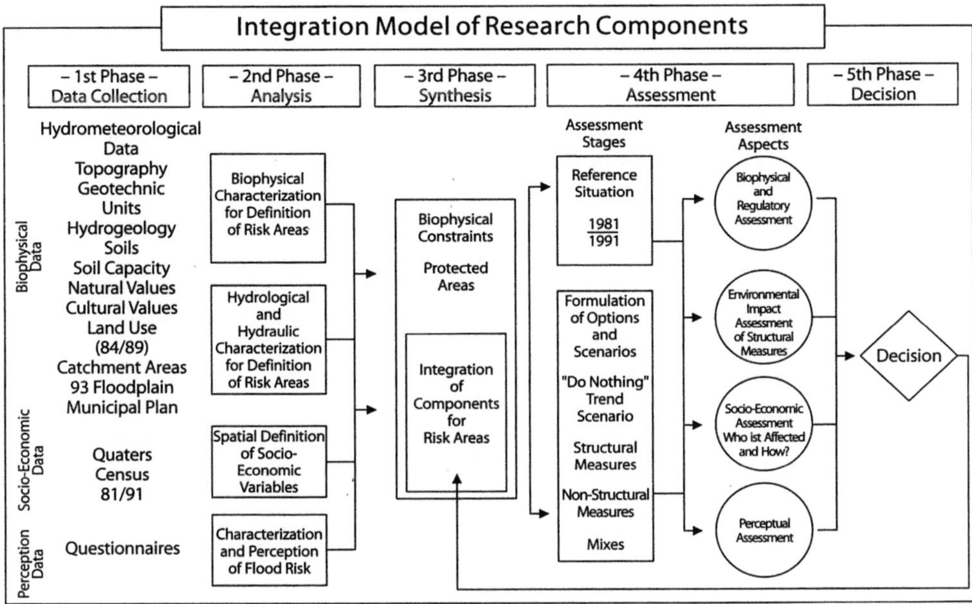

Fig. 13.2. Conceptual model for flood analysis and management

policy making. This global picture, displayed in Fig. 13.2, has been presented before by Correia et al. (1994) but still captures the most relevant aspects of flood policy formulation.

This figure presents a conceptual model with five basic stages of floodplain management policy making: data collection, analysis, synthesis, assessment and decision-making.

In the first phase, a digital data base for the catchment area is implemented, collecting and storing different types of data, namely biophysical, socio-economic and perceptional data, from the survey questionnaire.

The second step is characterisation. Main components of the research are related and analysed through the data base to select the main variables to assess the general situation of the catchment and floodplain in terms of biophysical and regulatory issues, hydrological and hydraulic regimes and variables, socio-economic assessment and characterisation of perception patterns and causal effects.

The next phase allows for the generation of a comprehensive synthesis of the catchment and for the integration of specific components in the risk areas, namely the floodplain. Public perception of flood risks, namely those of residents, shop owners, local authority politicians and professionals have been studied and understood, jointly with the physical and human processes that contribute to the increase of flood risk and vulnerability in the catchment.

The following two steps will lead directly to the decision-making process in floodplain management. It incorporates the formulation of scenarios and the assessment of the impact of various scenarios on flood effects.

The scenario formulation will be based on one hand, on urban growth, and on the other hand, on different options for flood alleviation measures. Four options can be considered in general terms:

- The "do-nothing" scenario, assuming that urban development will continue with few constraints as in the previous decades, and no structural or non-structural measures will be implemented;
- The case for structural measures for flood control, namely, building a dam in the headwater area and retention basins on the floodplain inside the city;
- The application of non-structural measures, namely floodplain regulation and zoning and regulatory constraints within the catchment, through the application of environmental protection regulations;
- The use of a mix of structural and non-structural measures.

For each of these scenarios, a comprehensive assessment can be made using the GIS and feeding a multicriteria decision process. A more simplified view of the policy making process is presented in Fig. 13.3. The use of GIS and multimedia techniques as supporting tools of the decision making process is emphasised in this figure.

In the present research work, two different Geographic Information Systems (Intergraph and IDRISI) are used in order to take full advantage of the specific tools and utilities available, thus making the processes less complicated. The only possible handicap in using different systems is in the area of data transfer. Many transfer formats have to be tried before the most adequate format is adopted insuring data integrity and avoiding data or structure losses.

The structure of these systems and the coupling between them are presented in Fig. 13.4. Further details can be found in Correia et al. (1995).

Social and economic characterisation of flood affected areas is essential and can be improved with the use of GIS and georeferenced database. An important aspect of the of the floodplain characterisation is to know who and what is affected, when a flood occurs. For this purpose it is necessary to evaluate the possible impact in terms of the

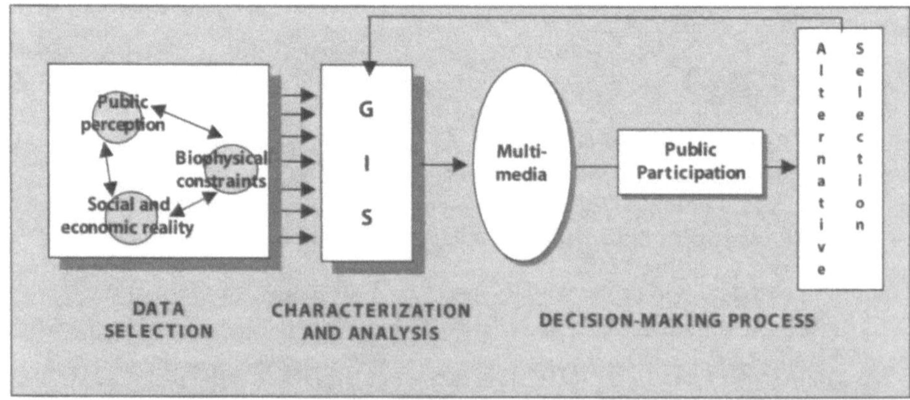

Fig. 13.3. GIS as a support tool for the decision making process

damage in homes, commercial activities and heritage. This information can also be very important for the preparation of flood emergency plans.

This aspect is particularly important in the case of Setúbal and many other coastal towns because the centre of the town spreads all over the floodplain, with very important commercial activities and historical values located in these areas.

Not only flood hazards can affect urban activities but also most of flood control measures can disturb existing situations. Thus, the great importance of finding out possible impacts on the residents, on the trading activities and on the historical heritage values. This information can be also very important to define emergency plans for flood hazards.

The characterisation of social and economic dimensions of flood affected areas can be done at different degrees of detail and accuracy. This is one of the first problems that need to be faced when dealing with flood damage assessment. In any case, it is adequate to have a data base with available information and be able to display such information graphically.

The objective is to create a georeferenced data base that integrates all the information available from the Census of 1991, concerning population and building characteristics, as well as, all the information that may contribute to the social-demographic and economic characterisation of the study area.

GIS, as an integration tool, makes it possible to combine the graphic elements with alphanumeric information. The alphanumeric information by itself is usually non-spatial data but it allows to interface with associated graphic features. The association procedure between these two kinds of information is made by the code that represents the geographic unit. Thus, our database have two types of information: on one hand the internal attributes of the software GIS that allow to make the univocal relation between the graphical elements and the alphanumeric data; on the other hand the attributes that allow to characterise the graphic elements.

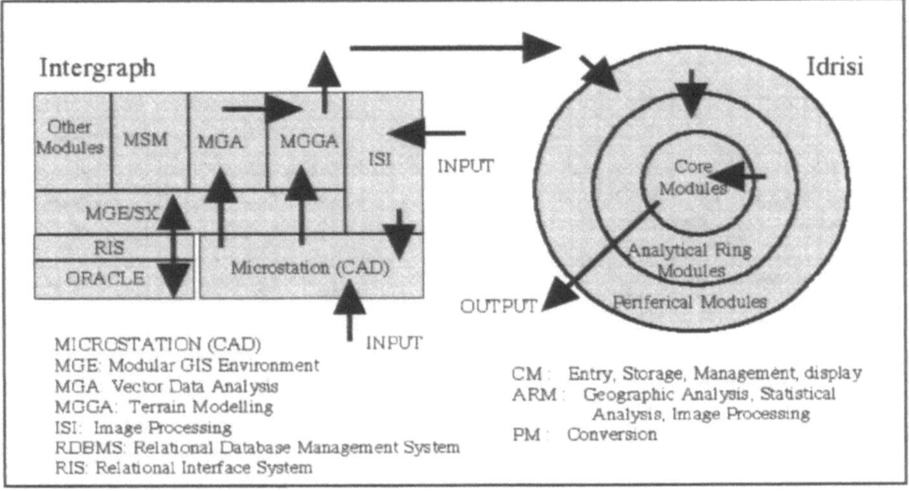

Fig. 13.4. Integration of Intergraph and IDRISI GIS in hydrological and hydraulic modelling

The linkage between graphic and alphanumeric data makes it possible for two types of analysis:

- To retrieve graphic elements and present the results in an alphanumeric way (graphic input → alphanumeric output);
- To retrieve alphanumeric attributes and represent them as graphic elements (alphanumeric input → graphic output).

It is possible to identify different steps in the use of social-demographic and economic data in the context of GIS-based flood analysis. The first step is to build the georeferenced database including graphic and alphanumeric information. The second step is the social and economic characterisation of the area in order to determine the possible impacts on the flood affected areas.

The construction of a social and economic atlas for the study area, that has a friendly interface when used by non-expert people, is one of the objectives that can be achieved with the GIS. However, it is necessary to have an advanced knowledge of a specific software language to manipulate the data. This is the reason why the construction of a user-friendly interface is an important step in order to make the information available without a large and specific knowledge of these kinds of technologies.

13.3
Land Use Management and Flood Impacts

13.3.1
Land Use Controls and Flood Policy

Integrated and comprehensive floodplain management and policy formulation can be considered as the partial overlap of four main areas of planning and management: water resources, land use, environmental management and emergency protection. Administration and institutional procedures for flood hazard management reflect the organisational structure and objectives of these fields of action at the national, regional and local levels.

An important area of research is to evaluate how well floodplain management is achieved through the land use planning process, namely the implementation of land use based non-structural measures. The assessment of the effectiveness of some specific regulations that have been adopted in the past by central and local governments is especially important.

In Saraiva and Correia (1993) a detailed review of the institutional, legal and regulatory framework of floodplain management and strategic land use planning for Portugal has been presented. A number of best practices related to land use controls have been identified, providing tools that can be used for the formulation of flood mitigation policies. Some of these best practices in Portugal are summarised as follows:

a   The concepts of "water public domain" and "adjacent zones", that allow the definition of flood prone areas on a very detailed scale with the definition of development regulations and restrictions. These concepts, and the zoning of the corresponding areas, are key instruments in terms of non-structural measures. The definition of the adjacent zones is the responsibility of the central Government after consultation with local authorities.

b The concept of "National Ecological Reserve" (REN), that establishes specific environmentally sensitive areas, subject to special regulations. In these areas, riverine systems are especially protected, namely rivers, river banks and floodplains. These areas must be identified and mapped within the local land use plans, with the involvement and collaboration of other regional and central agencies.

c Local land use Master Plans (PDM), prepared by local authorities. Legislation requires that these Master Plans include the definition of various land uses and establish development regulations, together with a network of protected areas based on environmental sensitive concepts, such as REN and other environmental and conservation resource regulations.

After the disastrous flash floods of November 1983, in the Lisbon area, a special Task Force for the assessment of flood problems in the Metropolitan Area of Lisbon was appointed. This Task Force conducted a very comprehensive and multi-disciplinary analysis of flood hazard problems in the area and formulated several policy recommendations largely based on land use controls and non structural measures (Cheias 1985, 1989).

Some of these policy recommendations were implemented, namely the official zoning of a few adjacent zones for flood protection in the most severely affected rivers. Approximately ten years after the implementation of these policies, it is very important to make an ex-post evaluation of the undertaken measures.

This is being done as part of this research effort and the Setúbal case study plays an important role because it corresponds to a critical situation. Other river basins in the Metropolitan area of Lisbon have been investigated.

13.3.2
Assessment of Land Use Controls in Floodplain Management

The importance of assessing the effectiveness of land use controls in floodplain management strategies has been suggested by several authors (Burby and French 1981; Handmer 1995; Muckleston 1972; Parker 1981; Fordham 1992; Saraiva 1995).

Burby and French (1981) argue for the verification of a so-called "land use management paradox"; showing that factors that stimulate the adoption of land use programmes in hazard areas can also promote encroachment, which in turn limits their effectiveness. Therefore, the local context is the best framework to evaluate the extent and aims of floodplain management programmes and their effectiveness, stakeholders involved and the emergence of conflicts related to development interests and regulatory restrictions.

In this context, an ex-post evaluation of the policy recommendations and land use regulations adopted in 1985/87 by the Task Force mentioned above, provides us with a view of the evolution of some critical areas in the past ten years. The recommendation of flood policies and flood mitigation measures have been prepared by central government agencies and proposed to local authorities as a *top-down* methodology. The general approach adopted by this Task Force is very comprehensive and can be considered as an interesting methodology to deal with flood problems.

During the last ten years, local authorities have been preparing their municipal Master Plans (PDM), that are already in force in most cases, after approval by central Government. It is very important to evaluate how floodplain management is considered in these plans and how non structural measures are used. This is a *bottom-up* approach, since measures have been considered by local authorities and approved by central Government.

The research in this area is based on existing plans and surveys, comparison of programme recommendations with land use plans and regulatory decisions, and interviews with local authorities and technical staff involved in floodplain management.

Four municipalities within the Metropolitan Area of Lisbon, seriously affected by the 1983 flood, have been selected – Setúbal, Oeiras, Cascais and Sintra.

An analysis of their Master Plans (PDM) has been done, in order to assess all references to floods and flood management measures. Spatial information and regulations over hazard areas is under analysis, as well as, potential conflicts with urban development.

More specifically, the situations of "strong enforcement", "weak enforcement" and "carelessness" over Adjacent Zones, legally established after the Task Force recommendations, have been looked at, through the analysis of maps and aerial photos.

The relationship between local initiatives for flood control and measures adopted by central government agencies are also being analysed, in order to evaluate the effectiveness of *top-down* and *bottom-up* measures.

13.3.3
Ex-post Evaluation of Land Use Controls

The evaluation of the main trends in floodplain occupation over the last ten years in the Metropolitan area of Lisbon leads to the following conclusions:

- The institutional framework for floodplain management is quite complex and co-ordination is lacking. In fact, there is little coordination between central departments and local authorities in defining strategies and tools for flood risk minimisation, namely with respect to non-structural measures such as land use controls.
- The use of the two main non-structural measures, namely the "adjacent zones" and the protection of rivers and floodplains included in the National Ecological Reserve (REN), although having similar zoning effects, are quite different in terms of regulation codes and types of enforcement. The prevailing measure that is being applied is the second one, because it is a requisite in the preparation of municipal land use plans (PDM). However, the REN is usually not defined inside urban areas, in which floodplain encroachment is more severe. In these urban environments, the implementation of adjacent zones is the most efficient tool because it is applicable in all kind of rivers and streams. Unfortunately, governmental agencies responsible for establishing "adjacent zones" have used this legal provision in very rare situations.
- The high value of land in peri-urban areas introduce complexities in the land use control decision process. The lack of zoning regulations, the complexity of preparing these regulations and difficulties in the consultation process, can lead to some undesirable situations. Illegal or inadequate development is still spreading along floodplains, despite legal and regulatory efforts, mainly due to lack of coordination and enforcement.

- It can be concluded that cooperation between central government and local authorities in terms of floodplain management needs to be largely increased, namely through a better coordination of overlapping duties and a more effective enforcement of the land use control measures that are available.

13.3.4
Urban Growth Modelling in Setúbal

13.3.4.1
Formulation Urban Growth Scenarios

The first step to define realistic scenarios for the future urban growth of Setúbal is the analysis of the evolution of the urban areas since the moment in which Setúbal started suffering the consequences of the metropolitan growth in the Lisbon region, that is, since the late sixties. This allows for the identification of the previous urban growth paths and trends and the quantification of the changes in the urban perimeter.

The urban areas in 1969, 1984, 1989 and 1991 were identified from the existing maps (1 : 25 000 and 1 : 50 000 scales) with the help of aerial photography. Using the GIS, variations were measured in Δ area values. Linking these values to the census data it is possible to get information about the evolution of urban density (inhabitants/ha). The geographic units for population and dwellings are the census blocks, which allow for linking to the attributes considered in the alphanumeric data base.

Relevant data for the characterisation of urban growth in Setúbal is presented in Tables 13.1, 13.2 and 13.3. The evolution of urban areas from 1969 until 1991 and the urban perimeter considered in the recently approved Master Plan are displayed in Fig. 13.5.

Table 13.1. Evolution of developed areas in Setúbal

Developed areas	1969	1984	1989	1991
S (urb) catchment (ha)[a]	610.0	1 180.0	1 240.0	1 275.0
S (urb) Setúbal (ha)	480.0	810.0	840.0	855.0
ΔS (urb) catchment (ha)	–	570.0	60.0	35.0
% ΔS (urb) catchment	–	93.4	5.1	2.8
ΔS (urb) Setúbal (ha)	–	330.0	30.0	15.0
% ΔS (urb) Setúbal	–	68.8	3.7	1.8
ΔS/year (urb) catchment (ha a^{-1})	–	38.0	12.0	17.5
ΔS/year (urb) Setúbal (ha a^{-1})	–	22.0	6.0	7.5
% S(urb) Setúbal/S(urb) catch.	78.7	68.6	67.7	67.1
N° of settlements in the catch.[b]	18	35	39	39
µS settlements (ha)	7.01	10.33	10.49	10.76

[a] The city of Setúbal is included;
[b] The city of Setúbal is not included.

Table 13.2. Population growth in Setúbal

Population data	1950	1960	1970	1981	1991
Population (municipality)	54772	56344	66099	98366	103634
Demographic growth rate (%)		0.30	1.60	4.00	0.50

Table 13.3. Evolution of settlements (isolated neighbourhoods) (ha)

Settlement N°	1969	1984	1989	1991	Settlement N°	1969	1984	1989	1991
1 (Setúbal)	482.4	809.3	841.6	855.3	24		9.90	9.90	9.90
2		7.88	7.88	9.90	25			2.70	2.70
3	1.23	1.30	1.30	1.32	26	4.24	14.60	16.43	16.45
4	2.11	3.73	3.80	3.80	27		19.40	19.42	19.42
5		6.00	6.00	6.00	28	1.52	5.20	5.18	6.40
6		2.45	2.50	2.50	29	1.04	2.50	2.60	2.60
7	2.83	2.86	2.90	2.90	30	8.19	15.80	17.60	17.60
8		3.50	3.50	3.50	31	47.68	96.50	97.30	100.60
9			8.20	8.20	32	1.90	3.10	3.11	3.11
10			1.60	1.60	33		0.80	0.80	4.50
11		3.94	4.00	4.00	34		1.30	1.30	2.20
12		1.40	1.40	1.42	35		0.80	0.80	2.20
13		10.50	10.60	10.60	36	34.70	62.20	62.34	62.70
14		1.00	1.00	1.00	37	2.70	5.60	7.80	9.40
15		0.60	0.62	0.62	38	2.00	6.20	6.14	7.10
16		1.80	1.90	1.90	39		1.20	1.20	1.20
17		7.45	7.40	7.50	40		0.50	0.50	1.03
18		4.80	4.80	4.82	41	1.83			
19			2.62	3.20	42	2.22			
20	6.96	28.90	29.30	33.15	43	1.36			
21		3.80	3.80	3.80	44	1.44			
22		2.40	7.33	7.33	45	2.18			
23		31.40	31.40	31.40	46				
Average area with the city of Setúbal not included						7.01	10.33	10.49	10.76
Average area with the city of Setúbal included						30.78	31.92	31.44	31.87

Four scenarios were formulated considering the previous trend of urban growth and several degrees of intervention from the local authorities:

- A "trend scenario" that consists in the continuation of the urban growth patterns of the seventies and eighties, without any enforcement of the recommendations of the Master Plan. This pattern reproduces the main characteristics of the urban sprawl of those days: discontinuous development nearby the existing urban perimeter, linear urbanisation along the main roads, random occupation of previous agricultural land. This can be considered a "do-nothing" scenario and corresponds to a pessimistic scenario from a land use management point of view.

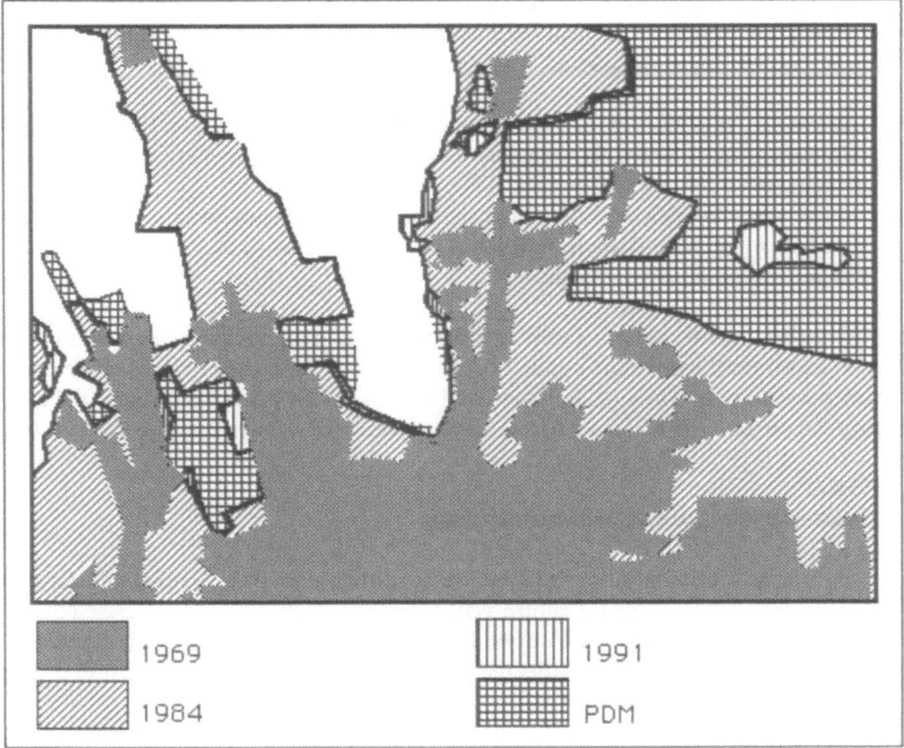

Fig. 13.5. Evolution of urban areas from 1969 until 1991 and urban perimeter considered in the recently approved Master Plan (PDM)

- An "official scenario" that considers the development on new urban areas as they are established in the Master Plan. It should be reminded that the Master Plan corresponds already to a compromise between the interests of the main developers of the area and the national rules for the preservation of natural and agricultural resources.
- An "optimistic scenario" that assumes a stronger intervention of the municipal administration. In this case the urban growth will take place essentially inside the existing urban perimeters, and new urbanisation fronts will occur only when the majority of the existing open land have been developed. These new urban areas will take place in the continuity of the existing ones filling up unacceptable voids of non used land.
- An "intermediate scenario" that results from a balance between the previous trends and the development apent proposed by the Master Plan. In this case, the urban growth will occur with the same pattern of the seventies but respecting the legal constraints regarding the agricultural activities and preservation of natural resources as identified in the Master Plan.

In order to calculate the growth rate and the importance of each growth typology, the population growth of Setúbal since 1960 was analysed, and three basic typologies of urban growth were considered. These typologies are the following:

- *Continuous,* when the urban growth takes place in the continuity of the existing urban areas, that is, when the distance between new developments and the existing urban areas is less than 25 meters.
- *Linear,* when the new developments are not linked to the existing urban areas but they follow the main roads at a distance of no more then 50 m.
- *Discontinuous,* all other situations.

The urban growth scenarios were established considering the following assumptions and constraints: the previous population growth rates; the rate of new developed land in the past (ha/year); the availability of land to be developed; and the natural and agricultural constraints established by the Master Plan in all scenarios except the first one.

13.3.4.2
GIS-Based Urban Growth Scenario Analysis

GIS can be used for analysing the dynamics of urban growth in Setúbal and for formulating new scenarios for the future. Naturally, these scenarios must be based on the assumptions and constraints identified for the situation being analysed and must reflect some basic options like the decree of conformity with existing plans and degree of intervention of local authorities.

The formulation of urban growth scenarios is a very important and useful tool in many areas of land use planning. The use of automatic procedures based on GIS is entirely justified because it makes possible the use of large amounts of data that can be easily related. It also provides an adequate means for visually displaying the future scenarios.

However, in the context of this research effort, the main objective is to evaluate the consequences of different land use policies on the flood regime. In fact, changes in the land use of the catchment area induce different flood conditions that can be predicted by the hydrological and hydraulic models as it is explained in Section 13.4.

The easiness of using GIS for this purpose should not be overstated. In fact it raises a lot of practical difficulties, even when one may think that all data are already available in the system. In fact, such data often have different sources and are presented in different scales making more difficult the process of joint manipulation.

In the Setúbal case study there was already a lot of information in the data base, collected in previous phases of the research work. However, the first step was to make an adjustment of classes and boundaries of different maps, harmonising classification criteria for the different land use units and assuming a level of accuracy never worse then the one given by the cartography at the 1 : 50 000 scale.

An additional problem is that the area relevant for the analysis is not only the Livramento catchment but all the urban area of Setúbal. In fact, it is only at this global level that the dynamics of urban growth can be analysed and alternative scenarios can be formulated.

There are two basic types of information needed for the analysis of urban growth dynamics: land use data and Census data. In order to make the manipulation of these data more useful, the following steps were followed:

a *Adjustment of the boundaries of the cells.* This was done in pairs, from the oldest to the most recent data: 69 and 84, 69 and 89, 84 and 89, 89 and 91.
b *Definition of urban areas based on land use maps.* The information on the urban areas contained in the land use maps was presented in separate data layers, one layer for each year. Each urban polygon was given a unique ID number in order to be able to follow its evolution over time. The urban polygons sometimes vanish, because they are absorbed by another polygon.
c *Data validation.* For using GIS, the graphic units need to be validated and depicted as features for future links to the alphanumeric database. This process is done for each year separately, corresponding to different layers of information.
d *Linkage of land use data to a simple database.* The graphic files related to land use were linked to an alphanumeric database with a simple structure, including only the label (land use), ID N° and area.
e *Linkage of Census data to a complex database.* The graphic files related to the Census data have to be linked to a more complex alphanumeric database, including data on dwellings and population. This information is only available for 1981 and 1991.

With all the graphical and alphanumeric data properly linked to the databases, some of the parameters relevant for the characterisation of urban growth dynamics, like area, perimeter and density values, can be immediately and automatically computed.

One important aspect that can be also immediately investigated is what kind of land use classes have been preferentially used for urbanisation. In the GIS this is computed by an overlay operation of the urban area at time (t) with the land use map at time ($t - 1$).

Another relevant aspect is the identification of the different growth typologies identified in Section 13.3.4.1. This identification needs to be done stepwise as shown in Fig. 13.6 and Fig. 13.7.

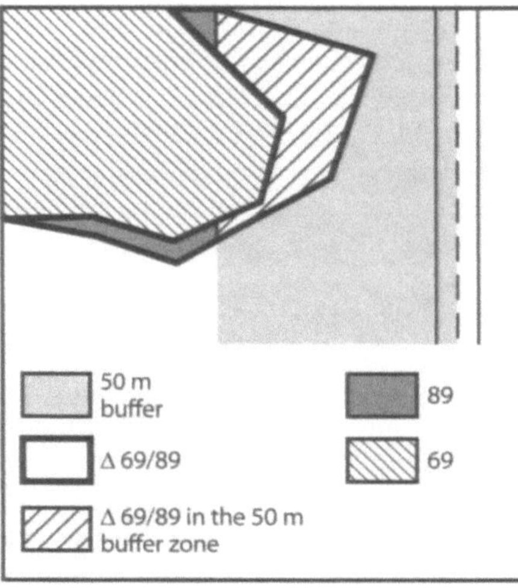
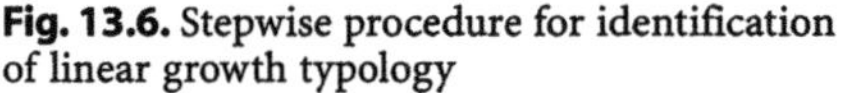

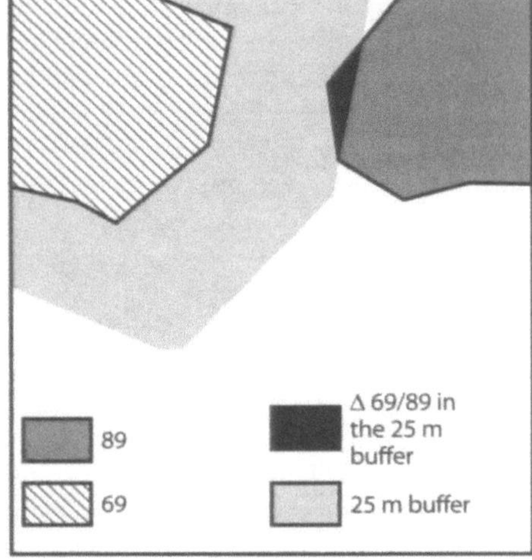

Fig. 13.6. Stepwise procedure for identification of linear growth typology

Fig. 13.7. Procedure for identification of continuous growth typology

For the linear growth it is necessary to built a separate layer only with the road system and afterwards use the buffer function included in the GIS to create a file with polygons that are overlaid with the result of the urban increased polygons (Fig. 13.6). These areas are measured directly through database operations in relation to the total growth and compared to the values for the different time periods.

For the continuous growth typology the procedure is similar, as it can be seen in Fig. 13.7. The buffer zone of a polygon at time $(t-1)$ is crossed with the urban area at time t. This type of occurrence is characterised in terms of frequency and area.

The analysis and quantitative characterisation of urban growth dynamics will allow the projection of scenarios for the future, according to the options presented in Section 13.3.4.1.

13.4
Hydrologic and Hydraulic Modelling

13.4.1
Lumped and Distributed Modelling

The mapping of flood affected areas is a basic tool for flood risk assessment. The most important hydrologic and hydraulic variables relevant for flood hazard assessment are: meteorological data, namely rainfall depths and intensities associated to different levels of risk, magnitude and frequency of flood peak discharges, characteristics of the flood wave, and location and size of flooded areas.

A simple representation of flood affected areas in a way that is easily grasped by the public, but still accurate and scientifically sound, is an important product of a flood study. With this concern in mind, there is much to gain in coupling the traditional and cumbersome hydrologic and hydraulic modelling with advanced and sophisticated computer tools like GIS.

According to this approach, a lumped rainfall-runoff model, XSRAIN, was previously used to describe the hydrologic behaviour of the watershed Verdin and Morel-Seytoux (1981). A physically based and fully distributed model, OMEGA, is being currently used to describe the flood generation processes (Correia 1984; Correia and Matias 1991). This model is much more sensitive to changes in the river basin, namely in what concerns land use practices. Different areas of the watershed can be modelled separately and subject to changes providing the ability for simulating scenarios and computing flood waves for these scenarios.

The use of the lumped model XSRAIN is described elsewhere (Correia et al. 1994; Correia et al. 1995). A brief description of the distributed model OMEGA follows.

OMEGA is as a package of fully compatible and consistent computer watershed models designed for simulation or real-time forecasting of river flows, and including complete calibration procedures. A more detailed description of the model can be found in Correia and Matias (1991), Correia and Morel-Seytoux (1985) and Correia (1984).

OMEGA can be used for event or continuous simulation of river discharges and can be utilised in a lumped or distributed fashion. When used as a distributed model the spatial variability of the input and the watershed response may be fully considered.

OMEGA incorporates to a large extent the physical conceptualisation of the hydrological processes. The description of the complex and highly non-linear processes of infiltration, ponding and redistribution of water in soil is addressed with special attention. The modern theory of infiltration provides us with tools that can be adequately incorporated in rainfall-runoff models enhancing the possibilities of these models and improving their results.

Three totally compatible versions of the model were implemented, namely the basic simulation version, the parameter calibration version and the real-time forecast version. This triple formulation adds a large degree of versatility. In previous applications, the model showed a remarkable ability to adapt to different hydrologic conditions, going from a mountainous and wet region to a semi-arid and flat one, and provided consistently good results in practical applications.

Naturally, OMEGA is much more demanding in terms of required data and much more complex in terms of parameter calibration, when compared to a lumped model like XSRAIN. This should be faced as a trade off for the better capabilities in comparing alternative scenarios for urban development. However, the semi-automatic calibration procedures incorporated in the model allow for a rather expedite calibration of the parameters, principally when plenty of information is already available in the GIS data basis.

The use of a rather demanding model like OMEGA directly coupled to the Intergraph GIS is not practical because of the complexity of the model and the need to make many experimental runs. For this reason, the relevant layers of information were transferred from the Intergraph equipment to a common 486 PC compatible system. The GIS IDRISI, developed by Clark University, in Massachusetts, USA, was used to compute most of the hydrologic parameters, creating input files that can be used by the model.

The transferability between Intergraph and IDRISI was found to be manageable and the most relevant hydrologic and hydraulic information, like Digital Terrain Model (DTM), type of soils, land use maps, hydrographic network, and hydrogeological maps, are available in both systems.

A general conclusion at this stage is that it is better to use a simple GIS if a complex rainfall-runoff model is being used, at least at a preliminary stage when many experimental runs are needed. As experience grows, good calibration of the model is achieved and a good control of all operations is acquired, then the coupling with a more complex GIS can be attempted.

The use of the OMEGA model is rather successful and a good agreement between observed and computed discharges are being achieved. Figure 13.8 displays the results obtained for the historical storm of November 1983, one of the most severe events in this century.

As with XSRAIN model, both a synthetic design storm and a historical record can be used with OMEGA. However, OMEGA is more flexible because spatial variability of the rainfall inputs can be also considered. This is particularly important when simulating the movement of a storm over a watershed.

The most relevant output of the model is a flood hydrograph expressed in terms of instantaneous flood discharges at specified time steps. This flood hydrograph, or simply the peak discharge, must be used as the input of the hydraulic model in order to evaluate the flood affected area and other relevant parameters, such as flow velocities at different points. The flood affected area can be transferred back to the GIS and represented as a flood map for the rainfall event under consideration.

The determination of backwater profiles for flood hazards of various frequencies is based on the hydraulic model HEC-2, developed by the Hydrologic Engineering Cen-

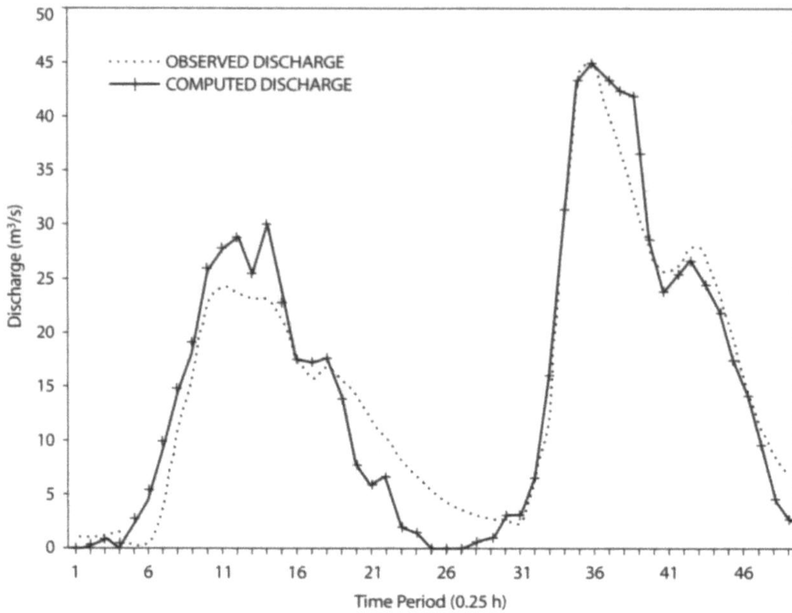

Fig. 13.8. Observed and simulated flood event of November 1983

tre of the US Army Corps of Engineers. This model provides the calculation and graphical presentation of backwater profiles for steady flow conditions. Supercritical or subcritical flows in natural or lined cross sections, for several flow rates, can be considered. Profiles with or without singularities such as bridges, culverts or structures in the overbank region can be also considered.

To use this hydraulic model the user has to input the following data: identify and describe a number of cross sections necessary to represent the river geometry and all singularities; roughness coefficients of overbanks and channel; water elevation for the first cross section (downstream section if the flow is subcritical or upstream section if the flow is supercritical); discharge value.

Here are some of the most relevant results after running the backwater program (HEC-2 or an equivalent model): the computed water surface elevations; the cumulative volumes of water in river since the first cross section; the mean velocity in the channel. This information can be transferred back to the GIS and flood affected areas can be displayed. A more complete description of these procedures can be found in previous papers like Correia et al. (1994) and Correia et al. (1995).

13.4.2
Coupling GIS with Hydrologic and Hydraulic Modelling

The weakness of GIS is the time based processes that are not easily accommodated in a GIS. To circumvent this limitation it is possible to couple the GIS with a peripheral computer in which the hydrologic and hydraulic models are ran. Results can be transferred back to the GIS and space based operations can proceed in this system. This proved to be a good approach for both hydrological models being used (XSRAIN and OMEGA) and for the hydraulic model used for flood mapping (HEC 2).

As described in Correia et al. (1994) and Correia et al. (1995), GIS have different backgrounds and different architectures. Even if results may be equivalent, the implementation process is quite GIS specific and demands an adequate methodological approach.

As a preliminary task, the existing cartography (source data) has to be converted from their existing form into one that can be used by the GIS at hand. This issue can be done by manual digitising or by map scanning. Both processes create a geo-referenced database files in digital format. In some cases, the agencies responsible for data publication can provide the user with cartography that is already in digital form. This is a tremendous breakthrough, since it is usually manageable to convert this source information to the system that is being used.

In the Setúbal case study, it was possible to purchase from the Portuguese Military Cartography Service the topographic map with some of the basic information at the scale 1 : 25 000. This map was used as the basis for superimposing other layers with all sorts of relevant information.

Various sorts of data were manipulated in order to produce the necessary input for running the hydrological model, such as watershed parameters and cumulative rainfall depths.

Precipitation values corresponding to design storms of a given return period and for a specified duration can be stored in the database and selected by the user for a specific run of the hydrologic model. As an alternative a historic storm can be also stored in the data base and used for model validation or to access the response to changes in the watershed.

The parameters of the hydrologic models are very much related to soil characteristics and land use. Soil characteristics can be derived from the existing hydrogeological maps. Land use maps can be established from aerial photography.

Urban developed areas play a crucial role in the hydrologic regime as they create large impervious areas. Since the existing and digitised land use maps are from 1984, it was necessary to update them with more recent information. Remote sensing data was used for this purpose. Aerial photographs from 1989, at a 1 : 15 000 scale, were used, taking advantage of image processing tools available with the Intergraph GIS system. After saving the geographically referenced images, it is possible to load and overlay the older land use map in vector format. The overlaid map can be manipulated and edited with the photography being a backdrop image to the vector design file. All data input and pre-processing are executed with the Intergraph GIS on a UNIX platform. Hydrologic and hydraulic models are run on a MS DOS platform and IDRISI GIS can also be used on this more common portable platform.

Vector-raster conversions were done within the Intergraph, taking advantage of the existing database linkage to the polygons with different database records for different cell values. The image size considered as satisfactory for the 1 : 25 000 analysis is a 15 × 15 m · meaning 500 rows and 550 columns. After the vector-raster conversion, all maps – including class of soil, land use and Digital Terrain Model (DTM) – can be transferred to IDRISI in ASCII files.

In order to create a CN map, the CN value is computed automatically for each cell by overlaying land use classification and type of soil.

Fig. 13.9. Buildings affected by the 100-year flood computed automatically by the GIS

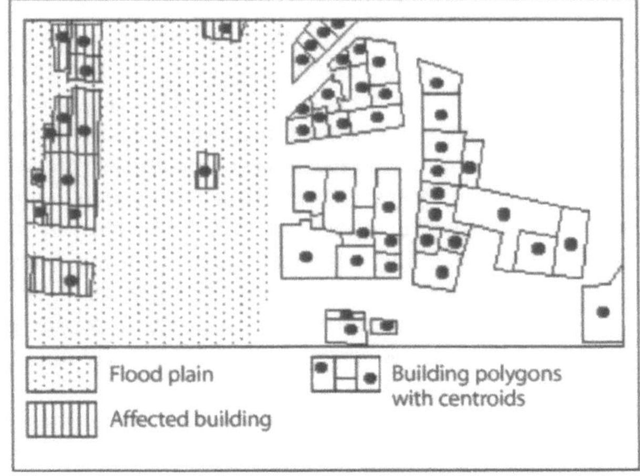

Table 13.4. Basic hydraulic data for the 1983 flood event and for the 100-years design flood event

Hydraulic data	1983	100-year flood
Volume of water (m^3)	598 000	1 413 000
Water level at an upstream reference section (m)	10.29	10.50
Water level at the beginning of the culvert reach (m)	6.71	7.20
Water level at a downstream reference section (m) (Luísa Todi Av.)	3.20	3.20

The terrain modelling software creates DTM maps and files automatically. Relevant watershed descriptors are derived from DTM. Therefore, slope, aspect and elevation maps can be very easily obtained in a GIS environment.

The final result that can be obtained by the methodology that was used is presented in Fig. 13.9. It is interesting to emphasise again that all sorts of information can be stored in the data basis for each building polygon.

Some relevant data for the 1983 flood, used for calibration, and for the 100-year flood are presented in Table 13.4.

13.5
Environmental Impacts of Flood Control Measures

13.5.1
Relevance of Environmental Considerations

Floods are one of the most devastating natural hazards, principally in urban and suburban areas. To control or mitigate flood hazard damages, several types of measures can be adopted. It is usual to establish two main types of flood measures: structural and non-structural.

Structural measures involve a set of works and structures for modifying flooding through the reduction of one or several hydraulic parameters that characterises a flood event, such as the volume of runoff, peak stage, time of rise and duration, depth, velocity or extent of the flooded area. Main structural measures include dams, reservoirs, channel alterations, dikes, levees and floodwalls, high flow diversions, spillways, retention basins and stormwater management facilities (FIMA 1992).

Non-structural measures include a wide range of prevention or adjusting measures to reduce flood risk through the modification of human activities vulnerability and susceptibility to flood damages in the floodplains. They can include flood forecasting and warning systems, disaster preparedness and emergency plans, land use regulations and development control, floodproofing and flood insurance.

Structural measures for flood control can cause significant environmental impacts, due to the modifications of the hydrological regime that affect the biotic and abiotic components of the river systems. The awareness of these impacts have been increasingly growing, calling to the need for establishing methodologies and processes for assessing and minimising them, through the implementation of these measures in the most environmentally appropriate manner. In fact, the Environmental Impact Assessment (EIA) became a general concern or even a requisite (Directive 85/337/EEC) for all major activities that change the biophysical systems significantly.

The evaluation of environmental impacts of flood control measures is a rather complex task and there is not much literature on this subject. There is a large variety of measures of different nature and the local context is always very important. Therefore, it is not easy to establish a general methodology.

This is particularly true for two preliminary steps of all EIA processes, both requiring important options to be made: the *screening*, which allows the definition of projects, plans or policies that needed to be submitted to an environmental study; and the *scoping*, which selects the components of the environment that will suffer significant impacts from the project, plan or policy.

GIS can be a powerful tool in the assessment of environmental impacts of flood control measures and flood management policies.

13.5.2
Environmental Impacts of Structural Measures

The structural approach in flood control policies has been traditionally predominant all around the world, until the emergence of its criticism in terms of the significance of environmental impacts and the raising of comprehensive non-structural measures for flood adjustment, during the sixties (White 1964). Since then, the assessment of environmental context and public awareness for flood defence and control has been increasingly growing. Otherwise, the ecological and cultural values and functions of wetlands and floodplains have been also highly recognised and valued. FIMA (1992) identifies, along the history of floodplain management in the USA, the following periods:

- The structural era, conducted by federal agencies with the aim of flood reducing losses mainly through structural measures.
- 1960s decade: A time of change and the raising of zoning and land use regulation programs and other non-structural measures for floodplain management.

- 1970s decade: The concern for environmental impacts and environmental led programs for water resources management; it is also the period of development of the National Flood Insurance Program which combines flood mapping and zoning with insurance programs.
- 1980s decade: Attention has been increasingly paid to comprehensive floodplain management, environmental concerns and post-disaster recovery programs.
- 1990s decade: Actual floodplain policies intend to consider a broader approach incorporating structural and nonstructural measures for flood loss reduction, together with the preservation and restoration of environmental resources.

Main structural measures, such as dam and reservoirs, retention basins, river regulation works and dike building, can introduce strong modifications in the flow regime, channel morphology, soil properties, groundwater circulation, riparian vegetation structure, habitat disturbance, landscape character and diversity, influencing natural functions and socio-economic uses along river corridors and floodplains. The conventional design methods used by public works engineers in the first half of this century have been mainly influenced on hydraulic models that had not taken into account in large extension the behaviour of natural streams and their geomorphological context (Williams and Swanson 1988). Some effects of structural works, such as flood defence measures, leading in many cases to river degradation situations, are summarised in Fig. 13.10 (Perrow and Wightman 1993).

Otherwise, it has been recognised that structural works can induce and encourage urban development in the floodplains, resulting in increased vulnerability and flood losses (Burby and French 1981; FIMA 1992).

Environmental impact assessment for flood control measures comprises a large set of concerns depending on the diversity and types of those measures. Dam and

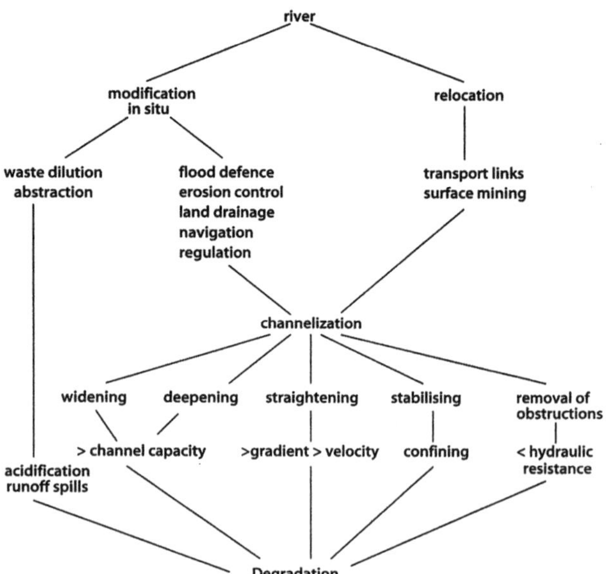

Fig. 13.10. The pathways of riverine modification (Perrow and Wightman 1993)

reservoir building are the most quoted in the literature, having been one of the subjects mostly studied from the environmental impact point of view within the water resources field (ICOLD 1982; Goldsmith and Hildyard 1984; OCDE 1982). Nevertheless, flood control is one of the multiple objectives considered in those projects and its specificities under that point of view are generally less considered in the methodology assessment.

A specific concern for flood defence works is contained in Mock and Bolton 1991, including a checklist for identification of possible environmental changes in hydrology, water pollution, soil properties, sediment transport, ecology, socio-economic aspects, health and imbalances caused by flood control, irrigation and drainage projects (Table 13.5). This checklist can be a useful tool for EIA scoping for flood control projects, after adapting it to the specific characteristics of each situation.

A more sensitivity and environmental concern has been growing in river engineering practices applied to river regulation and flood control, paying more attention to geomorphic processes and hydrological and biologic factors that influence stream

Table 13.5. Check list of environmental impacts of flood control measures (After Mock and Bolton 1991)

Type of impact	Impact	Yes/no	Type of impact	Impact	Yes/no
Hydrology	Low flow regime		Ecology	Project lands	
	Flood regime			Water bodies	
	Flow compensation			Surrounding area	
	Water table fall			Valleys and shores	
	Water table rise			Wetlands and plains	
Pollution	Solute dispersion			Rare species	
	Toxic substances			Animal migration	
	Organic pollution			Natural industry	
	Anaerobic effects		Socio-Economic	Population change	
	Gas emissions			Income and amenity	
Soils	Soil salinisation			Human migration	
	Soil properties			Resettlement	
	Saline groundwater			Women's role	
	Saline drainage			Minority groups	
	Saline intrusion			Heritage sites	
Sediments	Local erosion			Regional effects	
	Sediment yeld			User involvement	
	Channel regime		Health	Water and sanitation	
	Local siltation			Habitation	
	Hinterland effect			Health services	
	River morphology			Nutrition	
	Estuary erosion			Relocation effect	
Imbalances	Pests and weeds			Disease ecology	
	Animal diseases			Disease hosts	
	Aquatic weeds			Disease control	
	Structural damages			Cultivation risks	
	Animal imbalances				

hydraulics, together with social awareness of preservation of natural and cultural values of rivers and floodplains (Binder 1991; FIMA 1992; Gardiner 1991; Newbold et al. 1983; Williams and Swanson 1988). A new approach to flood defence should take into account the complexity of multi-purpose floodplain management within the local context, minimising or avoiding the adverse

13.6
Conclusions and Recommendations

Flood policy formulation and floodplain management should be based on rigorous information and accurate prediction capabilities. These activities are rather complex not only because they are technological demanding, but also because decisions must be made with the involvement of several actors and taking into consideration several points of view, i.e. in an open and participatory environment.

This participatory approach to flood problems is more than an attitude or a philosophical starting point. It requires specific technological tools conceived to be used by different actors, some of them non-experts in flood analysis. In fact, a participatory approach to flood policy formulation and floodplain management should be based on tools that help communicating with the public in a scientifically correct and yet are rather simple in manner.

GIS is indeed a useful tool to integrate and manipulate information which is useful not only for technical purposes but also to communicate more easily with the public. In order to take advantage of this tool it is important to integrate as much data and models as adequate for the purposes that are being pursued. Progresses that are being made in information technology show a clear trend towards the integration of various tools in one single system. This improvement may result in better interfaces, richer databases and more realistic analysis provided by a Multimedia GIS.

To take advantage of these features, GIS should be coupled with hydrologic and hydraulic models that are able of predicting the flooded areas for different scenarios of storm events and catchment land use. The use of GIS can be misleading if the models are not adequate.

The linkage of the hydrologic and hydraulic models to the GIS may be a quite complex task. Very powerful GIS have better capabilities but are less flexible and less portable. Simpler GIS have increased flexibility but are not so powerful. The advantages of both types of systems can be used if we take advantage of the conversion and interfacing capabilities that are increasingly available. It is possible and advisable to develop an architecture that allows for the joint use of different systems with different levels of requirements.

In our case study, heavy data and display capabilities are done preferably with Intergraph and complex hydrologic and hydraulic models are ran with IDRISI, a much simpler GIS. Using these two systems may substantially increase the flexibility of GIS as a tool for flood studies.

Some of the most serious inundation problems happen with flash floods in fast growing towns in which previously open land becomes developed. This is particularly the case of suburban areas in Southern Europe where people are attracted to coastal regions often located in floodplains. Two important considerations can be made under these circumstances. Firstly, non-structural measures for flood mitigation are essential and their effectiveness needs to be improved. Secondly, flood analysis need must explicitly consider the dynamics of urban growth in these regions.

Non-structural solutions to flood problems are essential in these areas to prevent the development of potentially flood affected areas. Unlike structural measures that become "instantaneously" ready and can be directly evaluated for their performance, non-structural measures are effective in the long term and can only be evaluated indirect-ly. For this reason it is important to monitor the performance of these measures in order to improve their effectiveness. GIS can also be a very adequate toll for this evaluation.

In fast growing suburban areas, the flood analysis should be based on a dynamic assessment of flood problems. Urban growth models are essential in order to predict the evolution of the river basin land use and the floodplain encroachment. GIS is a valuable tool for urban growth modelling.

Flood control measures, especially the structural ones, have impacts on the environment and must be subject to a environmental impact assessment. This impact assessment is mandatory in most European Member States and are a key component of the decision making process. The previously defined framework for flood analysis may help the analysis of the environmental consequences of flood control projects.

Participatory evaluation of flood consequences and the formulation of flood mitigation policies are rather complex and demanding activities. A participatory approach requires specific technologies and decision support algorithms. This research project was an opportunity of developing and linking some of these technologies with the general purpose of contributing for an integrated and comprehensive view of flood mitigation problems.

Acknowledgements

EUROFlood research project was made possible with the support of the European Community. Funding for this research project was provided by the EC under the Climatology and Natural Hazards component of the Environment Research Programme. We want to express our gratitude towards Prof. Edmund Penning-Rowsell, Co-ordinator of EUROFlood, and Maureen Fordham, his collaborator for most of the duration of the research activity. At an early stage of this research project, JNICT, the Portuguese National Board for Science and Technology, provided a very crucial support.

Several institutions and colleagues were very helpful in several phases of this research work. The authors wish to thank the main Portuguese institutions involved in the project, namely IST, ISA, CNIG and LNEC. At CNIG we want to especially acknowledge the collaboration of Fátima Bernardo and Isabel Ramos. At LNEC we want to acknowledge the collaboration of João Rocha.

References

Binder W (1991) Rivers and streams – conservation, improvement and restoration. Riverbank Conservation. Occasional Papers in Environmental Studies N° 11, Hatfield Polytechnic

Bowles D, O'Connell P (1991) Recent advances in the modelling of hydrologic systems. NATO ASI Series C, vol 345, Kluwer Academic Publishers, Dordrecht, Netherlands

Burby RJ, French SP (1981) Coping with floods. The land use management paradox. Journal of the American Planning Association 47(3)

Cheias GT (1985) Estudo das Causas das Cheias na Região de Lisboa. Relatório Síntese. Bacia Hidrográfica da Ribeira da Laje. (Study of the flood causes in the Lisbon region. Laje river basin). MPAT, SEALOT, DGO, Lisboa

Cheias GT (1989) Estudo das Causas das Cheias na Região de Lisboa. Relatório Síntese. Bacia Hidrográfica do Rio de Loures e Ribeira de Odivelas. (Study of the flood causes in the Lisbon region. Loures and Odivelas river basins). MPAT, SEARN, DGRN, Lisboa

Correia FN (1984) OMEGA, a watershed model for simulation, parameter calibration and real-time forecast of river flows. Ph. D. Dissertation, Civil Engineering Department, Colorado State University, Fort Collins, Colorado, USA

Correia FN, Matias P (1991) OMEGA: Impact of spatial variability of infiltration parameters on catchment response. In: Bowles D, O'Connell P (1991)

Correia FN, Morel-Seytoux H (1985) USER's manual for OMEGA – A package of FORTRAN 77 programs for simulation, parameter calibration and real-time forecast of river flows. Technical Report Nº ITH 18, Laboratório Nacional de Engenharia Civil, Lisbon, Portugal

Correia FN, Saraiva MG, Rocha J, Fordham M, Bernardo F, Ramos I, Marques Z, Soczka L(1994) The planning of flood alleviation measures: Interface with the public. In: Penning-Rowsell E, Fordham M (ed)

Correia FN, Saraiva MG, Rocha J, Bernardo F. Ramos I (1995) Public perception on flood risk and flood defence policies. In: Gardiner et al. (1995)

Correia FN, Saraiva MG, Silva FN, Costa CB, Ramos I, Bernardo F, Antão P, Rego F (1996) Innovative approaches to comprehensive floodplain management – A framework for participatory valuation and decision making in urban developing areas. Technical Annex 12, IST Contribution for EUROFlood Project, Instituto Superior Técnico, Lisboa

FIMA (1992) Floodplain management in the United States: An assessment report. vol I, Summary. prepared for the federal interagency floodplain management task force and The Natural Hazards Research and Applications Information Center, University of Colorado at Boulder

Fordham MH (1992) Choice and constraint in flood hazard mitigation: The environmental attitudes of floodplain residents and engineers. Ph. D. Thesis. Middlesex Polytechnic, School of Geography and Planning, in collaboration with the National Rivers Authority Thames Region

Gardiner John L (ed) (1991) River projects and conservation. A manual for holistic appraisal. Wiley & Sons, Chichester

Gardiner J, Starosolszky O, Yevjevich V (1995) Defence from floods and floodplain management. NATO ASI Series vol 299, Kluwer Academic Publishers, Dordrecht, Netherlands

Goldsmith E, Hildyard N (1984) The social and environmental effects of large dams. Wadebridge Ecological Centre, UK

Handmer J (1995) Cooperation, coercion, capacity and commitment: Key concepts in floodplain management. International Conference "Flood protection of towns – ideas and experiences. Krakow, Poland

ICOLD (1982) Dams and the environment. International Commission on Large Dams, Bulletin 35, CIGB/ICOLD, Paris

Mock JF, Bolton P (1991) Environmental effects of irrigation, drainage and flood control projects. checklist for environmental impact indication. Revised draft for discussion by the ICID Working Group on Environmental Impacts of Irrigation and Drainage, Report OD/TN 50, HR Wallingford

Muckleston K (1972) Problems of ordering flood plain occupancy. Oregon State University

Newbold C, Purseglove J, Holmes N (1983) Nature conservancy and river engineering. Nature Conservancy Council, Peterborough

OCDE (1982) Guide pour la conception et l'évaluation du point de vue economique, social et de l'environnement des projets hydrauliques à buts multiples. Organisation pour la Coopération et le Developpement Economique, Paris

Parker DJ (1981) Flood mitigation through non-structural measures: A critical appraisal. International Conference on Flood Disasters, New Delhi, India

Penning-Rowsell E (ed) (1996) Improving flood hazard management across Europe. Final report of EUROflood project, Middlesex University, Enfield

Penning-Rowsell E, Fordham M (eds) (1994) Floods across Europe. Flood hazard assessment, modelling and management. Middlesex University Press, London

Perrow MR, Wightman AS (1993) The river restoration project. Phase 1: The feasibility study, Final Report. ECON, University of East Anglia

Saraiva MG (1995) O Rio como Paisagem. Gestão de Corredores Fluviais no quadro do Ordenamento do Território. (The river as landscape: management of river corridors in the framework of land use planning). Ph. D. Dissertation, Instituto Superior de Agronomia, Universidade Técnica de Lisboa, Lisboa

Saraiva MG, Correia FN (1993) Floodplain management and strategic land use planning in Portugal. Case studies and identification of Best Practices. National Rivers Authority R&D Project 426, Review of Best Practice in European Strategic Land Use Planning, Final Report, Lisbon

Verdin, Morel-Seytoux (1981) User's manual for XSRAIN-a Fortran IV Programme for Calculation of Flood Hydrografic for ungaged Watershed. Federal Highway Administration. DC, USA

White GF (1964) Choice of adjustment to floods. The University of Chicago, Department of Geography, Research Paper Nº 93, Chicago

Williams PB, Swanson ML (1988) A new approach to flood control design and riparian management. California Riparian Systems Conference, Davis, California

Chapter 14

The Economical Consequences of Floods and Landslides

H. Tiedemann

14.1
Introduction

It is very often not appreciated by the academic community that those affected by floods and landslides, or by other natural perils, are not or at least primarily not interested in academic research and the results produced by it but by the consequences of such disasters for them and in preventive steps. As the money spent by research is produced by the taxpayers and as one of the main aims of research in such fields should be a pragmatic service to the public, it is felt that more attention should be paid to such issues.

In view of the much greater importance of the economical consequences of floods, this paper will deal primarily with such issues and only in a secondary manner with damage and losses due to landslides.

The approach described in this paper has been developed by the author nearly four decades ago for risk assessment and rating in the field of natural perils when introducing scientific, quantifying assessment methods when working for the two largest reinsurance companies in the world. The approach described in this paper can nevertheless, be used not only by the insurance community but by all parties who are interested in quantifying, for instance, the exposure to floods.

14.2
General Aspects

When studying the exposure, that is the risk, or when dealing with it, we should bear in mind and quantitatively consider the following correlation whenever possible:

$$I = LE\,p \quad . \tag{14.1}$$

That is the impact of an accident (I) is in general the product of the loss expected (LE) and the probability of its occurrence (p). This is in fact a shorthand version of a equation for rating risks proposed by the author several decades ago, which reads:

$$X = \frac{10\,LE\,f\,u\,P\,p}{SI} \quad . \tag{14.2}$$

A different and even older equation does not employ the annual probability p but the return period R in years of an accident which leads to a loss equal to the LE. This equation reads:

$$X = \frac{1000\, LE\, f\, u\, P}{SI\, R} \, .$$

(14.3)

The above symbols mean:

- X = Gross rate in ‰ for the period P (for a single event.) It represents the risk rate, irrespective of who is shouldering the exposure.
- LE = Loss expected from an event corresponding to the return period R used in the calculation. Calculations should always be double checked, first with a lower loss figure, or percentage, for a shorter return period, and then with a higher loss estimate for a longer return period. Normally the two rates obtained should not be too far apart. When the LE is inserted in percent, the Sum Insured (SI) will have to be inserted as 100%.
- f = Factor covering all overheads like deductions, costs, margin, profit. Deductions affect the cost of insurance noticeably. If cost, margin and profit are taken as 10%, and if deductions are 15%, or alternatively 30%, this factor would be 1.31 or 1.59 respectively.
- u = Correction factor allowing for the uncertainty in the determination of the return period/expected loss complex and the confidence desired.
- P = Period of exposure. In the classical branches of insurance P is generally equal to 1 year. Not so in engineering covers. If a cofferdam has to protect a site during, e.g. 3 flood seasons, P would be 3; if an excavation is exposed for 6 months, P would be 0.5, the return period R being given as the annual probability converted to years of average recurrence of the event.
- 1 000 = Factor to obtain the rate X (‰) of the insured value.
- SI = Sum Insured, either as an amount if the loss (LE) has been expressed as a sum of money, or as 100% if the LE is given in percent. The sum insured corresponds to the new replacement value of the element at risk.

  Whenever possible the sum insured (value of the element at risk) and the expected loss should be inserted in the corresponding currency, as this will generally produce a more precise rating than if percentages are used for LE and SI.
- R = Return period of the event liable to cause damage equivalent to LE. The probability of the damaging event could be used as well, as shown in the earlier equation. If, for instance, the return period R is 20 years the annual probability (p) is 5%.

The above equation is quite correct if the return period is long compared to the period of exposure. If this is not so, the following Eq. 14.4 may be used.

$$P = 1 - qN \, ,$$

(14.4)

where q is the probability that the rainfall, inundation or flood will not occur in a particular year. To calculate the probability of occurrence, q is arrived at by deducting the probability of occurrence from 1. If e.g. a heavy rainfall may be expected once every five years, its probability of occurrence is 1/5. If the period of exposure, for instance, the construction period of a factory is 3 years (N), one would calculate $1 - 1/5 = 4/5$, and then $P = 1 - (4/5)^3 = 61/125$, i.e. the chance of a loss would be somewhat less than 50%.

As regards the LE, the damage caused to the elements at risk, does not need emphasising that we are talking about damage and/or losses which will probably arise, the probability depending on the factors enumerated in the respective sections. But the above discussions suggest to keep also the probability or return period of an even liable to cause the LE in mind because only this will show where to start investigations, research and loss minimisation.

To assess the probable extent of any loss, it is first of all necessary to study the essential details of an element at risk, including the site and its "vicinity". Someone may feel that statistics on losses are indispensable for ascertaining the loss which may be caused by an event of a certain magnitude. Although such records are helpful, in particular in the assessment of the loss distribution function in general, which produces a better estimation of very large or borderline losses their value for individual projects is generally overestimated for the reasons discussed in connection with damage parameters and trends.

Since damage depends on a very large number of parameters related to rain and the ensuing flood, the general topography and the site, vegetation and subsoil, the vulnerability of the elements at risk, contributing external parameters, like failure of a dam or slope, contamination of the flooding water with dangerous or noxious substances and damage reduction possibilities, and since for each homogeneous subsample literally hundreds of observations are required to guarantee a given reliability of the results, an extremely large number of observations would be necessary. The only pragmatic way out of this dilemma is a thorough familiarity with the essential parameters and their proper application to each case studied. Such an approach is also needed because it is not fair to deal with good and bad risks alike and to use statistical general data indiscriminately like statistical "Irish stew", even if the sample is large.

Before dealing with specific elements at risk and their vulnerability, let us remind ourselves of some essential extraneous parameters, because even an intrinsically very vulnerable element at risk poses no problems if safely sited. Conversely an element with a low vulnerability per se can become a total loss if it is located in a very exposed place.

The factor which shapes topography, especially in the mountains, but even in the desert, is water. Water gnaws away at what tectonic movements, volcanism, or sedimentation have built, and occasionally even transports huge masses of rock. The author vividly remembers what imposing masses of rock were washed downhill and downstream by the runoff from the sparsely overgrown mountains south of Alma-Ata, Kazakhstan. For good reason the government had constructed a large dam to retain the bedload and, in the valley below the dam, a sequence of massive structures to stop any excess bedload from entering Alma-Ata.

As many maps, documents, inspections and photos of damage teach, streams and rivers have a tendency to meander not only in flat regions (flood plains!) but also in the mountains. This is due to the "centrifugal force" of flowing water. More precisely, unless acted upon by external forces, water flows in a straight course. In a riverbend the outer bank acts on the flowing mass of water, forcing it to deviate. According to the principle of action = reaction, force = counterforce, the water exerts an effect on the bank, undercutting it or carrying it away, sometimes according to the proverbial principle of "constantly dropping hollows the stone". Any element at risk sited on such an exposed bank must therefore, be viewed very critically because not only inundation with water but scouring of foundations determines the *LE*.

This also holds for the general course of rivers flowing more or less parallel to the geographic longitude, i.e. north-south or vice versa. The reason for this lateral deflection is the coriolis force resulting from the different circumferential velocities of the earth's surface at different latitudes. Because of this slowly acting force the course of several rivers has changed considerably over time. The coriolis force has made the Brahmaputra, for instance, change its course in Bangladesh to a bed which is up to 130 km west of the old one. This process will continue. Because of the immense discharge of the Jamuna, Ganges and Meghna and the very deep scouring of these rivers, projects to force this river system to keep its present course are bound to fail in the long run. The same applies to the lower range of the Mississippi. It may be added here that this lateral progression of the river bed is one of the reasons that one bank of a river is higher than the other one. This can, for instance, be easily observed at Baghdad for the Tigris or at Kiev for the Dnjepr, where the western banks are at a substantially higher level. The bluff at the western bank of the Volga played an important role during the battle of Stalingrad during World War II.

Such changes are likely to occur during very large floods. One must therefore, not only consider the exposure along the present course of such rivers but the additional damage if they should leave their present bed.

If in particular the threatening parameter described earlier cannot be ruled out, the general topography must be studied carefully. Both aspects must be considered in connection with maximum loss assessments.

Most exposed are elements at risk which are located in depressions. This can lead to frequent and lengthy inundation and the accumulation of silt. To determine depressions one must have good small scale maps or resort to extensive levelling. Levelling of an adequate accuracy can today be achieved quickly and economically because of the availability of high-precision altimeters.

Elements at risk next to rivers whose level is known or likely to rise often and dangerously, or in the low region of a normally dry valley or ravine, can also be exposed if cloudbursts cause flooding at such sites. Beds or valleys of streams in arid regions (wadies, oueds) are notoriously dangerous in this respect.

But even flat areas far from bodies of water can be inundated if very strong rain (cloudbursts) occurs. The drier the region, the higher the risk in general because such rains tend to fall within a short span of time and on ground with very low infiltration rates.

And as a last general rule, the elevation of the site above the surrounding topography is a very decisive factor determining exposure. As maps which show such details are in general unavailable, except in a few countries, a site inspection is needed if substantial values are exposed.

Flood and inundation can cause tremendous direct, indirect, and consequential damage or losses because of the sensitivity of many items to water, mud, and silt, or because of damage caused by erosion or scouring. As regards indirect losses, if one considers only interest on investments during the interruption of production or of specific activities caused by the inundation, this loss alone can be graver than the direct damage.

Moreover, one must consider that a region which has experienced one flood in the past is generally exposed to repeated flooding unless special steps have been taken to protect it. It therefore, pays rich dividends to assess this exposure carefully. It is unfortunate that hydrologists are at present very rarely consulted to determine and optimise this risk. Moreover, even if advice is sought, the limited value of *past* water-level records is sometimes not even realised by hydrologists. If the use of the land in the catchment area has changed much, or if the characteristics of the river discharge have been altered, or if infiltration in an area of ponding, i.e. a depression, has been affected, or if the climate has changed for natural reasons or due to human activities, the value of the records is particularly dubious.

It is therefore, recommended to study the following questions very experty and carefully. It must, however, be stressed that the aspects discussed here can only briefly cover the broad spectrum of hydrological and meteorological issues.

The simplest information is the "total loss" or "damage" from a given event, as frequently presented by the media. Some may think that such a statement is of some use for general national economics, but even there its usefulness is extremely limited – even if the figures reported were correct. Such information neither differentiates between direct and indirect damage nor does it provide any information on the relative contribution of the essential groups of elements at risk. Moreover, and this should be borne in mind, nearly all such figures, even if released by governments are mostly back-of the envelope estimates.

From such very crude data one could derive "average" damage and losses caused by floods or expressed in a somewhat more refined manner as the sum of money per inhabitant in the area. One could even involve the GNP or GDP of the region and express the losses as percentage of the per capita GNP or GDP. However, in order to perform such calculations the number of residents per region experiencing flooding of given magnitudes must be known.

Moreover, as such estimates produce only crude generalisations, a more refined assessment is desirable. This could and should be based on damage parameters. So after considering some important parameters for selected categories of elements at risk, we shall discuss some specific examples to show in which way assessments can be improved economically.

The essential aspects which require checking and/or allowance can be listed as follows:

1. If the area under study is near to a body of water:
   a Water level and discharge records of rivers, etc. in the catchment area;
   b Rainfall records for the catchment area;
   c Infiltration data for the catchment area;
   d Effect of changes in land use in the catchment area on infiltration and runoff;
   e Trends due to climatic change;
   f Effects of projects affecting the river proper;
   g Data on siltation of rivers (changed discharge capacity);
   h Topography of the site;
   i Stability of banks;
   j Stability of slopes;
   k Danger from dams upstream of the site;
   l Risk of the river changing its course during a severe flood.

2. If the area under study is at a safe distance from rivers or other bodies of water:
   a Rainfall records for the region;
   b Records of earlier inundations of the site (consider ponding, study profiles of deposits);
   c Infiltration and runoff data for the region;
   d Effects on infiltration from deforestation, loss of topsoil, changed farming methods, construction activity affecting infiltration and/or runoff;
   e Topography of the site and of the surrounding region which can lead to inundation of the site through normal runoff and/or ponding based on precise levelling.

We shall now deal more specifically with the consequences of floods, viz. the direct and indirect damage caused by them and the direct and indirect consequential losses. As it is obviously impossible to treat the extremely large number of elements at risk in such a paper, we shall restrict the discussion to common classes of elements at risk.

14.3
Parameters Influencing Damage by Category of Elements at Risk

14.3.1
Residential Buildings

This category of risks is, as a rule, the most ubiquitous one contributing therefore, the largest share to the total physical losses due to floods. Moreover, it can represent a colossal exposure in aggregate. These are good reasons to start the discussion with this category of elements at risk.

Generally speaking any element of risk incorporates parts whose vulnerability ranges from zero or very low to extremely high. This rule also holds for buildings. To assess the possible degree of overall damage the relative importance of the respective elements, parts or sections must be known.

Table 14.1 illustrates this aspect and allows us briefly to discuss the vulnerability of the components.

Table 14.1. Average investment percentages for various components of ordinary Central-European houses

Item	% of value
Earthwork	2
Concrete and reinforced concrete	17
Brickwork	18
Plaster and paintwork	12
Roof structure	3
Roof cover and accessories	5
Floors, tiles, natural stonework	11
Joinery (doors, windows, stairs)	12
Electrical installation	4
Heating and warm water	9
Sanitary installations	7

The earthwork of a house under construction can be fairly exposed to flooding of the pit as a result of erosion and slides. This is generally not the case for completed buildings unless they are located where heavy scouring is possible.

Concrete and reinforced concrete are generally immune to water unless the foundation is undercut to such an extent that the building collapses partly or completely. This is, however, only probable if the building is sited on an erosion-prone bank (Tiedemann 1988). If there are many buildings along an endangered bank of a ravaging river, the aggregate exposure can be considerable.

What has been said for concrete applies as a rule but not always to brickwork and plaster, but not to paintwork and other sensitive materials covering walls, like wallpaper, panelling and so on. Paintwork and the other even more sensitive elements mentioned above must be considered total losses in the storeys likely to be flooded.

Roof structures and tiling are only damaged by partial or total collapse, which has been discussed above.

Natural stonework and tiles generally pose no problems. This holds for floors too as long as the material is stone or terrazzo. However, wooden floors or wall to wall carpeting must as a rule, however, be considered total losses. This is also true of wooden doors, windows and stairs.

Heating and warm water systems must be segregated into vulnerable and non-vulnerable components. Boilers can generally be recommissioned after cleaning and drying and replacement of sensitive insulating material. Burners, blowers, and pumps with their electric motors are, however, very vulnerable. In countries with high wage costs it may not be economic to dismount, disassemble, clean, reassemble and reinstall such items. In such cases replacements will be made.

Control and switch panels will in general have to be replaced if they were immersed in water for some time. This brings us to electrical installations, including telephone systems. The wiring system may be serviceable after proper rinsing and drying. Switching systems and telephones will, however, be replaced in general.

Sanitary installations will mostly only require cleaning, which is not very costly. If the flood water, however, had a chance to take mud, silt and other objects into the sewer system, this can create very costly repair and even replacement if the pipes are clogged.

This shows in general, that is if structural damage can be ruled out, only approximately 35% of the total new replacement value of such houses is exposed, provided the flood affects single storey buildings. If the elements at risk are buildings of two or more storeys the percentage must be reduced proportionately, as in most cases it is very unlikely that the flood level would reach the second floor. One must, however, not overlook the exposure of basements in this respect.

This discussion also shows in which manner this general approach can be refined, viz. with the help of an experienced architect or engineer who can establish the values of the components more precisely and also determine whether particularly vulnerable components are present.

14.3.2
Commercial Buildings

The above approach can also be used for commercial buildings. As a general rule it should be remembered that the more luxurious such buildings are the higher the proportion of non-structural elements, i.e. those which are mostly more vulnerable. If such buildings not only have expensive panelling, floor covers and doors and concealed wiring in places where water and mud may accumulate but also air-conditioning plants in the basement, etc. the damage can be very substantial.

14.3.3
Business Interruption

As regards indirect losses, e.g., business interruption or loss of use, two things must be noted:

General construction experience teaches that the time needed to build or install the non-structural components takes much longer than the construction of the structural elements. This also applies to repairs or replacements after flood damage and it will be realised that particularly those elements whose construction is time-consuming belong to the high vulnerability category.

Secondly, constructing the original building is one thing but repairing flood damage is a different issue.

If many elements at risk were damaged by the flood, contractors will be heavily booked, i.e. a "natural" element of delay must be considered, in particular since these firms or people were not sitting idle at the time of the accident but busy with contracts in hand.

Moreover, it can be taken as a rule that repairs cannot be effected as time- efficiently as original construction and, in addition, the water must recede, access to the site must sometimes first be reestablished, the affected part of the building must be cleaned and dried to some extent and the damaged items removed before repair or replacement can take place. Such delaying parameters should be carefully assessed.

14.3.4
Theatres

In theatres a substantial component are the chairs, which are generally very vulnerable. It must be stressed that the more luxurious variety of chairs can produce total losses. Many other elements resemble those found in buildings discussed earlier.

Theatres with complex stage machinery pose a particular problem if part of the installation is below ground level. Such machinery will often contain many vulnerable elements. Moreover, it is custom-made, which implies very costly repairs and replacements. This also signifies that interruption periods can be long.

14.3.5
Factory Buildings

Severe damage to factory buildings by flood is only to be expected if collapse occurs. Walls, floors, doors, etc. are generally not vulnerable. One must, however, consider underground trenches and ducts for cables, tubes, etc. Such trenches and ducts are practically never watertight. If flood water passes through such plants for some time, one will have to reckon with deposits of mud, etc. and thus with costly and time-consuming cleaning.

Cubicles and switchgear are considered separately under the equipment and machinery found in industrial plants.

14.3.6
Plants and Industries

The main parameters which determine vulnerability are:

- A design which permits the easy entry of water, mud, dirt and foreign matter;
- A design which renders repairs impossible or difficult;
- Parts and components which are sensitive to water, dirt and corrosion and/or difficult and/or costly to dismount, repair or replace, and to reassemble;
- The location of elements at risk during pre-storage and construction, and after commissioning;
- Raw material, feedstock, semifinished and finished products which are sensitive to water and dirt;
- Substances which are dangerous, noxious, contaminating if released during a flood.

As it would not be appropriate to reiterate information on the vulnerability of specific elements at risk, we suggest to consider this entire section carefully, even if interest exists only in the vulnerability of one given project or element at risk in it. This will help to develop a comprehensive view and avoid tunnel-vision.

1. Thermal Power Plants. Such plants are, because of their cooling water requirements, often located next to rivers and other bodies of water. This implicitly means exposure to floods. On the other hand some of the very sensitive equipment, like that in the switch and control rooms, is installed at a level which is generally safe. This does, however, not apply during the pre-storage period, construction and erection. Damage to such equipment may be severe during this phase.

Sensitive equipment at a level which can be flooded are motors to drive pumps, fans, etc., transformers and open-air switchplant.

If turbo-generators are not immediately put on their foundation but stored at a lower level because of delays or complications during construction, this may pose a grave flood risk. Even if the packing appears to be tight, it is very likely that water and dirt will enter, cause damage and corrosion, and therefore, considerable cost.

2. Nuclear Power Plants. The remarks under thermal power plants apply to the conventional part of such plants. The nuclear part, that is the reactor building and its contents should in theory present no problem as the reactor building is gastight and therefore, watertight. There are, however, some problems associated with NPPs which are more substantial than with conventional power stations.

Such plants cannot be "switched off" like the engine of a motor car. The shut down of a steam turbine, whether in a conventional or nuclear plant, requires power for turning the shaft as it is cooling down; the nuclear section must similarly undergo a lengthy shutdown for which power and other services are required. If such services break down, the consequences for the turbine are costly but not alarming. The issue is graver in respect of the nuclear element.

It is interesting to note that the earthquake design of NPPs takes account of earthquakes with overall return periods of several thousand years. Such safety considerations are, however, surprisingly not applied to hydrological and meteorological exposure.

It requires no explanation that, for instance, a 5 000-year flood (and/or a similar windstorm) can cause immense problems and endanger the continuation of the services needed to avert serious danger. It therefore, makes sense to scrutinise the damage and loss potential accruing from these natural perils as carefully as for earthquake exposure.

3. Gas-Turbo and Diesel Power Plants. Contrary to the types of plants discussed above, these utilities can be located far from bodies of water. As pointed out earlier, this does not mean, however, that the flood risk can be brushed aside, a factor particularly true of diesel power plants where most of the equipment is installed at ground level. If the engines should have crankcase openings, e.g. for venting, water can even enter the diesel engines and create substantial losses. The generator and the electric switchplant are at any rate very vulnerable.

4. Hydro-Electric Power Plants. The exposure of such plants depends very much on their location and it is therefore, not possible to present general rules. Such power plants have been flooded and not only because of penstock fracture. The water turbines are not vulnerable but generators, transformers, and switchgears are.

5. Power Distribution/Switching Facilities. We have already alluded to such equipment when discussing power plants, but as regards to transformer banks, grid stations and switchyards, some additional information is required.

The transformers used in such stations are in general tall and the depth of the water inundating such sites must therefore, be considerable to reach their upper part. Such transformers do, however, sometimes have fan-operated oil-coolers at a lower level and the electric motors powering the fans are generally sensitive.

High tension switches and similar apparatus are not only tall but often erected on high supports. In general this equipment is beyond the reach of floods.

This can unfortunately not be said for control rooms or switch and control panels. This equipment is mostly found at ground level and which means it is highly exposed, in particular as it is vulnerable.

Moreover, below ground level one finds cable ducts which are prone to flooding and filling with silt and dirt.

6. Water Works. During construction exposure can be very high because the electric motors, switchgear and especially the tanks and accelators are vulnerable. The problems with electrical equipment have already been discussed in connection with power plants.

Tanks and accelators deserve attention for several reasons. Flooding of the excavations can cause damage, but this damage is dwarfed by that likely to occur, if tanks or accelators are constructed below ground level or on the surface. In the former case the damage may be particularly severe. If the ground water level had been lowered, e.g. by well-point pumping, failure of the pumps will in due course cause the vessel-like structures to pop out of the ground because they are buoyant. This is also likely to occur if the site is flooded. Damage will be severe to total.

Pipeline trenches under construction are also likely to suffer badly if flooded before refilling and compaction of the fill.

Completed tanks can be floated and torn from their foundations if they are empty.

Water treatment plants today usually include chlorination and so storage tanks for chlorine. This gas is very poisonous and such facilities therefore, require particular attention. We have seen heavy, very large caissons being taken downstream for many miles by a river in spate, and it is therefore, reasonable to expect that chlorine tanks will suffer a similar fate if they are torn lose by the flood. If pipe connections should be broken during such accidents, the consequences could be grave.

7. Sewage Treatment Plants. The equipment, which is similar to the one mentioned under 6. above, can be assessed according to the information given in the preceding. Sludge digestors under construction behave like the vessels described under 6.

8. Radio and TV Stations, Telephone Exchanges. The electric and electronic contents of such facilities are very vulnerable and those sections of these stations which are installed at a level that can be reached by water must be considered a total loss in most cases.

9. Plants for the Storage, Handling, and Processing of Liquid and Gaseous Hydrocarbons. This heading covers a large number of plants, which cannot be discussed here. However, as they incorporate equipment and machinery which has comparable vulnerability properties, we shall discuss all of these plants jointly.

During the construction and erection phase one has principally to worry about two things. Those plants where excavation and foundation work represents a significant portion of the total cost and projects with an endangered pre-storage site.

The first problem generally only applies to tank farms and the relevant sections of chemical plants.

The latter peril has to be assessed on a case by case basis and it is stressed that the pre-storage risk should not be underestimated, as it appears to be an unfortunate habit to select the most exposed areas of a project for prestorage, the safer places being reserved for offices and production facilities.

In line with what has been pointed out earlier, electric equipment and machinery permitting the entry of water and mud pose grave problems. If there should be machinery with important sensitive parts like polished rollers in calenders or spinner heads with fine polished nozzles, damage can be severe.

Large piston compressors are to be viewed with concern because most of them will not be watertight and, moreover, the high-quality surface of bearings, journals, liners, pistons, valves and their seats can be damaged. At any rate, such machinery must be disassembled, cleaned meticulously and reassembled by experts who may come from far away countries or those where wages are high. All this can generate expenses equal to a significant percentage of the total value of such equipment, not to speak of the delays in completion.

Centrifugal compressors can be better risks if, but only if, all openings have been properly sealed. In general, water and dirt entering such equipment will do less damage to this variety of compressors than to piston compressors.

Once the machinery is shifted to its final position, the risk may improve considerably if installation is effected on a high foundation pedestal or a raised or upper floor.

Some of the equipment, in particular pipes and cables in trenches and tanks as well as reforming furnaces, distilling columns, heat exchangers, cooling towers, and a lot other equipment is at or in part below ground level.

The vulnerability of such items must be assessed during an inspection of the respective site, where particular attention must be paid to:

- Trenches, which may not only hold sensitive items like pipe systems whose vulnerable insulation can get damaged by water and silt, or wiring systems, which are not immune to water.

   Moreover, trenches are in general exposed because mud, rubbish and other objects are likely to be deposited in them. The repair of such damage can be quite costly.
- Some of the items mentioned above can have sections which are below ground level. They are exposed in the same way trenches are. We have even found switches and electric panels in such dangerous locations.
- Tanks or other items insulated with certain materials are vulnerable and one must consider that repair can involve sections above the flood level. At any rate most insulating material is not water-repellant and therefore, becomes soaked with water. This reduces the insulating properties and these may be affected even after the drying of such layers, if they have been compacted by the accident. Tank farms are discussed under a separate heading.

10. Tank Farms. Tank farms require extensive excavation and earthwork for the foundations of the tanks and their dikes, and for embedded pipe systems. This type of work is very vulnerable during construction. It must be pointed out that massive floods are not needed to cause heavy damage, alone rain can cause costly losses. Some types of foundations are also vulnerable.

After completion, tanks present another problem because of their buoyancy when they are more or less empty. For large tanks adequate anchoring is extremely difficult or impossible to provide. If they are surrounded by dikes built for reasons of fire exposure this can provide protection, as long as the dikes are well-built, impermeable, and protected against scouring at places where the velocity of water many be high, and are high enough to cope with floods.

Horizontal tanks are often placed on foundations at some distance above the ground. Their safety is a function of this distance and of the quality of anchoring, which can compensate for at least some of their buoyancy.

The risk is not restricted to damage to the tanks and the pipes but can extend to contamination of the soil, groundwater and entire river systems. Tanks which float downstream can bump into other elements at risk and cause further damage.

11. Steel Mills and Foundries. Those not familiar with such plants may think that they are not vulnerable. This is, however, incorrect because steel mills in particular have many vulnerable items at floor level or in pits. If water and silt penetrates the driving equipment of roller trains or endless casting facilities, damage can be costly.

If the characteristics of a site are such that deep flooding can occur, damage to furnaces and coke oven plants is possible. It should be noted that in the past many floods have occurred even in developed regions which immersed parts of towns and developed land in water many metres deep.

If sites exist where more than nominal flooding is possible, reliable flood warning is of paramount importance. If furnaces and ovens cannot be shut down early enough before the flood reaches furnaces or interrupts the power supply, damage can be very great.

12. Engineering Plants. The general exposure of such production facilities is high to very high. In such plants one finds many vulnerable machines like sensitive machine tools with their electric motors, gear boxes, switch and control panels, and even computer assisted machine tools.

Machine tools, moreover, have many polished parts and others finished to a high degree of precision, like gears, spindles and beds. All such equipment is today built for high precision and output. If such equipment is immersed in water, extensive disassembling is often required to permit internal cleaning and replacement of affected parts, which is then followed by reassembly again. Such work can in general not be done by the maintenance team of the plant proper and within a short enough time to avoid corrosion and deterioration.

13. Galvanising and Plating Plants. In part, the problem is similar to that discussed in connection with steel mills if heated galvanising tanks are employed. All such equipment is installed at a low level, i.e. no deep water is needed to cause problems.

Electro-plating incorporates another risk, viz. that is associated with the contents of the plating tanks. These liquids may be toxic and extremely dangerous when mixed. For this reason such facilities are often governed by stringent environmental laws. On sites where flooding cannot be ruled out, it is essential that adequate emergency plans and special facilities exist.

The vulnerability of electrical equipment in such plants is not different from that found in other industries.

14. Cement, Chalk, Gypsum Plants and Quarries. In general, the exposure is moderate because the proportion of vulnerable components is small. Such items are in the main electric motors, gears, switching and control panels. Some items are installed above ground, like the drive of rotary kilns, and as a result are only exposed if deep water can accumulate.

We have, however, seen such plants where the conveyor system taking raw material to the plant was installed in such a manner that heavy runoff undercut some foundations of the conveyor system, causing the collapse of the supporting structure and a lengthy down-time.

15. Glass Factories. Glass furnaces, machinery to make plate glass, bottles, tumblers, etc., and annealing furnaces are installed at ground level. The furnaces must be shut down in time to prevent costly damage but not much can be done to protect the glass making machinery and the often extensive conveyor systems. These items will require cleaning and repair as previously described.

16. Saw Mills. That the tree trunks and the cut lumber will be swept away by the flood water requires little imagination. We therefore, only need mention that the saws and the other machinery, including electric motors, will require extensive cleaning, in part after disassembly. Some items may have to be replaced because of corrosion, depending on the period of immersion. In some countries many lumber mills are next to rivers.

17. Wood Processing Plants and Furniture Factories. Plants producing plywood, veneer, and chipboard, as well as furniture factories contain a lot of machines of an open design and with many polished and unpainted parts. Flood water will therefore, enter such equipment and deposit silt and dirt in all openings. Moreover, extended exposure is likely to cause corrosion, e.g. of knives and other wood-cutting tools.

Plywood and chipboard plants use large multi-stage presses. Although such presses appear to be sturdy, they are nonetheless vulnerable if immersed in water. The water may penetrate into the tanks holding the liquid used in their hydraulic systems, water and silt will contaminate the hydraulic pistons and seals, and may also affect the guides and plates. Electrical equipment has already been discussed elsewhere.

18. Paper and Cardboard Factories. Practically all valuable machinery in such plants is of an open design and contains vulnerable elements. Extensive disassembling, cleaning, repair and replacement of parts followed by reassembly and re-commissioning will therefore be required. This will have to be done by the manufacturer's experts, so one must expect substantial losses. Moreover, business interruption can last for a long time, for instance if rollers have to be shipped back to manufacturer's workshop for repolishing.

19. Printing and Stationery Plants. What has been said under 18. applies here as well. In fact, vulnerability is even higher since modern printing machines are built to very exact standards and run at high speeds. Moreover, n printing press is in general far more complex than a paper machine. Photographic equipment and computers must as a rule be considered a total loss if they have been immersed in water.

20. Leather Goods Factories. These plants resemble engineering plants (12) and textile mills (21).

21. Textile Mills and Ready-Made Clothes Factories. General vulnerability can be compared to paper mills (18) and printing plants (19). The opening and carding section is somewhat less vulnerable than spinning and weaving. In the latter case, and if machinery in ready-made clothes factories are to be assessed, one would do well to allow for complete disassembling, cleaning, etc. if such plants are flooded.

Spinning and weaving machinery is not as tall as paper machines, so a lesser flood suffices for complete immersion and damage at all levels. This not only increases the mean damage ratio (MDR) but also shortens the return periods of dangerous floods.

22. Factories Producing Plastic and Rubber Goods, Including Tyres. The vulnerability is comparable to that of engineering plants (12).

23. Factories Producing Cables and Wiring (electric). The vulnerability can in general be equated to that of engineering plants (12).

24. Production of Pharmaceuticals and Cosmetics. Although the vulnerability of the machinery found in such plants is per se lower than that in printing plants (19), the extreme cleanliness required compensates the lesser cost of repair and replacement.

25. Bakeries, Biscuit and Candy Production. Although the machinery is generally less vulnerable than machine tools (12), cleanliness requirements are more stringent.

26. Breweries, Soft Drink Plants. The equipment in the bottling section is very vulnerable and requires thorough cleaning. In breweries part of the installation can be below ground level. This poses particular problems from siltation and the particular exposure of tanks found there (buoyancy). On the other hand, the brew-house proper is sometimes a multi-storey building, which protects the items found high above the ground.

27. Tobacco Processing Plants. Cigarette factories can be considered to be as vulnerable as weaving mills. The vulnerability is high as the machinery is very complex, consisting of many elements, into all of which water, silt and foreign bodies can easily enter. Corrosion can be substantial.

The general pre-processing section is less exposed.

28. Sugar Factories. In sugar factories part of the sensitive equipment, like centrifuges, is often installed at such a level that contact with floodwater is unlikely. Otherwise an inspection should be considered because layout differs from plant to plant.

One must also differentiate between mills processing sugar cane and those handling sugar beet. The cane crushers and their drives render these plants more sensitive.

If large packing plants are found at ground level, care is needed because such machinery is vulnerable,

29. Dairy Plants and Ice Cream Plants. Such plants are in general very vulnerable because the machinery, although mostly of stainless steel, must be disassembled and cleaned meticulously before recommissioning, This also holds for refrigerating machinery and equipment. Sensitivity of filling and packing machinery is similar to the corresponding sections in bottling plants.

30. Canneries. The vulnerability of such plants is similar to those described under 29, If the cans are made in the same plant which processes and packs the food, such sections are slightly better risks.

31. Laboratories and Research Facilities. As there are many different kinds of such facilities, only general statements are possible. Most of the equipment in modern laboratories and research facilities are very sensitive to water and dirt or requires extensive cleaning. If repairs are possible, they can in general only be made by manufacturers whose bills are likely to be high.

Many laboratories cannot tolerate contaminating dust and cleaning operations are therefore, not cheap.

As labels on chemicals and reagents are likely to disappear or become unreadable and as material which is not stored in water-tight containers will be contaminated or destroyed. This part of the overall damage must also be viewed with concern.

32. Elements at Risk during Erection. The exposure to hydrological perils is generally high during pre-storage. As more and more equipment is installed the exposure often decreases. The reasons have been given earlier.

The considerable probability of flooding of sites which are only a few metres above the normal river level and their often substantial vulnerability means that larger projects should be scrutinised (Tiedemann 1976).

It must be remembered that not only sites next to bodies of water can be exposed, but also areas which can be invaded by runoff from large slopes in the vicinity.

14.3.7
Civil Engineering Projects

Concerning these elements at risk, some general rules can be stated. We will deal with the construction phase of such projects as information on the vulnerability is either contained in earlier sections or can be inferred. Whenever required some additional information is provided hereunder:

- Any project incorporating extensive excavations which represent a significant proportion of the total value is potentially vulnerable and requires analysis by an expert.
- If such projects extend over considerable distances and if there is a chance that runoff collects in such excavations and flows in them, they must be considered very vulnerable.
- If there are elements at risk which are below ground level, the potential accumulation of silt is cause for concern, in particular if the silt can accumulate in conduits.

1. Towers, Chimneys, and Silos. If such sites are flooded, the resulting damage will in general be only a small fraction of the total value. If water can, however, trigger a slide endangering the project, a careful analysis is warranted.

2. Roads, Railways, Airport Runways and Taxiways. Road construction projects belong to the category of civil engineering projects, which must be approached with care (Tiedemann 1976). Roads under construction are far more vulnerable than completed ones and the runoff from rain, likely to fall, for instance, once every five years, can already cause serious damage. The extent of damage depends a good deal on the design of the work progress schedule, the capacity of culverts and the care taken by the contractor.

Railway construction is generally rare today but modern high-speed trains may revive such activities. In principle, the risk is similar to road construction.

Airports are mostly built in flat areas, which reduces the risk of strong runoff eroding fills and causing damage to other elements at risk. There is, however, the risk of ponding, which must be considered. The run and taxiways of operating airports are much less exposed. For airport buildings the earlier sections should be consulted.

3. Bridges and Overpasses. General bridge design often does not protect such projects against discharges in rivers which have return periods longer than 50 or 100 years. As damage is bound to be severe if abutments, piers or pillars are affected by a flood, a careful analysis is nearly always warranted, in particular, as bridge designers are mostly not very experienced in the field of hydrology and climatology.

4. Subways, Tunnels, and Galeries. One must analyse the chance of water entering tunnels, galeries, and their approaches. The water flooding such projects may come from runoff rivers, or the ocean. Also mountains can hold considerable volumes of water which may accidentally be tapped by the construction work. Since the damage is often substantial, an expert analysis is warranted.

5. Irrigation Canals. This element at risk can be likened to road construction. If more than nominal volumes of outside water can enter canal construction sites, this is cause for concern. Excavations for such projects not only collect water but conduct it over long distances. This can lead to extensive damage.

6. Dams and Barrages. The fact that these projects generally incorporate a hydrological study is no reason to view these risks through rose-coloured spectacles. These studies are in general made to ensure the economic viability of the project, but not to ensure safe construction. Moreover, these studies have so far never allowed for adverse trends in runoff and climate. Cofferdams and diversion facilities generally afford much less protection than assumed (Tiedemann 1976). Floods are the most important single hazard in dam and barrage construction, in particular if one considers earth dams. As both damage probability is high and the values involved tend to be considerable, the help of experts should be sought. As regards completed dams one must consider that although the probability of catastrophic damage is small the consequences including loss of life can be appalling. Moreover, the exposure can be substantially larger than estimated in the original design. For such reasons careful expert studies are warranted, in particular if one is dealing with earth dams.

14.4
Examples of Flood Assessments

The following examples are designed to illustrate the different approaches which may be selected to assess the exposure. It goes without saying that no detailed examples can be presented in a publication of this kind. Such an analysis would require not only many pages of general text but also a detailed description of each case and stock-taking of all essential elements at risk and their specific vulnerability. Moreover, many maps, illustrations and graphs would be needed. It must be left to the reader to expand the example(s) choosen according to his requirements.

Similarly, the values used in the examples must be adjusted to suit the particular case of study. It should again be stressed that proper allowance must be made for trends and an adequate safety factor (Tiedeman, Floods: A guide to risk assessment; under preparation).

The Australian example has been included to show that a first hand assessment can be based on simple information, and that people in all continents are confronted with similar problems.

14.4.1
Flood Exposure in Sydney, Australia

The purpose of this example is to illustrate a general approach to estimating accumulative exposure to direct flood damage.

There are approximately 1 149 000 dwellings in the Sydney region. It is assumed that approximately 10% of them are flooded.

In this case, a more precise estimate is made feasible by "flood maps" developed for Sydney many years ago. Employing a hydrological and hydraulic calculation and, as a further refinement, an actual model test in a hydraulic laboratory, one can estimate the areas flooded, say, once every 100, 250, 500, and 1 000 years, and the depth of water per area.

It must, however, be appreciated that even the most refined approach cannot precisely predict what effects a future flood may have, even if many different scenarios are used in model tests. Therefore, one must work with a safety factor.

As the average value of one dwelling is about $A 60 000 according to a statistical yearbook, the approximately 100 000 dwellings cost roughly $A 6 milliard.

The mean damage ratio (MDR) used in all examples represents the damage as a percentage of new replacement values. It would therefore, be wrong to apply the MDR to book or market values.

The next important question to be answered concerns the average damage to be expected. For the sample used here (mostly single storey, bungalow-style residences) the MDR is approximately 25%. A precise assessment necessitates an analytical approach which considers the composition of the sample and the vulnerability of homogeneous groups of buildings, requiring much experience gained during actual inspections of flood damage and some form of stock-taking.

Applying the MDR of 25% to the total value (*tv*) of the dwellings of $A 6 milliard the property damage potential can be estimated at $A 1.5 milliard. The next question relates to damage to contents. Again, this is considered in summary form.

Based on statistical information we assume that approximately 400 000 people reside in the 100 000 buildings at risk. The value of the exposed property per person which cannot be removed in time is taken to be $A 10 000. This is certainly a very optimistic estimate because it presupposes that practically all motor cars and some valuable property can be removed before flooding occurs. If this should not be so, for instance because extreme short-term rainfall disrupts traffic very early, because the residents are not warned in time, or because part of the flooding is caused by the failure of an earth dam, the damage to motor cars alone would amount to approximately $A 8 000 per inhabitant. Anyone who has seen flood disasters first-hand or even on television knows that large numbers of motor cars are often damaged. As an indication one could assume one car per two inhabitants in the region and an average value of $A 15 000 per car. This crude approach immediately suggests in which way it can be refined, viz. by superimposing a demographic map on the zones likely to be flooded and by ascertaining actual values exposed by proper sampling or stock-taking.

According to the earlier assumptions the total value of exposed private movable property is approximately $A 4 milliard. The vulnerability of the private property considered here (furniture, clothing, electrical and electronic equipment, books, stores, etc.) is very high and the MDR stands at approximately 75% in the case of single-storey dwellings. A total loss of $A 3 milliard is therefore, likely.

To the total loss of $A 4.5 milliard resulting from dwellings and contents one must add that sustained by shops and their contents, warehouses, factories, machinery, semi-finished and finished products, and the indirect losses caused by business interruption, etc. As a very crude indication, this could raise the total loss in this mostly residential area to approximately $A 6 milliard.

On the basis of detailed flood maps and an actual inspection of the exposed regions, one can similarly estimate the damage to be expected from more severe events which engulf, for instance, 20% or 30% of the buildings and which flood a high-value commercial and/or industrial area.

Let us now look at the second example, which includes *LoP* (Loss of Profit or Business Interruption, in insurers terminology).

14.4.2
Flood Exposure in the Rhine valley, Germany, South of Mainz

The general observations related to the earlier example hold, mutatis mutandis, here as well.

An estimate has produced a population of approximately 3 million people and approximately 1 million dwellings in the general area. Each of these buildings has a new replacement value of at least US $200 000. The total new replacement value of these buildings is therefore, about $200 milliard.

To illustrate that the assessment of economical consequences must be correlated with probability aspects we assume a flood with an annual probability of about 1% or somewhat less. Based on a very tentative estimate this flood would probably affect approximately 10% of the buildings, viz. a value of $20 milliard.

A MDR of 25% is assumed, that is an average loss of $50 000. These are the items which contribute most to the loss: the boilers (heating and warm water) including accessories, doors, floors made of wood or wall to wall carpeting and such stairs, electrical distribution which is often installed in the basement, paintwork and wall papers. The total loss will therefore, amount to at least $5 milliard.

The 300 000 persons living in the flooded zone own approximately 150 000 cars. At an average value of $20 000 the total value of these cars is therefore, about $3 milliard.

If we assume that each person owns only $10 000 of personal property $3 milliard have to be added. It is suggested at this place that interested readers prepare an inventory of their own belongings, as practical experience teaches often more than theoretical considerations. They should list all important items and the respective approximate new replacement values. The following enumeration can serve as a guide. In general, the following items are found in households:

Furniture, curtains, carpets, paintings and other pictures, books, radios, television set, electrical household appliances (vacuum cleaner, washing machine, dish washer, refrigerator, deep-freeze, toaster, coffee machine, shavers, hair dryer, telephone, telefax machine, personal computer, lamps), clothing and textiles of any sort, fur coats, shoes, musical instruments and the contents of refrigerators, deep freezes, and store rooms.

If such items cannot or are not removed to a safe place the MDR is invariably very high and we therefore, assume an average of 75%. The optimistically estimated total loss on personal property and cars would therefore be $4.5 milliard.

Turning now to the commercial and industrial sector we assume that the working force represents one third of the population, viz. 100 000 persons. In this region the average investment per working place is substantially higher than the German average of very approximately $100 000. As a more precise figure would, however, require detailed stock taking we shall work with the above figure. This produces approximately $10 milliard of exposed values.

At an average MDR of 50% the total loss would be $5 milliard. In a complete assessment the value and MDR of raw materials, semifinished and finished goods, packing material, etc., would have to be added.

From the total GNP of a country the GNP can be calculated per person. In this example we have assumed 300 000 persons in the flooded area. At a GNP of $20 000 per person, which is again on the low side, and an average interruption of 3 months, a loss of approximately $2 milliard must be reckoned with.

Adding the respective total losses per sector a figure of $16.5 milliard results. This is still not yet the aggregate loss of this hypothetical flood because damage to utilities (life lines), infrastructure, agriculture, horticulture, etc., will arise. Moreover, it is very likely that substantial damage to the environment will be caused, if only because many tanks holding fuels, oils and chemicals will be torn from their foundations and spill their contents. It is therefore, not impossible that the actual total loss from such an event in the region south of Mainz amounts to approximately $25 milliard. Such a flood with water levels exceeding those of the floods at Christmas 1993 and early in 1995 by a considerable margin will obviously also cause considerable damage in the valley of the Rhine below, not to speak of that in the valley of tributaries.

A flood having at present conditions (climatic trends, infiltration rates, lack of retention basins, build up areas and obstacles like road embankments, etc.) an annual probability of occurrence of one permille would immerse places which are several metres above those on which the above example was based.

This would result in vastly larger inundated regions, much higher damage to multistorey buildings and in particular catastrophic damage in some industrial facilities, in particular chemical plants, and also in the valley downstream of Mainz. Such a gigantic flood would, moreover, not only destroy many bridges but tear many ships from their anchoring, make them collide with other ships, bridge piers and other elements at risk and sink them spilling or destroying their contents.

14.5
Landslides

Whereas floods present problems as regards correct assessments of event probability and MDR, the situation is different for landslides. In this case the damage is in most cases total, that is one has to reckon with a MDR of 100%. The assessment of the probability of landslides is, however, substantially more difficult if we exclude those cases where a situation has to be assessed which has visibly deteriorated during the recent past.

As regards commercial consequences of *slides*, in principle, the methods can be applied which have been described above. To estimate basic direct damage one will need an inventory of the elements at risk in the area which is probably covered by the next slide and the respective values.

In connection with indirect losses one will mostly have to estimate how long it takes to rebuild the destroyed commercial facilities and factories at a new site, provided there are survivors, that is persons capable to rebuild. This is mostly the case if the commercial or industrial facility which was destroyed was a local branch of a company having its headquarters somewhere else.

As regards more remote causes of indirect damage, the potential loss of tourism may have to be considered.

Another aspect which may require attention is the chance that the slide dams a valley and that very heavy rain which falls at the time and/or after the slide fills the reservoir generated by the slide in a short time. An overtopping of this earth dam can cause its sudden failure and a disaster in the valley below in particular if timely warning and evacuation is impossible (Tiedemann 1988).

In conclusion of this brief chapter it should be mentioned that large slides can also be triggered by earthquakes. In this case no warning of the population on the slopes or in the valley below is possible. The number of victims can be appalling. The earthquake-triggered an ice and rock avalanche on Mount Huascaran, Peru, in 1970 killing approximately 18 000 people in Yungay. This brings us to the question of how to assess such losses.

The cost of loss of life to the national economy can be estimated by assuming, for instance, a median age of the population of 30 years and based hereon a loss of 30 years of contribution to the GNP or GDP until retirement. At a per capita annual productivity equal to about $20 000 the loss per person killed would therefore, be approximately $600 000. In the case of crippled people, the persisting medical care and other indirect expenses would have to be considered.

This is admittedly a very simple method but it suffices to show that casualties weigh heavily even if considered on the basis of the loss to national economy only, that is neglecting the unmeasurable personal grief and drama caused by such a disaster.

14.6
Concluding Remarks

The most obvious lesson from the information presented in this paper is that a professional assessment of the economical consequences of a disastrous flood must be based on detailed flood maps, proper stock-taking of the exposed elements at risk, and the establishment of mean damage ratios per category of elements at risk and depending on the water level, that is on various return periods or event probabilities.

Such a detailed analysis will immediately show not only the accumulative losses but the relative importance of groups of elements at risk and of outstanding individual risks. This in turn is the basis for cost-efficient risk minimisation which is today another badly neglected issue. We know from decades of international experience that damage can be avoided or substantially reduced if proper steps are taken in time. Such measures cost in general a very small fraction of the exposed values and are therefore, worthwhile taking.

Only once protective devices have been prepared, will flood warning serve its full purpose. There are more very important issues which are like risk optimisation beyond the scope of this paper, for instance, catastrophe plans and disaster management.

If one considers the total consequences of a "one hundred year flood" and of cataclysmic floods corresponding to a one thousand year event it becomes evident that research into the issues discussed here have so far not received the proper attention and funding.

References

Tiedemann H (1976) Rain, inundation and flood in C.A.R and E.A.R Insurance and Reinsurance, Swiss Reinsurance Co, Zurich
Tiedemann H (1988) The force of water. Swiss Reinsurance Co, Zurich
Tiedeman H, Floods (to be published) A guide to risk assessment. Partner Reinsurance Co, Bermuda

Socio-Economic Perspective of Developing Country Megacities Vulnerable to Flood and Landslide Hazards

L. Solway

15.1
Introduction

The paper is based on the United Nations – International Decade for Natural Disaster Reduction (UN-IDNDR) pilot study project "Mitigation of the effect of natural hazards on urban areas particularly developing country megacities" for which the writer was the project leader.

The project was carried out during 1993–95 and considered the effect of six natural hazards – tropical cyclones, floods, tsunamis, earthquakes, landslides, and volcanoes with reference to three case study cities – Karachi (Pakistan), Jakarta (Indonesia) and Metro Manila (The Philippines). The project included several visits to the case study cities and many meetings with government and city departments, commerce and business, NGOs, public works and service authorities, health care units, emergency services and many other interest groups. The United Nations Scientific and Technical Committee (UNSTC) invited the Union des Associations Techniques Internationales (UATI), the World Federation of Engineering Organisations (WFEO) and the Institution of Civil Engineers (ICE) to undertake the project funded mainly by the Overseas Development Administration (ODA) and to produce a guidance manual. This manual was published Autumn 1995 and is available from the I.C.E., One Great George Street, Westminster London SW1P 3AA.

15.2
Analysis of the Phenomena of Megacities

The megacity, defined by the UN as an urban development with a population of over 8 million, is perceived as significant not only because of its concentration of population, but also because of its inherent strategic and economic role. For a variety of reasons, some of which are considered in the paper, the megacity of the developing world are seen as being particularly vulnerable to the impact of natural hazards.

Over the past 20 years, the population of megacities in developed countries has tended to show either a steady rate of growth, a stabilisation or indeed a decline in size. (Los Angeles is the one exception to this generalisation, as it has expanded rapidly over the last 20 years). By comparison, the population of megacities in developing countries has increased rapidly over the same period. United Nations predictions suggest that as the world population grows, urbanisation is increasing faster and the number and the size of megacities in developing and newly industrialised countries will increase considerably. This comparison is illustrated in Fig. 15.1, which shows the population for 1970 and forecasts population for the year 2000 of the developing worlds' largest cities.

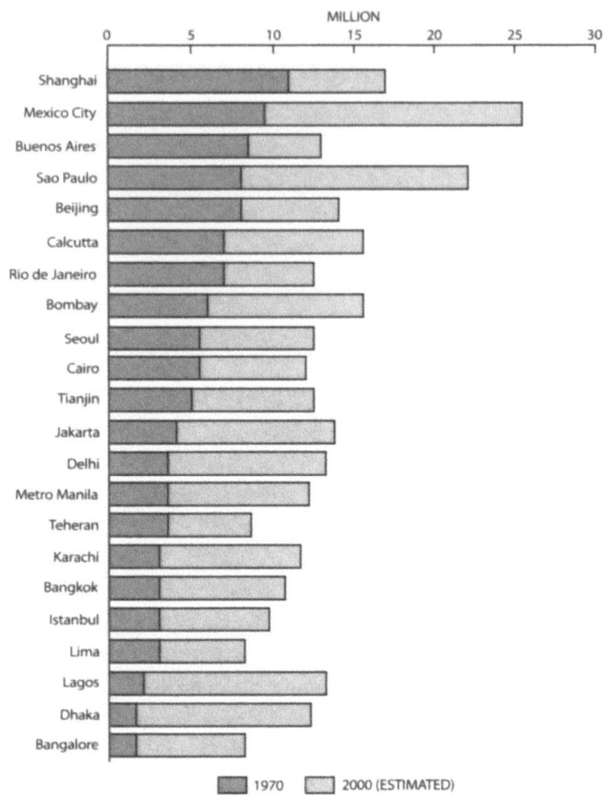

Fig. 15.1. Population growth of major cities in the developing world from 1970–2000

It is predicted that within the developing world there will be 20 megacities by the year 2000 compared with only 3 in 1970. This pattern of enlarging cities will be repeated on a wider scale with an estimated 67 cities in the developing world with populations of between 2.5 and 8 million. The growth of the megacity in developing countries is supported by many factors.

Megacities are regional and often national capitals, major centres of investment in industry, commerce and finance and frequently the major point of entry and exit for overseas trade. These characteristics all tend to attract further growth. By their nature, these cities account for a disproportionately large percentage of the country's economic activity and GNP. They act as magnets to the rural population encouraged by the perception that the city offers greater education, training employment and leisure opportunities. The possibility of earning money instead of the certainty of unemployment in rural areas. High birth rates coupled with improving health facilities in urban areas brings increasing life expectancy and lower infant mortality rates and these have a marked influence on the rates of growth. The occurrence of natural disasters in rural areas often leads to a large-scale migration to urban areas which are seen as offering relative safety.

The rapid growth rate of megacities is a significant factor in their development and compounds their problems. Very few can be said to have been planned. Most began as small planned historical towns or cities but around these old planned centres, unplanned or informal residential areas have burgeoned, overwhelming the capacity of both the cities planning processes and the infrastructure. Thus, the typical urban characteristics of pressure on land and accommodation are exacerbated by unreliable and inadequate supplies of water, electricity and transport and little public health provision. The urban poor and informal communities which can account for 30–70% of a megacities population, are often concentrated on marginal land (unstable slopes, flood plains) and have no land rights and therefore, little incentive or opportunity to improve living conditions and reduce their vulnerability to natural hazards.

All megacities exist with a degree of equilibrium, (it could also be described as perpetual crisis) which can be disturbed by a number of circumstances including natural hazards. The extent to which a hazard disrupts the equilibrium of a megacity depends on the stability of the city's infrastructure and on the nature of the hazard. Poorly planned and constructed buildings and infrastructure will be directly vulnerable to the impact of natural hazards. Damage to infrastructure in turn leads to secondary effects such as the disruption of transport, water and power supplies, food shortages and health hazards. These have major social and economic consequences for the city and the problems are compounded where there is a poor institutional framework and already fragile economy.

With reference to the three case study megacities, under the influence of three different colonial powers, three distinct cities have developed. However, all display in various degrees, those features already outlined which could be described as typical of developing country megacities.

15.3
Identification of the Socio-Economic Consequences of Hazard Impact

15.3.1
Effects of Natural Hazards on Cities

An understanding of the characteristics of natural hazards and of how they release energy is an essential precursor to predicting their effects on the city (Fig. 15.2) and an analysis of how the effects can be mitigated. Each hazard has characteristic features – probability, frequency, intensity, coverage and duration. Their impact also depends on factors such as location, topography, geology and soil characteristics (see also Table 15.1). The probability of natural hazard events occurring in a particular region can be estimated with varying degrees of accuracy. The formation and track of tropical cyclones and floods can be forecast and it is possible to predict the onset of some types of landslides. With limited exceptions, it is not yet possible to prevent the onset of natural hazards because of the enormous energies released. The main exceptions are ground stabilisation to reduce ground movement and embankments to prevent flooding.

Evidence suggests that in a developed country, the existence of sound infrastructure, forecasting capability, evacuation procedure and strong institutional organisation, prevent the escalation of major problems even though natural hazards may disturb transport, electricity, water and public health services. Thus, despite major economic losses and some health problems, the death-tolls are generally not large, but the economic losses can be substantial.

Fig. 15.2. The extent of flooding within the city can be predicted

Table 15.1. Potential effects of natural hazards

	Social or human effects	Physical effects	Economic effects
Primary effects	– *fatalities* – *injuries* – *loss of income* or employment opportunities – *homelessness*	– *ground deformation* and loss of ground quality – *collapse of and structural damage* to buildings and infrastructure – *non-structural* damage loss of ground quality to buildings and infrastructure	– *interruption of business* due to damage to buildings and infrastructure – *loss of productive workforce* through fatalities, injuries and relief efforts – *capital costs* of response and relief
Secondary effects	– *disease or permanent disability* – *psychological impact* of injury, bereavement, shock – *loss of social cohesion* due to disruption of community – *political unrest* where government response is perceived as inadequate	– *progressive deterioration* of damaged buildings and infrastructure which are not repaired	– *losses borne by insurance industry* weakening the insurance market and increasing premiums – *loss of markets and trade opportunities* through short-term business interruption – *loss of confidence* by investors, withdrawal of investment – *capital costs* of repair

In developing countries, perceptions of acceptable living conditions tend to be very different. Many people, particularly those within the informal sector, experience a very low or zero quality of services on a daily basis. Where adequately constructed buildings and infrastructure are gravely deficient or non-existent, the physical vulnerability to natural hazards can be low. But despite such perceptions, the actual effect of a major natural hazard upon over-crowded and impoverished areas of megacities can be devastating. The sheer density of population in the path of a flood results in very heavy death and injury levels, but the economic losses can be low by world standards.

Megacities as they expanded have had to look to the region to provide their water, gas, electricity and food. So when considering the effect of natural hazards on megacities, the effect on the supply lines and effects of upland catchments must also be considered.

Water and power supplies are frequently conveyed tens or hundreds of kilometres, for example a very small landslide or flood could disturb the main power supply to Metro Manila and after the 1991 Mount Pinatubo eruption, mud slides block the road north from the city every year, which blocks one of the main food supply routes to the city. Landslides within the three case study cities are insignificant although there is always the possibility of a critical section of the infrastructure being effected. Landslides in South American cities have had substantially more disastrous consequences.

All three of the case study cities suffer regular flooding. Jakarta is facing a disaster with falling land levels, rising sea levels and increased runoff from upland catchments so that some two thirds of the city is vulnerable. Metro Manila has large areas near the port which flood regularly. But it is Karachi, which in the 80's lost 20 000 inhabitants who had settled in the flood plain of one of the city rivers. Now the authorities have installed a warning system and still have to regularly remove dwellings from the vulnerable areas in the flood plain.

15.3.2
Vulnerability of Physical Infrastructure

Infrastructure is the physical framework of the city. It plays a vital role in economic and social development, and ultimately in providing improved standards of living for the population. Because the various elements of the physical infrastructure are essential for human activities, they are often described as 'lifelines'. This description is particularly appropriate in the context of disaster mitigation, as in the aftermath of a hazard impact, infrastructure plays a crucial role in emergency operations. Reducing the vulnerability of the physical infrastructure networks will therefore, effect a reduction in social and economic vulnerabilities within the city. The primary elements of a cities infrastructure are:

- Transportation: surface transport, ports, airports;
- Water and Public Health: water supply, waste water, solid waste, storm water;
- Utilities: gas, oil, electricity, telecommunications;
- Buildings and Structures: dwellings, offices, public buildings, bridges, stadia, etc.

The vulnerability of each element needs to be assessed for each potential natural hazard that can occur in the city and preferably for different magnitudes of hazards. Only then can assessments be made of the risk and this compared with the cost of mitigation. To take a few examples:

Highways and railways. These are crucial for commerce and industry and for the emergency services. They rely on the provision of electricity supplies, communication networks and surface water drainage. This interdependence means that a failure in one lifeline can effect others. Highways also suffer like river valleys, from intrusion of property, particularly from the informal community.

Hazard events will affect the most vulnerable parts of a highway network; damage to a small percentage of the network could paralyse large areas of the city. The failure of one bridge, embankment or tunnel could isolate large areas and prevent movement between the city and the surrounding area.

Watersupply. Interruption of water supplies could be tolerated for a short time, as many people in developing country megacities have become accustomed to routine shortages and interruptions. But a lack of clean water for more than a few hours could be a public health hazard and the health risk would increase if the sanitary systems also broke down. Protection of water supply systems is hence of the highest priority; emergency repairs must be made within hours and the overall system back in operation within days. Water supplies are essential to the operation of hospitals and emergency facilities.

Flooding can damage treatment works and reservoirs. Landslides and lahar flows have caused disasters by precipitating dam failure or causing surge waves from reservoirs over dams. But it is more likely that a reservoir embankment fails or ground movement disturbs supply pipelines and pumping mains which can be hundreds of kilometres long. Supply canals and pipelines are vulnerable to ground movement and flood erosion.

Solid waste. A solid waste disposal system depends on supplies of fuel for collection vehicles and for disposal sites and on efficient transport networks. The system is necessary to maintain standards of hygiene and to minimise the spread of disease, but of equal importance to discourage illegal disposal methods which are frequently into drainage canals. In Jakarta for example, only 40% of the waste reaches official disposal sites. The remainder is dumped into drains and rivers causing pollution, flooding and nuisance. It is the vulnerability of the highways and fuel supplies which could put a solid waste disposal system at risk.

Electricity generation and distribution. The components of a power generation and electricity distribution system for megacities are frequently located both within and hundreds of kilometres remote from the city. A reliable supply of electricity is essential for the megacities water supply – (pumping) gas supply, telecommunications network, transportation systems and to hospitals, industry and commerce. Generation stations, transmission line towers, sub-stations and transformers can all be vulnerable to the effects of floods and landslides. Where electricity supplies are frequently interrupted, called 'brown outs' in Metro Manila, industry, commerce and the population will have adjusted to the situation and may therefore be less inconvenienced by disruptions caused by hazard events. Informal settlements rarely have access to mains power supplies although some make illegal connections, some will use small generators, the majority will go without.

15.3.3
Vulnerability of Economic Assets

The relationship between a national hazard crippling a megacity and the effect on its economy is complex (Fig. 15.3). A hazard may damage a major part of the city assets; industrial systems, the infrastructure, important export. It may deter investment in the region and it will certainly set back the rate of economic growth. But it will generate a subsequent boom in construction work with beneficial effects upon the labour market and GDP.

Poorer countries may not have the resources to repair damaged infrastructure; this would result in further losses in production time and industrial earnings. Losses suf-

Fig. 15.3. Prompt payment by insurance companies can reduce the impact of natural hazards, but is not a substitute for mitigation

fered by workers with specialist skills could delay industrial recovery. The prospect of civil unrest after the event could lead to further economic deterioration.

The economic impact of a major disaster striking a city such as Tokyo would be widespread, affecting commercial activity throughout the world. The country's asset base would dwindle, due to the vanishing confidence of banks and investors, and its overseas investments would have to be withdrawn in order to support the domestic economy and to provide funds for recovery and rehabilitation. International investment in the emerging markets of cities in the developing world is increasing considerably, a trend which is likely to continue as the economic significance of these cities grows. The economic impacts of natural disaster in these cities would therefore, carry increasing financial implications for the international financial markets. An economic appraisal of the effects of natural hazards needs to consider the following categories.

Direct economic effects. As urban settlements have grown, so have increased industrial and production assets developed. The risk of direct damage is correspondingly greater. Industry and import and export facilities for raw materials and finished products are usually concentrated in megacities in developing countries. A natural disaster would cause disruption of production, communication and transport networks, and energy supplies and could have a devastating effect on the national economy. Components of infrastructure that are susceptible to damage by natural hazards; these include all types of buildings and their contents, utilities and services, transportation systems, vehicles, raw materials and manufactured goods, the social sector, and leisure facilities.

Estimated costs of the direct effect of hazard events of different magnitudes are relatively simple to calculate.

Indirect economic effects. The loss of productive workers (through death, injury, bereavement or involvement in relief efforts); and lost production time where supplies of electricity, gas or water are disrupted. This in turn could lead to company failure and job losses, and therefore, to a reduction in personal expenditure on goods and services, setting off a chain reaction of reduced profits and increased unemployment. The loss of export markets and overseas trade, and the decline of tourism are other potential indirect effects. Indirect economic effects will multiply as trade slows; the limit value of a disaster could be an economic depression. Theoretical assessments can be made for all indirect losses from natural hazards.

Secondary economic effects. These could include:

- Failure of potable water supplies or of public health facilities, causing epidemic diseases;
- A shortage of food and other commodities, causing increased prices, and therefore increased interest rates and inflation;
- Reduced tax income and the need to fund rehabilitation and reconstruction could reduce funding for social, educational and economic activities;
- Property values could fall, restricting the scope to raise capital for repairs;
- High demand for building materials and skilled labour could increase costs (particularly as large projects attracting overseas aid can be more attractive to the construction industry).

Long term economic effects. Ultimately nations and their cities will be judged by their preparations for and reactions to potential events. It will therefore be necessary to study the city's response to past events in order to estimate the potential long term effects of a major natural disaster. The primary economic losses will be the loss of foreign investment, the loss of overseas markets and the collapse of particular industries. The long term economic effects can perhaps be best displayed in terms of the amount of time needed for the city's economy to return to a fixed level, rather than in terms of the actual costs to the city of the event.

The issue of government and international aid needs careful study and application. There are many examples of post-disaster relief aid distorting local markets – a short term economic effect. Concentrating aid on rehabilitation and reconstruction can discourage the construction of new, and potentially safer, buildings; an opportunity for mitigation can therefore be missed. In some cases, damaged buildings have been replaced by new constructions of the same type in the same locations; this is often based on a belief that government or international aid will provide protection from future disasters.

15.3.4
Human Dimensions in Disasters

15.3.4.1
Social Perspective

> "All disasters are failures on the part of human systems. In every disaster, the physical and social infrastructure fails to protect people from conditions that threaten their well-being. At times, the infrastructure itself creates conditions that result in extensive social disruption. To reduce the vulnerability of people to disasters, social and technical systems must adapt to their changing physical and social environment."
>
> International Sociological Association, Denver, Colorado

It is people who suffer as a direct result of disasters and as a result of a lack of preparedness and disaster mitigation. It is also people who will prepare for and mitigate the impact of disasters. It is, therefore, vital to develop a social perspective on disaster mitigation and preparedness (see also Table 15.2).

Table 15.2. Human reactions to natural hazards

Stage	Time relative to event	Reaction	
		Positive	**Negative**
1	Before event	Understand warnings, preparedness	Panic, fear
2	During event	'What should I do?'	Fatalism: 'Act of God'
3	1 min to 1 day after event	Response by survivors, initial search and rescue	Looting; sight-seeing adding to traffic chaos
4	1 day to 1 week after event	Community effort, search and rescue by emergency services	Increase in price of basic food and commodities
5	1 week to 1 month after event	Provision of temporary camps, burial of dead, analysis of problems	Provision of wrong food and medicine, disposal of dead
6	1 month to 1 year after event	Clear up debris, commitment for major rehabilitation projects international funds	Provision of unsuitable and unacceptable temporary accommodation, allocation of blame, corruption and misuse of emergency funds
7	1–5 years	Some mitigation work, some progress on rehabilitation	No visible evidence of damage, bureaucracy and aid-fund accountability delays rehabilitation and reconstruction
8	5–30 years	Increased mitigation works, review preparedness and response programmes	Conflicting objectives lead to relaxation of building regulations to reduce cost of social housing, smaller event produces same effect on larger city
9	Until next major event in 100, 200 or 500 years	Steady programme of mitigation works	'Not happened in my lifetime', no votes for mitigation, funding reallocated
10	Next major event	See stages 1–9 above	Urban area much larger, effect much worse

Megacities possess a network of experts in planning, architecture, engineering and science, with ready access to hundreds of other experts world-wide. Knowledge on the effects of earthquakes, landslides, strong winds, sea surges and volcanoes and the planning and engineering solutions to mitigate the effect of these hazards are well established. So why does instituting mitigation measures remain a problem? To some extent this is because of the priority accorded to mitigation vis a vis the other needs of the city; but to some extent it is also because mitigation plans often ignore the people and their needs when planning and setting priorities and also when considering implementation. At best mitigation plans treat people as if they constituted one homogeneous group with the same needs, abilities and aspirations.

For example, it is easy enough to see how people with different levels of income are likely to be affected differently by the same disastrous event. If we assume that the investment that households make in their housing and other property varies directly within their income level, then it is likely that the damage suffered is likely to be positively related.

On the other hand, the vulnerability to disaster of different income groups reduces with increasing income, partly because the more expensive structures are more likely to have been engineered and therefore, incorporated mitigation measures in their design and construction, but also because it is more likely that the better off households would be located in planned areas that should be less vulnerable to hazards such as flooding, land-slips or mud-slides, whereas the poorer households (and certainly the poorest households) are likely to be located in marginal locations whose low value to large extent probably reflects their very real vulnerability.

However, for the poorest households, the greater concern has not so much to do with how to mitigate against the sort of disastrous events that are the focus of mitigation planning, as it is on how to cope with the more immediate need to survive, to ensure an adequate livelihood, to prevent being evicted. These are much more real and pressing than those events that perhaps may happen at some distant point in the future. In many situations, even where a household knows that the land it is occupying is liable to be inundated every year, as happens in many parts of Bangladesh, it nevertheless affords a living, and there is always the hope that the next year's floods may not be as catastrophic as last year.

Income has been used as one variable to suggest how different groups in the same city are likely to react differently to different hazards and therefore, to the kinds of mitigation effort they are likely to welcome. Similar analyses for the different social, ethnic and cultural groups will show a similar variance. As well, there is likely to be differences in the perception of hazards and willingness to mitigate based on gender or age and other such differences. To the extent that some of these groups are threatened by and concerned with mitigating against every-day disasters, as against disastrous events, urban planning is of crucial importance.

Understanding the social situation involves identification and characterisation of the range of coping strategies and giving reasons for any differences, such as the ethnic, religious, cultural, economic or political context. The extent to which people are integrated into coping strategies should be assessed. Are women and children, for instance, more vulnerable to disasters than men because their needs are not as well catered for? In Islamic societies for example, are women more likely to be trapped in the home? Are women's clothes more likely than men's to cause them to drown? Are the elderly or disabled more vulnerable? *Poverty and gender* are likely to be significant determinants of vulnerability to disaster in much the same way as they are of other aspects of development and access to opportunities and benefits. Consequently, the most vulnerable groups should become the focus of mitigation and preparedness planning.

But in the case study megacities people were asked if they were concerned about the effect of a natural hazard and they replied that they are concerned about having a supply of water, about health facilities and education for their children and they are frightened about the possible effects of fire. But natural disasters, what can we do as they are acts of god.

Hospitals are a focal point for any community because of the trust placed in them to care for the sick and injured. In the event of a major emergency caused by a natural hazard, they are required to play the pivotal role in saving lives and provide immediate care to the injured. However, all too often, hospital facilities and buildings are frequently as susceptible to the effects of natural hazards as any other part of the city. So with the exception of the dedicated staff, they behave no differently from other city institutions and contribute to the number of fatalities rather than being in a position to carry through their objective of saving lives. Schools are also a focal point for education of parents in addition to children, as communal places, providing shelter and as an indicator of a return to normality when they re-open after the passing of a hazard.

The veterinarians also have a critical role to play in the control of disease vectors, the welfare of animals and maintaining *safe food supplies* after a disaster. All of which contribute to Human welfare. Whatever the nature of the disaster several features are common. Natural disasters typically strike suddenly and often with little or no warning. People lose their homes and possessions, many lose relatives and friends and are thus faced with disaster at a time of bereavement. Many will be in shock. Needs in the immediate aftermath are great and resources are almost invariably unavailable. It has been said many times that following a natural hazard event, such as a flood, it takes time for the authorities *to make a disaster* out of the event.

Planning for the recovery, identification, burial and bereavement of dead is frequently overlooked. Bereavement is universal yet there are many problems in working through it. Hodgkinson and Stewart in their studies identified *four main* factors which contribute to the recovery of the bereaved:

- Type of death;
- Characteristics of relationship;
- Characteristics of the bereaved;
- Social circumstances.

15.4
Discussion of Possibilities for Mitigation of Risk

15.4.1
Phases of a Disaster Event

Disaster Management is a continuum of interlinked activity. Within the process the main phases of activity are the natural hazard impact or disaster event, followed by response, recovery, reconstruction, mitigation, and preparedness. These phases are inter-related and activities within them are all necessary for effective disaster management. The process of disaster management is often referred to as a cycle linking the main activities and is illustrated in Fig. 15.4.

The onset of a major natural disaster attracts national and international attention. There is considerable emphasis upon the phases of *Relief, Recovery* and *Reconstruction.* International, national and city resources are concentrated in an effort to alleviate loss of human life and suffering, to re-establish the basic requirements of human existence in the affected areas and to restore the means of livelihood and commercial activity. The competence of national and local government is judged by the ability to react to such major disasters. Media are drawn to portraying the aftermath as major

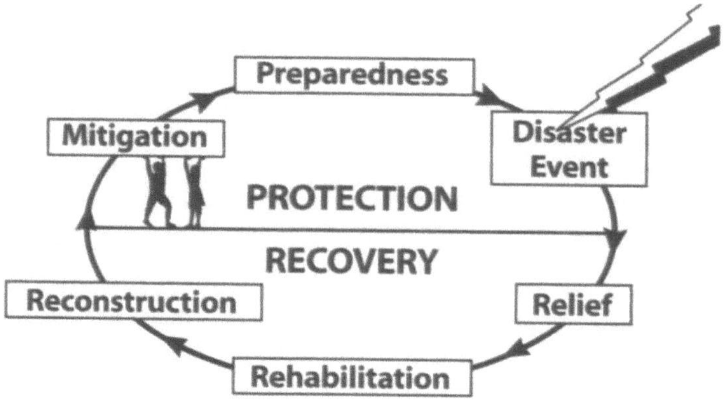

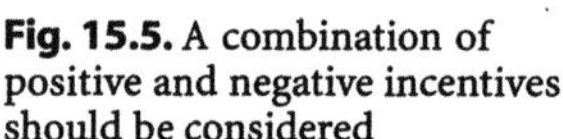

Fig. 15.4. The disaster management cycle

Fig. 15.5. A combination of positive and negative incentives should be considered

stories. There are considerable incentives for deploying urgent assistance and major resources (Fig. 15.5). The city mayor will get more votes if he is seen to be rushing around spending money after an event has occurred, rather than spending the city's money on preventing the possible occurrence of damage from an event.

The approach to the "looking forward" phases of mitigation and preparedness is of a different order. Public demand for these functions in preparation for a disaster, at a time unspecified in the future, will only be as great as the perception of the risk, and the availability of resources to be committed to competing priorities. The government of cities, particularly those that are rapidly expanding, impose immense demands upon those in responsibility. Just catering for the month on month requirements, finding a place on the city agenda for mitigation and preparedness is one of the fundamental problems that the city administration has to overcome.

15.4.2
Strategy for Action

Procedures for reducing vulnerability are already well established in most cities. These procedures are of little effect without properly resourced planned and coordinated implementation by trained personnel. It is in this respect that most developing countries remain weak. Information, education, communication and incentives are the keys to implementation.

One of the conclusions of the study project was that considerable progress could be made at very little cost simply by raising awareness amongst key individuals who can then begin to consider disaster mitigation issues in the course of their work to influence others to do the same. More, fundamentally, megacities at risk can be encouraged to gradually incorporate disaster mitigation into their existing planning process, to improve the economy and maintain sustainable development.

What is an acceptable level of risk? What is an acceptable level of natural hazard impact? With a few limited exceptions, nothing can be done to avoid the occurrence of a natural hazard but the level of impact in an urban area can be controlled. Setting an acceptable level for mitigation will set the requirements for the city's preparedness and response plans as no developed or developing country city can afford mitigation to the extent that a natural hazard has no impact. The cost of some mitigation is the minimum price an urban area must afford. The zero mitigation is not relevant today and will be even less relevant in the 21st Century as the cost in physical, social and economic terms will be too high.

One of the early activities which should be undertaken is an initial assessment of hazard vulnerability and risk in the urban area. The assessment should be completed fairly rapidly (say within 3 months) in order to identify priorities for future action. Using existing sources of information and with the cooperation of the leaders of the major sectors of urban infrastructure the following parameters should be identified:

- The nature of potential hazards, their predicted frequency, intensity and duration.
- The areas of the city which are most vulnerable including areas outside the city which could effect the city e.g. power transmission lines.
- The communities business sectors and infrastructure components which are most vulnerable.
- The estimated losses which would result from hazard events of different magnitudes.

An illustration of this process is given in Fig. 15.6.

In order to define the vulnerability of the physical infrastructure, it is necessary to classify components into distinct categories and sub-categories as outlined in Section 15.3.2 Paragraph 1.

Classification of the physical infrastructure should include the age of the component, its design, construction materials and type and its location (areas exposure and ground condition). Buildings should also be classified in terms of their height and occupancy. Once the classification has been made it can be refined and updated at intervals of, for instance, every three years. Eventually all information should be held on GIS with immediate access.

Model for Assessing Probable Maximum Loss

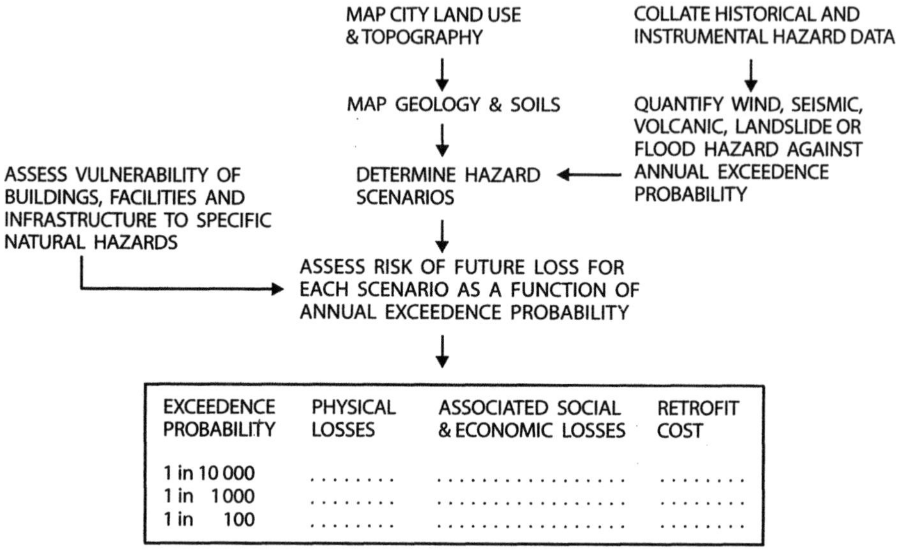

Fig. 15.6. Model for assessing probable maximum loss

15.4.3
Mitigation of Risk

The next step after assessment of the risks and potential losses is to establish the range of possible actions to mitigate the risks together with the estimate of costs and benefits of these actions. Mitigation can be implemented in one or more different ways with different objectives, including:

- *Preventive Measures,* aiming to minimise the physical damage created by hazard events. e.g.: Review design standards, specifications and good practice guides and revise them according to the priority of the component and the level of finance available.
- *Spread the Risk,* aiming to reduce the effects of physical damage by ensuring a range of alternative facilities e.g. introduce redundancy or reinforcement into the distribution system for re-routing operations.
- *Spread the responsibility,* aiming to provide more than one responsible authority (public + one or more private + community) e.g. encourage informal communities to install and manage local systems, subject to regulations on minimum standards for security and quality of supply.
- *Cover or minimise the impact,* aiming to manage potential effects e.g. provide disaster insurance for physical losses particularly mechanical and electrical plant and equipment to facilitate rapid recommissioning of the system.

- *Plan disaster management,* aiming at long term resilience e.g. arrange regular workshops and training programmes for the continuing education in hazard preparedness and mitigation for staff involved with lifelines.

Mitigation is not just about strengthening physical infrastructure. It does not mean immediately retrofitting or rebuilding the whole city. It does mean starting with actions which *do not* require major investment and *do not* require a change in the political system. It *does* mean hundreds of activities all over the city, including education, public awareness, investigations, planning, design and community activities. It has to mean strengthening strategic infrastructure for example hospitals to ensure floods don't damage equipment.

Informal communities are severely disadvantaged in a large urban area. For the urban poor it is a question of who owns the house, who owns the land and what is their right of tenure before progress can be made with the mitigation of their physical vulnerability. But with international funds, low cost labour intensive schemes are attractive to both the donor and community as community involvement and support would be high and results achieved quickly. It would be important to work through an established NGO to ensure value for money and standards of workmanship. In this respect, the very large informal community of 'Orangi' in Karachi has made substantial progress by helping themselves assisted by a low budget NGO on water supply, sewerage, house construction, health and social matters. However, even with this community, district offices have been pressurised to allow building in areas that should have been kept clear for surface water floods and a hazard event dramatically highlighted this mistake by damaging many houses.

15.4.4
The Scientists Contribution

The answer to the question what is the role of a landslide and flood scientist in a city to reduce the impact of hazards, is actually wider than at first thought. They must obviously press for budgets so that *'adequate'* monitoring and warning equipment is installed and they must provide usable hazards maps for the city planners and public work department. But bearing in mind the limited city budget they must identify the potential hazard effect and risk of damage to critical facilities. Critical facilities can be defined as those facilities which must be in operation immediately after a hazard event e.g. hospitals, schools, highways. A very few chief executives control most of cities physical infrastructures. Helping them to understand the problem of natural hazards and the benefits of mitigation could have much wider benefits for the city.

Similarly there is a major public awareness role required in developing country cities where landslide and flood scientists and engineers can provide guidance to specific interest groups such as hospitals, schools, industry and particularly informal communities. But perhaps their main role is to try and persuade the city to put mitigation of potential hazards as an item on their agenda. Mayors, governors, and administrators who live day to day with running urban areas and their many conflicting problems will need a lot of persuasion to spend money on mitigation. Cooperate with the media, give them access to your work and provide them or spoon feed them with good copy, is part of the answer.

References

Asian Development Bank (1991) Disaster Mitigation in Asia and the Pacific. ADB, Manila
Centre For Advanced Engineering. Lifelines in earthquakes – Wellington Case Study. Published by CAE University of Canterbury, New Zealand
Hodgkinson PE, Stewart M (1991) Coping with Catastrophe. A handbook of Disaster Management. Published by Routledge, London
Horlick-Jones T (1994) Planning and coordinating urban emergency management; some reflections on New York City and London Journal of Disaster Management vol 6
Horlick-Jones T, Amendola A, Casale R (1995) Prospects for a coherent approach to civil protection in Europe. Natural Risk and Civil Protection E & FN Spin London
Solway LM (1994) Urban Development and Megacities: Vulnerability to Natural Disasters. Journal of Disaster Management vol 6
Solway LM et al. (1995) Megacities – reducing vulnerability to natural disasters. Thomas Telford Publications, London

The Potentialities of a Risk Disaggregation Between Vulnerability and Hazard: Example of the Relative Stability Induced for Flooding Risk Alleviation

G. Oberlin · O. Gilard · P. Givone

16.1
Introduction

It is assumed here that the land use planning practice is mainly an action of negotiation between land private owners and/or land public authorities. The ultimate target is to improve the localization of the land uses, taking in account the natural hazards, so as to be able to alleviate the risks. An intermediate target is to improve the civil and hydraulic structures and the water bodies maintenances undertaken. Alleviate means not only a global reduction of damages, but also a progress in social equity toward the residual risks. These lasts remain always present, even when an actual application of strong land planning measures, and/or of protection structures, is undertaken and respected.

16.2
Economical or Social Target?

To improve land uses means here to lower the damages (economical aspect) among, and to equilibrate the remaining risks (social aspect) between, the land users and the concerned populations. Due to the difficulty to estimate the damages (size of the investigated area, short- and long-term evaluations, scale of the analyses, non-direct damages, non-monetarizable damages, benefits induced by risk acceptance, bad side effects of some protection measures, etc.), the social target of land use planning can be considered as very important.

A better negotiated balance of the remaining risks could effectively be a powerful measure to alleviate the numerous bad effects of natural hazards, and perhaps allow a better taking in account, by the concerned people, of the protection measures and of the land use planning decisions. A better equity on (residual) risks could also be a powerful way for both a social stabilization and an encouragement for sustainable development. In addition, a presentation of the risk management tools under a negotiable way, is able to develop the fruitful links which should exist between neighbouring communities. For flooding waters, this concerns also the complementarity between urban and rural areas. In fact, rural areas are frequently more or less relatively overprotected toward the flooding risks (see further: a negative sign for the risk definition proposed here), once accepted that a level of protection toward flood is necessarily limited (see Section 16.7: the concept of a reasoned maximal risk acceptance level), and urban ones are frequently under-protected (positive sign for risk). So, a tool which makes this potential market of negotiation and collaboration arise, is probably a very efficient and sustainable one.

16.3
A Manageable Definition of the Risk, Disaggregating Vulnerability and Hazard

For such a negotiable approach of land planning, it seems compulsory to disaggregate the risk into its two components, more or less independent: the vulnerability (the sensitivity of the land use and of the population) and the natural hazard. This leads to a more adequate, flexible and manageable definition of the risk (Vie Le Sage 1985). Then, by crossing locally vulnerability and hazard, it becomes possible to estimate a more realistic level of the local risk. For the areas with both high vulnerability and high hazard, the benefit of such disaggregation seems not very relevant: just able to qualify and classify them in different levels, all very severe. But for other areas, the disaggregation is very interesting: it makes appear other classes, like areas with a significant level of risk in spite of a moderate vulnerability or of a moderate hazard, and also areas with a large credit of relative security (relative to their vulnerability) (see Fig. 16.1).

16.4
The Crossed Map as a Basic Tool for Negotiation

It is already very interesting, for many purposes, to map clearly, and as detailed as wanted, these local vulnerabilities, hazards and the results of their crossings. It is particularly powerful to map clearly the areas qualified as "relatively over- protected". Their maps give first a clear perspective for future implementations of more vulnerable land uses on them, exploiting their credit of security by increasing their vulnerability, obviously in the limit of this credit, i.e. respecting the hazard present. But these "relative security credits" can also be exploited by increasing their hazard, if it is possible and relevant (so is the case for flooding waters hazard), i.e. if it can be useful for other land users and other areas elsewhere, in particular for areas with a large deficit of relative security (a much too large level of risk).

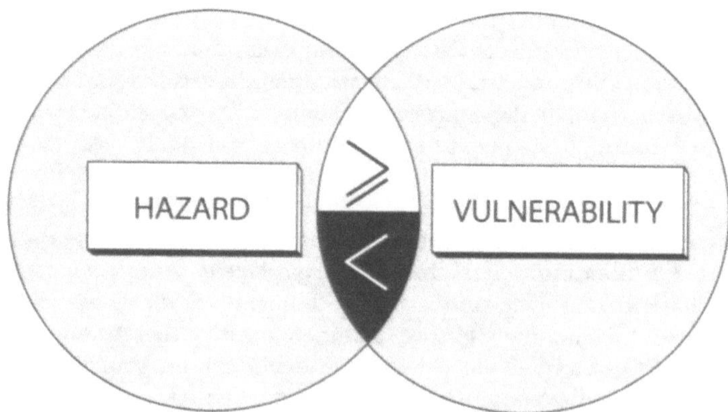

Fig. 16.1. The disaggregation of risk between its two independent components, and the result of their crossing: a risk with a sign

16.5
A Relevant Application Case: The Flood Hazards

The previous considerations are particularly relevant for the case of flooding hazards: it is technically possible, and already in practice since a long time in many countries, to manage the flood volume so as to make it more present (more duration, more depth, more velocity, higher probability, etc.) in a given area which has a large credit of relative protection (a low level of risk), and in compensation to have this volume less present (less duration, less depth, less velocity, lower probability, etc.) in an other area with a severe deficit of relative protection (a high level of risk). It seems that it is now the adequate moment to improve structurally the management of this very widespread natural risk (Rasmussen 1994).

16.6
The Relative Stability of Such Risk Maps

In addition, such research of a better balance between the remaining risks taken among local communities, is in fact more leaded by the sign of the risk (positive = actual risk; negative = credit of relative security), and by its rough estimation (low, medium, high), or its relative level (lower, higher), than by a precise economical and monetarized estimation. In fact, land use allocations and choices result mainly from a political behaviour, sustained but not fully constrained by monetarized estimations. So, if the maps of such disaggregated risk levels are clear, and if possible relatively stable, at least on the signs displayed (area clearly and simply mapped as having credit, or deficit, of relative security), then they could become a powerful and respected tool for land use planning, and could progressively lead to an actual and respected risk alleviation policy.

16.7
An Example of Model Exploiting these Concepts: Inondabilite

The hydrological integrated INONDABILITE model was launched to exploit the above mentioned concepts of vulnerability, hazard (in a way adequate for crossing it with vulnerability) and risk, and to be able to map them at any scale, with a display devoted to the management of water excesses (flooding volumes) through negotiated land use planning.

16.7.1
A Preliminary: The Synthesis Hydrological Model *QdF* as Basic Tool

INONDABILITE exploits a synthesis hydrological sub-model (acronym: QdF) which displays explicitly (a discharge Q for a duration d and a Frequency F), or gives implicitly (for depth p and velocity V, when completed by the local mean rating curve $Q(z)$, z being the water level), the hydrological knowledge needed. The needs concern the quantification of both the social requests toward flood risks (vulnerability) expressed in terms of equivalent flood hazards characteristics (see further: the quadruplet (d^*, T^*, p^*, V^*) interfaced with socio-economical characteristics), and the actual hazards characteristics themselves. These QdF models are formalized, either in curves $F(Q(d))$ with physical values of Q and d, or through mathematical models (nested hyperbolic functions) with dimensionless variables d and Q. The durations d can be continuous or cumulated ones. The discharges Q represent mainly thresholds of discharges overpassed during d (continuous or cumulated), and if necessary volumes, i.e. means on d: for extrapolating the $F(Q)$ curves, for rainfall (P)–runoff (Q) $Q(P)$ models, etc. They are extracted from discharge chronics $Q(t)$, either observed or simulated (through $Q(P)$ models), and they summarize efficiently the flood regime.

These QdF models are continuous, at least until the second order, monotone (decreasing with d), have a large range of validity on d, F and basin areas, can be aggregated on basins having different flood regimes, and have very relevant properties of regional transfer (from a basin to an other). They seem to be able to display some of the very rare arising properties in hydrology, here on regime features and for them-

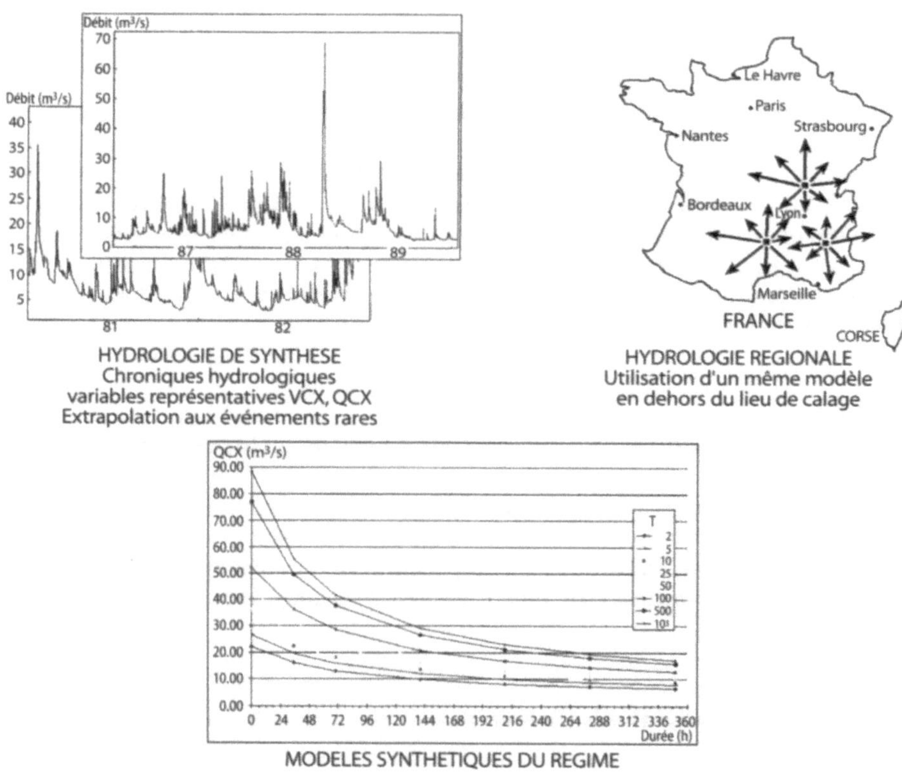

Fig. 16.2. Synthesis flood models QdF: principles, schemes of elaboration and of use

selves as such. In spite of their hierarchical definition through $F(Q(d))$, they are used, as it is common in hydrology, in any direction between their 3 variables d, Q and F (or T: mean return period, as an equivalent to F).

For the following points, it is assumed that a local QdF model is available at any section of the concerned river (basin). Such assumption is now frequently feasible (Galea and Prudhomme 1993), obviously with the classical set of uncertainties always present in hydrology, but also enough relevant for the following algorithms (see Fig. 16.2).

16.7.2
An Interfacable Definition of the Vulnerability Toward Floods

This section deals with the local unicity of a synthetic variable TOP, summarizing a vulnerability expressed through a (complete or uncomplete) quadruplet (d, T, p, V).

The social sensitivity mentioned above (flood protection level request) is estimated here through a so-called vulnerability, represented by the vector (d^*, T^*, p^*, V^*), and being the hydrological interface toward any social, economical or policy criteria. The x^* elements of this vector are local flooding water characteristics (duration, Frequency, depth, Velocity), but not the actual ones expressing the flood regime of the place: they are the upper limit of flooding water characteristics which could be accepted if, due to natural constraints, or to a planified effort (flood alleviation, damage reduction, equity, negotiated and accepted solidarity, …), a minimum of flooding should be accepted (with or without compensations). The x^* values are quantified (by ones, by couples, by triplets, etc.) through socio-economical, or a priori, criteria, and they express either a tolerable (maximum) level of hazard, or a demanded (minimum) level of protection, both linked to the local land use vulnerability and globally designed as a maximal risk level acceptance (French acronym RXA).

It appears effectively (but this has still to be generalized and validated) that any socio-economical information, or modelization about flood hazard damages, or drawbacks, can lead, after having taken in account the actual processes in floods, to the following quadruplet (Desbos 1995) defining, either a maximum flood acceptability, or a minimum of protection need, in any given place (actually floodable or not):

- no water velocity $> V^*$ …
- and/or no water depth $> p^*$ …
- and/or no duration $> d^*$ …
- … more frequently than F^*, i.e. with a mean return period smaller than T^*.

Notice that, for both levels and durations, the linked discharges are thresholds ones (acronym $QCXd$, in QdF modelling), no mean or volumes (Q integrated on d; acronym $VCXd$, in QdF modelling) values (see Fig. 16.3).

Remark. To respect the numerous water resources aspects always present in waterways, even the flooding ones, this vulnerability is presently extended to water ecosystems needs, which include minimal water level aspects, in addition to classical flood protection ones as presented here.

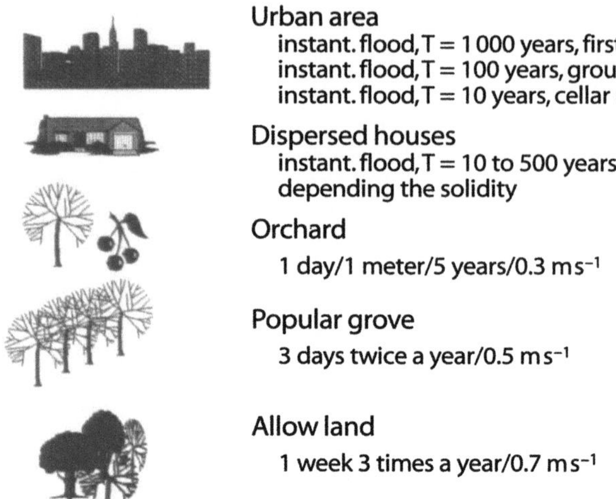

Fig. 16.3. Examples of quadruplets (d, T, p, V) expressing, in hydrological variables, the reasoned vulnerabilities toward flooding water hazards

16.7.3
The Principle of the Transformation Needed to be Able to Cross and to Map

Such vulnerability will then be crossed with the local hazard. To prepare this crossing, it has been chosen to transform the quadruplets (or their reduced/simplified form) in a variable representing an equivalent peak discharge ($d = 0+$) $QIEO$, or its mean return period ($TOP = T(QIEO)$). The acronym TOP comes from the French terms "Taux d'Objectif de Protection".

The reason of this peculiar choice, which can appear here as artificial, is the following: such an instantaneous variable is easy to map as it represents a flood-peak (here symbolic, not actual), more or less mappable in its flooding area effects by the maximal extension of this area (with some rating curves representativity problems, but resolvable, even in case of a high degree of hysteresis: see further). The hazard aspects, dealing actually with flood peaks and the spatial extension of their flooded areas, are treated beyond. The following section details this way of transformation between the vulnerability quadruplet (of x^* values) and its symbolic equivalent peak discharge $QIEO$ (i.e. its mean return period TOP).

16.7.4
The Local Unicity of the Variable *TOP* Displaying Such Vulnerability

To explain the bases of the algorithm (Oberlin et al. 1993), it is better to start with a simpler vulnerability reduced to a *couple* (d^*, T^*). A synthesis QdF model gives a discharge Qe for such couple: $Qe(d^*, T^*)$. The triplet (Qe, d^*, T^*) is a solution of the QdF model. There is theoretically always a solution: mathematically, with the forms chosen for these QdF synthesis models, Q is defined for any d or T (positive real numbers). In some rare cases, QdF could be not calibrated, or not validated, but only for some very peculiar values of d and T.

This discharge Qe is now interpreted, inside the QdF model, as linked to an other duration artificially (see above) chosen as $d = 0+$ (the duration of a peak discharge), i.e. Qe is interpreted as the $QIEO$ (see above). Changing the duration d but not the discharge Qe, needs to change the Frequency T into Te, as the triplet (Qe, $d = 0+$, Te) has to remain a solution of the QdF model. So, TOP is just equal to such Te, and is unique, due to the univocity of the QdF model. The following basic equation can be used:

$$TOP = Te[d = 0+, Qe = Q(d^*, T^*)] \ ,$$

with

- $Qe = QIEO$.

This is the basis generic relation. Once the QdF model is well defined, and would it be under a mathematical, or a table, or a graphical form, the practical solution of this equation is very easy to solve (implicit, when mathematical form) (see Fig. 16.4).

Notice that, with such a "quadruplet" reduced to a couple (d^*, T^*), the solution for TOP is only hydrological (through QdF). No rating curve $Q(z)$ is needed for such simplified vulnerabilities. This simplified vulnerability form is already very useful: a lot of actual sensitivities toward flood are sufficiently defined by such (d^*, T^*) couples. The $Q(z)$ curves will obviously appear, and be compulsory, further: for the hazard estimation and mapping.

For a triplet (p^*, d^*, T^*), as it is given on vulnerability considerations, more or less independent from the flood regime features, there must be a gap toward QdF: such triplet is normally not a solution for the given local QdF model, because it is generally overdetermined, except the very rare case where it falls quite as a solution. Like for the basic couple (d^*, T^*), it is necessary to attain the target of a Frequency representative of an equivalent peak discharge $QIEO$. Several ways (all equivalent) are mathematically possible. The best is to exploit first what is always defined and easy to understand, i.e. d^* and p^*.

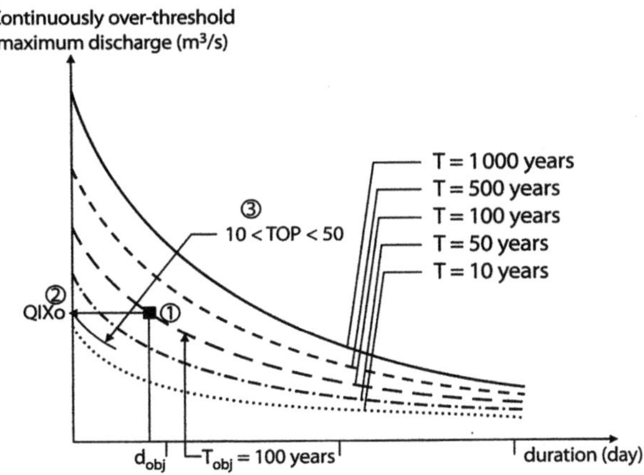

Fig. 16.4. The univoque determination of a mappable TOP in the simple case of an hydrological quantification of the vulnerability through a *couple* (d^*, T^*)

So, the first step proposed is to find a first intermediate variable, corresponding to the couple (d^*, p^*). A local rating curve $Q(z)$, taken as a mean representative one if there are difficulties (see further), gives a discharge $Q(p^*)$, because the local depth p and the section level z are closely interrelated (by the local topography of the river section, including the concerned local place where the vulnerability is here treated). The *QdF* model gives an intermediate equivalent Frequency *Te* by the following equation:

$$Te = T[d^*, Q(p^*)] \ .$$

The basic input data d^* and p^* have now been exploited, and a local (mean) rating curve $Q(p)$ was needed (see Fig. 16.5).

The second step is now to try to approach a peak discharge situation for the analysed place. This is made by a second intermediate triplet where p is chosen as equal to $p = 0+$, corresponding to a just flooding discharge $Q(p = 0+)$ for the given place analysed (recall the basic rule: approach an equivalent peak discharge), with respect to the other local data: *QdF* model, rating curve $Q(p)$, *Te* and T^* (p^* and d^* are already exploited). This choice will introduce here a second intermediate variable, a duration de, respecting these conditions, and solution of the following equation exploiting again the *QdF* model for *Te* and for the new discharge $Q(p = 0+)$ given by the mean rating curve for the chosen depth 0+:

$$de = d[q(p = 0+), Te] \ .$$

The remaining basic data not yet exploited is T^*. It allows us to find a third intermediate variable, the discharge Qe, solution of the following equation:

$$Qe = Q[de, T^*] \ .$$

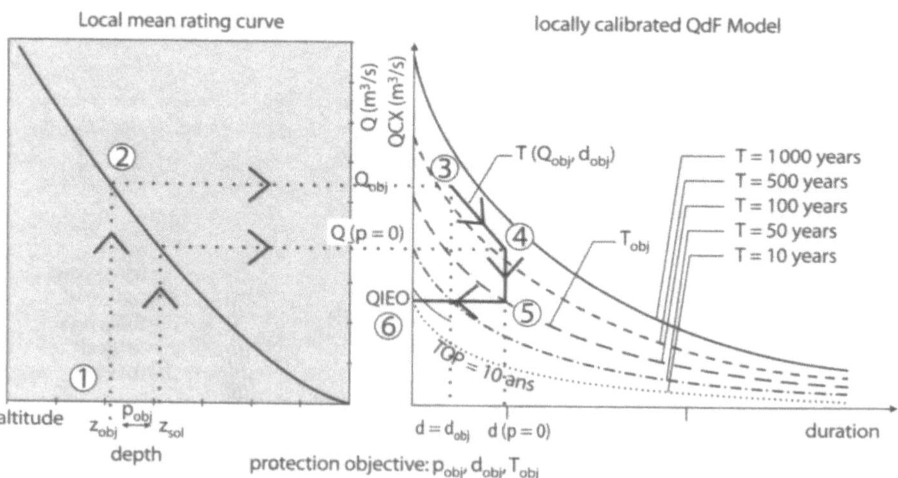

Fig. 16.5. The univoque determination of a mappable *TOP* in the general case of an hydrological quantification of the vulnerability through a *triplet (p^*, d^*, T^*)*

The case is now the same case than previously presented for the simple couple (d^*, T^*), and it is just necessary to interpret this discharge Qe as a *QIEO*:

$$TOP = T[d = 0+, QIEO = Qe] \quad .$$

For a *quadruplet* (V^*, p^*, d^*, T^*) the present rule is to exploit a relation between V and p, like $V(p)$, eventually a mean representative one if local scattering or turbulences. Such relation gives a second depth $pV = p(V^*)$. Because there is a redundancy between p^* and pV, and because the target of reasoned security exploited here (see the basic RXA concept above), it is allowed to take the most exigent, i.e. the lowest p, without bad side effects. So the basic quadruplet is reduced to the following triplet, to be treated as above:

$$(\min(p^*, pV), d^*, T^*) \quad .$$

Due to the continuity and inversibility of QdF synthesis models, all the previous determinations are solvable, and locally unique, so allowing a clear mapping of these vulnerabilities (see Fig. 16.7a).

It is frequent that local vulnerability cannot immediately, or simply, be defined by an unique quadruplet (triplet, couple, ...). For instance because there are actually different levels: in a house/building, the vulnerability is different for the different floors (cellar, ground-floor, 1st one, etc.). So is the case for a fruit-garden between the soil/ grass level and the branches/leaves/fruit one. Difficulties can also derive from socio/ economic/policy discrepancies, or from different hypothesises not yet selected (Desbos 1995). In such undetermined cases, the best way is to estimate all the local *TOP* variables representing these different options, and to choose the most exigent (the highest value of the different local *TOP*s). Here also, due to the basic effort (see RXA concept) of a security request level which is reasoned and moderated, there is no bad side effect to make such relative over-security choice.

Summary. It can be useful to summarize here the conditions for obtaining such mappable (through *TOP*) vulnerabilities:

- Define the local vulnerability through an hydrological interface using one or several among the following flooding water characteristics: Velocity V, the depth p, the duration d and their Frequencies T, all quantified as "reasoned maximal acceptable" ones;
- Have a local synthesis model representing the flood discharge regime (normally: thresholds discharges);
- Have locally a mean rating curve $Q(p)$ if p is used, and a mean $V(p)$ relation if V is used.

Remark. As presented further (see next Hazard § and *TAL* definition), the use of maximums for both p (equivalent to level z) and Q, allows the feasibility of a representative and adequate rating curve $Q(z)$ or $Q(p)$ in a very large majority of cases. The situation is less comfortable if a relation $V(p)$ or $V(z)$ is needed, and basic research is still ongoing for that velocity point.

16.7.5
The Local Unicity of the Variable *TAL* as Representative of the Local Flooding Regime

This point is now very easy to understand and to treat. Following the efforts made previously to have a hydrological equivalent of the vulnerability, equal to the mean return period (*TOP*) of a symbolic "just flooding discharge" *QIEO* represented by a symbolic relation $Q(pe = 0+, de = 0+)$, it should appear obvious that the flood regime will be represented by the actual just flooding discharge $Q(p = 0+, d = 0+)$ of the analysed (local) place. And the variable to be crossed with the *TOP* will be the mean return period of this discharge, named *TAL* (French acronym coming from alea, which is the French translation of hazard). The algorithm can be summarized by an unique equation (Oberlin et al. 1993), solving the synthesis model *QdF* formulation, constrained with the local mean rating curve $Q(p)$. This synthesis presentation can be detailed as follows.

For any place in a flooded area, whose altitude is *zo*, it is possible to affect a $Q(zo)$, discharge corresponding at $p = 0+$, $(z = zo)$ and $d = 0+$, i.e. corresponding to the lowest Q that begin to flood the place (a "just flooding" peak discharge *QIX*), In case of rating curve problems, due to unstability, or non-bijectivity, of $Q(z)$ (or $Q(p)$), the following properties are exploited: if one considers only the flood maximum levels $zx(t)$ and discharges $Qx(t')$, generally non simultaneous $(t > t')$, the pseudorating curve $Qx(zx)$ where the current time is eliminated, is bijective by definition, and generally more stable than the current $Q(z)$ ones. For the *TAL* and *TOP* applications, devoted to *QxdF* and *zxdF* curves (models), the adequate curve is $Qx(zx)$, and then used in case of failure for the basic $Q(z)$ one. In case of stability and bijectivity, the two are equivalent. This can be summarized as follows: for peak problems, the rating curves can be simplified. In any case, the application of hydraulic transient models allows the control of the $Q(z)$ curves and the consistency of the $Qx(zx)$ ones when needed.

So, using again the *QdF* models, the local flood hazard of a given place (whose level is *zo*) is here resumed by the Frequency, *TAL*, of the actual just flooding discharge of the place. *TAL* is the solution given by the following *QdF* unique equation:

$$TAL = T[Q(p = 0+, d = 0+)] \ ,$$

with
- $Q(p=0+, d=0+)=Q(zo)$.

This *TAL* variable is obviously mappable. If observations (photos, …) and historical inquiries could supply these hazard treatments, and even allow a rough mapping, the most adequate and recommended way is here to exploit hydraulic models, permanent or transient. In that last case, the *QdF* models have to be completed by an hydrograph representing the flood regimes. This point uses the concept of HSMF (Mono-Frequency Synthesis Hydrographs), not detailed here but validated (for such frequency mapping applications, not for general flood routing), and operational. Some probabilities composition tasks at confluents must here been assumed (not detailed here, research on-going, provisional solutions operational). In addition, such hy-

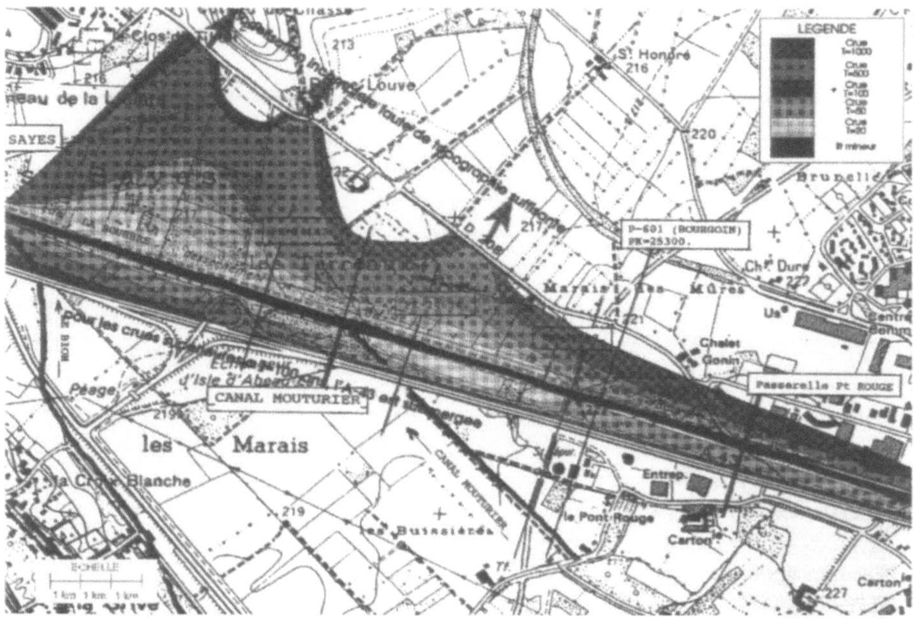

Fig. 16.6. Hazard mapping through the synthesis *TAL* variate, expressing the detailed hazard characteristics (d, p, V, T) synthesized by the *TAL*. (the original maps are in colours, so more legible)

draulic models give rating curves ($Q(z)$ or $Qx(zx)$ ones) in any cross sections (see Fig. 16.6).

Intermediary conclusion. The transformation of the local vulnerability and of the local hazard in two comparable and mappable varieties is now finalized. The relations used are easy to computerize as far as the following materials are available: local QdF (for a river reach, affluents included: transferable QdF), local rating curves $Q(z)$ or $Qx(zx)$, local topography (zo known, and related to the water levels z), and the vulnerabilities expressed as previously detailed: hydrological variables validated as interfaces with socio-economical data, representative of reasoned social requests.

16.7.6
The Synthesis Variable for Risk, *Delta*: Definition, Use, Mapping and Relative Stability

The crossing between vulnerability and hazard is now very simply expressed by the variable *Delta* (Oberlin et al. 1993), which is an efficient quantification of the risk level, expressed here basically in probability units (through mean return period units):

$$Delta = TOP - TAL \quad .$$

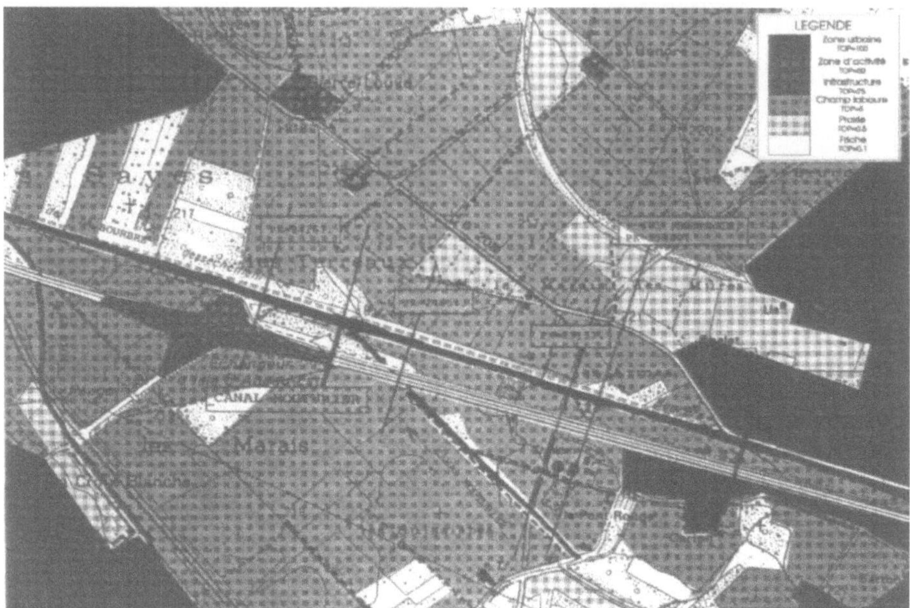

Fig. 16.7 a. Example of an independent vulnerability map expressed through the synthesis *TOP* variate, to be crossed further with the hazard map of the Fig. 16.6. (the original maps are in colours, so more legible)

Considering the *QIEO / QIX* and the *z(QIEO / QIX)* physical variables that *Delta* represents here, this synthesis risk variable *Delta* can be expressed (locally) in discharge units *DeltaQ*, or in level units *Deltaz*. Its continuous mapping allows to integrate the *Deltaz* form in water volumes *Deltav* (flooding waters volumes).

A *positive Delta* indicates a positive, i.e. an actual, level of risk: the request of relative security is not satisfied, because the actual flood regime is worse than those considered as acceptable. This is illustrated (but only illustrated: the complete meanings are deeper and more detailed) by a *TAL* smaller than *TOP*, indicating that the just flooding discharge is more frequent than the *QIEO* equivalent to the vulnerability.

A *negative Delta* indicates a negative, i.e. a non actual, level of risk: the request of relative security is satisfied, because the actual flood regime is better than those considered as acceptable. This is illustrated (but only illustrated: the complete meanings are deeper and more detailed) by a *TAL* larger than *TOP*, indicating that the just flooding discharge is less frequent than the *QIEO* equivalent to the vulnerability.

It is easy, and powerful, to map this quantified risk variable *Delta* (from a weak regional scale, to a demanding individual parcel owners one). The first and main results which appears from such mapping is a clear display of the two categories of areas: over- and under-protected ones (considering the RXA basic concept, in fact compulsory, even if it could be socially, but theoretically, rejected). A marginal category has no sign, or a less significant one: *Delta* has a small absolute value. Such limited cases represent a (hazardous or planned) quite balanced situation between vulnerability and hazard. This

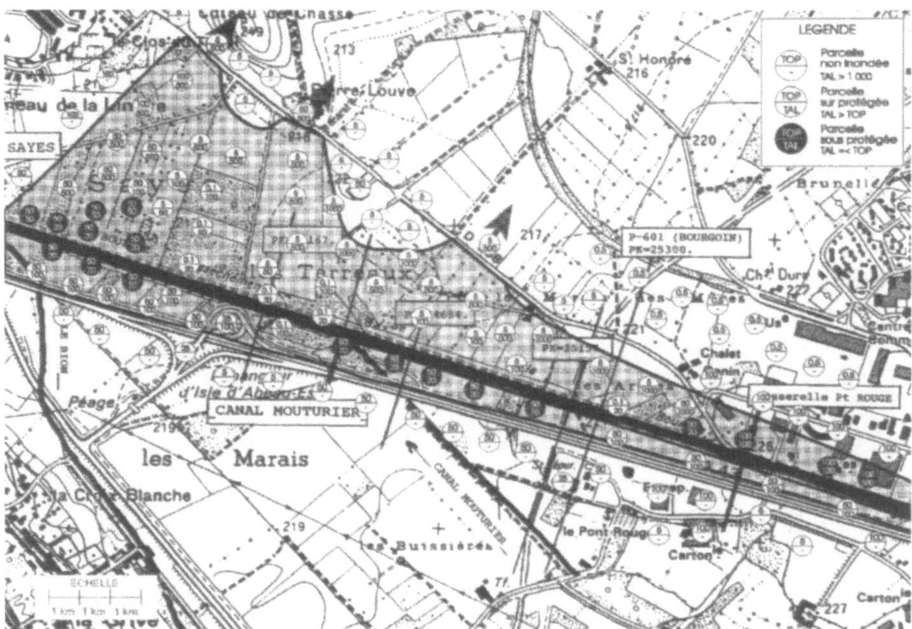

Fig. 16.7 b. The pedagogic display of the synthesis crossed risk maps, devoted to Launch the local land use and water management negotiations, with a target of risk alleviation and of better equity between Local communities. (the original maps are in colours, so more legible)

small category can be considered, either as already well-planned, or to be pushed in one of the two main others, depending the on policy undertaken (to develop or reduce the security requests levels) and the level of uncertainties of the analyses having led to such balanced *Delta* values. Apart from this marginal category, the two others have generally a significative position, largely over- or largely under-protected: a sort of stable and solid status, in spite of the common hydrological (and other) uncertainties. The main decisions to take (develop the protection for the one, eventually exploit the credit of protection on the other) are now relatively stable, and this is very important to sustain a risk alleviation policy, and a land use evolution, both politically very difficult.

So as to improve the understanding of these diagnostics, the maps are generally in colours (Oberlin et al. 1993), with a green for the over-protected areas and a red for the under-protected ones. If needed, the basic map of vulnerability (yellow) and hazard (blue) can be displayed, but it is facultative, the crossed map being able to display also the basic terms *TOP* and *TAL* (in small local screens) (see Fig. 16.7b).

It must be noticed that such synthesis representation (in *T*) can be disaggregated, for pedagogic purposes, under a $f(t)$ representation, *t* being the ordinary current time (chronics). With *TAL* and *TOP* placed on a discharge ordinary chronic $Q(t)$, the green areas should appear as mainly green in current time, but sometimes red (and sometimes yellow: not flooded at all). The red ones should appear as mainly red in current time, but sometimes green (and even sometimes yellow) (see Fig. 16.8).

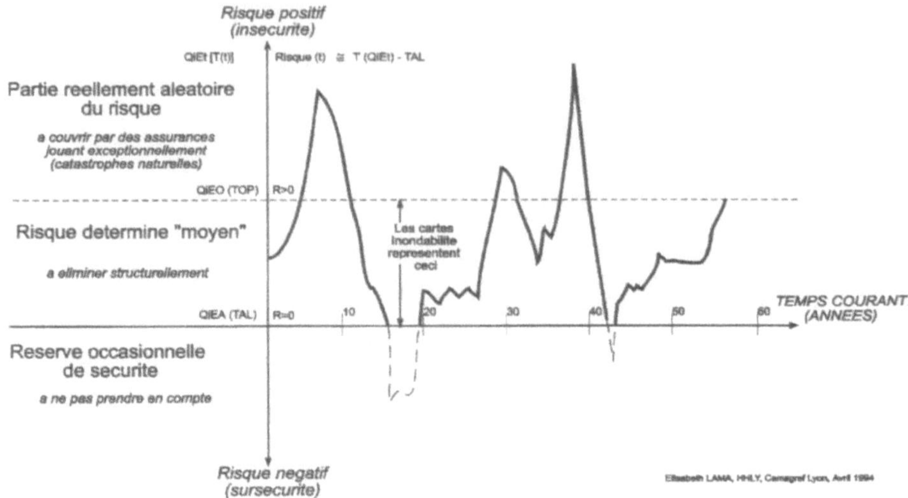

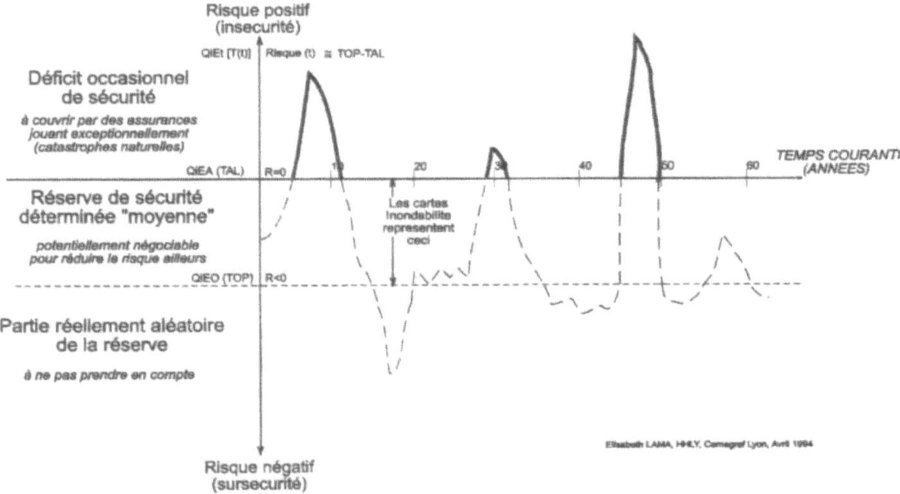

Fig. 16.8. The symbolic current time realities which are represented by the synthesis crossed risk maps, with the two main cases (over- and under-protected areas)

16.7.7
The Management Exploitation of the Risk Variable *Delta*

The absolute values of *Delta* measure the gap between the actual flood regime beared by the concerned parcells, and the socio-economical wishes or rules that they represent. *Delta* can be expressed in water volumes *Deltav*, by spatial integration of the *Deltaz* values (*Delta* transformed in water level units):

- Direct local integration for under-protected areas, with excess flooding volumes; they give positive volumes;
- Remote integration for over-protected ones, remote because their potential storage capacities are below their *zo*, i.e. in a lower part of their river cross section or reach; they give negative volumes, i.e. unemployed storage capacities.

These volumes give the basic data for eventual hydraulic interventions: derivate excess waters from the under- to the over-protected areas.

But such civil works interventions are not compulsory, or not essential, and in any case not the priority ones. The first ones are to inform the concerned people, then to launch the negotiations between the representatives of the two main categories. For that, the relatively stable sign of *Delta* is essential. In addition, the reduction of the positive *Delta* values can also be managed by lowering their vulnerabilities (their land uses), an important alternative (when possible), or a complement (more easy) to water structures management or implementation.

An other efficiency of the single sign of *Delta* is linked to the time needed by any decision to be taken in land use planning. As the land use negotiations are running, and they can be launched with just the sign of *Delta* as soon as it is known, the hydrologists have time to improve their basin modelization, so as to reduce the uncertainties on the absolute values of *Delta*, and to have more precise maps and design figures for the near future, when costly civil works, or when severe land use decisions, will be finalized.

Naturally, once some land uses change, or some hydraulic works are done, all these diagnostics materials have to be periodically up-dated, or at least controlled. There is a first step of control and eventual up-dating to assume, if strong actions are immediately undertaken after the diagnostic phase, which induce changes or trends. With QdF and hydraulic models, this is feasible, and in many cases quite cheap and easy.

16.8
An Encouraging Conclusion

The first results, and the first already operational aspects of these new risk approaches, are encouraging for the risk management in general, and for hydrology and its destiny in the particular case of flooding water risks used here as illustration.

Such flexibility on the risk management (diversified diagnostic, negotiation potentialities, alternative or complementarity between vulnerability or hazard treatment, etc.) is mainly the result of the basic disaggregation, which displays at this management level its power and efficiency (Gilard and Givone 1996). The emphasis of these advantages for the peculiar flood risk is mainly linked to the arising properties observed in regime synthesis modelization (not only due to the QdF form), compared to the chaotic behaviour of the basic chronics $Q(t)$. It is also linked to the specific and powerful role of the continuity equation in hydrology, which allows an actual flooding volume management situation with probably no equivalent in other natural risks (except for snow launches?).

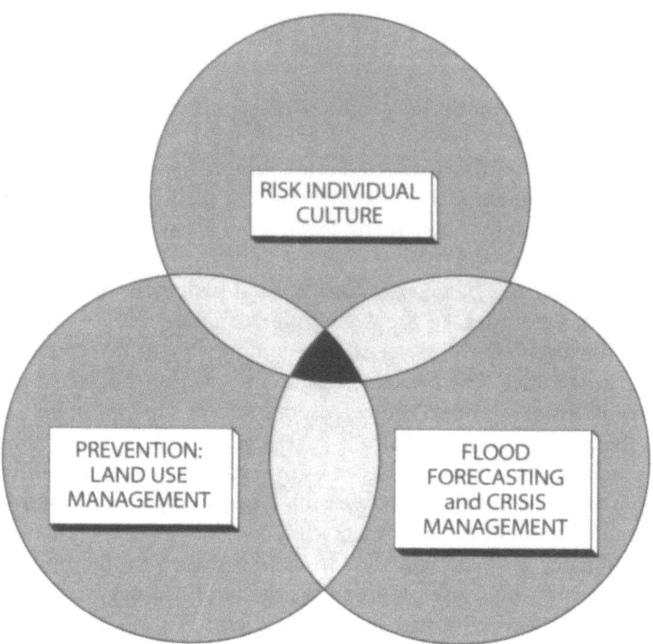

Fig. 16.9. The three compulsory poles of a successful risk alleviation policy

Beyond the case presented here, it seems that specific estimators, dealing with comparisons between natural hazards and social requests, and having simultaneously, on one hand a quasi-stability or a limited uncertainty (like for the sign of *Delta* in our example), and on the other hand a moderate accuracy (like the absolute value of *Delta* presented here), could help risk managers (and hydrologists for flood cases) to both exploit rough regional knowledge and hurry for more rigour.

The first point is compulsory, for society recognition and sustainability.

The second point is compulsory, for progress in risk knowledge (and in hydrological sciences, for flood cases). These two points are not contradictory. So, inside the large uncertainties which still remain in natural risk assessment (and in hydrology), it is possible to, simultaneously, already display stable results, and also sustain basic research, the first action being not disarming for the second.

The practical interest for the society is important. Using a last time here the INONDABILITE illustration, it is hoped, and practically scheduled, that these approaches will progressively see (Fig. 16.9):

- Change the culture of people toward risk aspects, mainly thanks to adequate and up-datable risk maps;
- Allow a larger use of the flooding areas with less flood damages and with more basin solidarity, instead of the classical behaviours (nothing; too small area devoted to river with high dangerous dykes; too large area devoted to river, which is missing for essential human implementations);

- Improve the quality of the water bodies by reducing the choices of channelization works;
- Develop the water resources (seepage);
- Develop a new approach of risk negotiation at any levels, which could definitely establish a permanent risk management (instead of the present scattered crisis practice).

As final conclusion. a challenge to reduce natural risk? Yes, but reduce the risk in its genuine meaning: not only by lowering the hazard, but also by developing reasonable and realistic vulnerabilities, and objective quantitative crossing, and maps allowing permanent negotiations between concerned people, and more equity between people suffering the compulsory remaining levels of risk.

Acknowledgements

This paper has already exploited the very first results produced by starting an European project titled FLOODAWARE (Environment and Climate, DGMI, IVth RDT Programme).

References

Desbos E (1995) Qualification de la vulnerabilite du territoire face aux inondations. Memoire de DEA, INSA Lyon, Cemagref (ed), Lyon, 44 pp

Galea G, Prudhomme C (1993) Characterization of ungauged basins floods behaviour, by upstreaming synthesis models QdF, 2nd International Conference on FRIEND, Braunschweig Oct. IAHS serie N° 221, pp 229–240

Gilard O, Givone P (June 1996) Flood risk management: new concepts and methods for objective negociations.. Conference on Destructive Water, Anaheim (California), 13 pp

Oberlin G, Gautier JN, Chastan B, Farissier P, Givone P (1993) Une methode globale pour la gestion des zones inondables: le programme INONDABILITE. Secheresses,4–3:171–176

Rasmussen JL (1994) Floodplain management into the 21st century: a blueprint for change-sharing the challenge, Water International 19:166–176

Vie Le Sage R (1985) Mise en oeuvre des plans d'exposition aux risques naturels previsibles, Delegation aux Risques Majeurs (DRM), Premier Ministre (ed), Paris, 14 pp

Part IV
Case Studies

Landslide Hazard Investigations in the Dolomites (Italy): The Case Study of Cortina d'Ampezzo

M. Soldati

17.1
Introduction

Many researchers who deal with mass movements feel that landslide hazard is generally neglected, more emphasis being given to other types of hazards such as seismic and volcanic hazards. It has been estimated that every year about 225 000 lives are lost because of natural events in general (Burton et al. 1978), among which several mass movements causing casualties are included (for a more comprehensive review of these events see Eisbacher and Clague 1984; Hansen 1984; Brabb 1991; Gares et al. 1994). The fact that landslide hazard is usually underestimated is even more unfortunate, since slope movements are usually more easily predictable and manageable than earthquakes, volcanic eruptions or hurricanes. Actually, besides high-magnitude mass movements which occur quite seldom, there is a huge number of medium to small sized landslides which are so widespread that the related cost for human society is even higher than that of catastrophic events.

This is also the case of the north-eastern Italian Alps where beside some historical catastrophic events, such as the Alleghe landslide (1771; 49 casualties), Antelao landslide (1814; 256 casualties), Vaiont landslide (1963; about 2 000 casualties), many medium- and small-scale mass movements have taken place recently and are still active nowadays.

The losses due to low-magnitude, high-frequency events, mainly consisting of earth and debris flows in the Dolomites, are also generally increasing because of human activity which, on the one hand, tends to increase landslide hazard and, on the other hand, favours vulnerability situations. Jones (1992) stated that landsliding is generally poorly understood by non-specialists and for this reason underestimated in urban and land planning; it may be added to this that sometimes even if the importance of landsliding is understood there is a tendency to minimise its eventual effects because of economic interests, especially where tourism activities are well developed.

The correct interpretation of landslide hazard and consequently the optimisation of the interventions to be carried out in order to mitigate the loss of human lives and economic assets, depends on some general considerations. First of all gravitational phenomena should be considered as an effect of natural landscape evolution. It is, in fact, the interaction of these phenomena with man's activities and structures that causes casualties and damage. In the Dolomites the increase of tourism activities has determined a greater need for space available for residential buildings, ski pistes etc.; as a consequence, natural events such as landslides have had increasingly serious consequences on man's activities.

According to Varnes (1984), the first principle on which landslide hazard assessment is based is that "The past and present are keys to the future". This means that, according to uniformitarian principles, areas characterised by geological, geomorphological and hydrological conditions which favoured or triggered landslides in the past, or are doing so in the present, are very likely to be subject to failures also in the future. Nevertheless, the absence of past and present failure does not mean that landslides will not take place in the future. Thus, detailed investigations on the physical environment and on the spatial and temporal occurrence of landslides are of primary importance in terms of landslide hazard assessment. However, the effects of human activities and climatic changes should always be taken into account.

17.2
Landslides in the Dolomites

Gravitational phenomena are widely disseminated in the Dolomites since the Würm Late-glacial period; at that time huge rock masses, deformed by ice pressure on the slopes and jointed by ice pressure release, fell on the retreating glaciers which redistributed the debris. Morphological evidence of this is observable in many sites and a few events have been ^{14}C dated back between 11 000 and 9 000 years B.P., especially by means of wood samples found in the area of Cortina d'Ampezzo (Eastern Dolomites).

The frequency and magnitude of gravitational phenomena is proved to be very high in the last early Post-glacial period when slopes, no longer sustained by ice masses, were affected by many large scale landslides. In particular, these landslides seem to be concentrated downstream of the confluences of glaciated valleys where "glaciopressure" might have been more intense, like downstream of the confluence of the Costeana and Boite valleys (south of Cortina d'Ampezzo) where some large landslides occurred (Panizza 1973). Huge complex mass movements took place, eventually obstructing valley floors with the formation of barrier-lakes. Many of these landslides have been further remobilised during the entire Holocene, probably with periods of higher intensity linked to climatic variations (Panizza et al. 1996a).

In addition to glacial implications, other significant factors have contributed to the development of mass movements throughout the entire Holocene. First of all the spatial distribution of geological formations showing different mechanical characteristics must be taken into account. In particular, the occurrence of landslides is high where rigid and resistant rocks with a brittle behaviour, overlie (plastic) rocks characterised by a ductile behaviour. This is the case, for example, of the Badia, Boite and Fiorentina valleys. Furthermore, also where the effects of tectonics are more intense, in correspondence of faults or overthrusts, mass movements have been favoured.

Gravitational landforms are also connected with the existence of deep-seated gravitational slope deformations, which have been only recently recognised in the Dolomites (Pasuto et al. 1994) and particularly in the area of Cortina d'Ampezzo (Soldati and Pasuto 1991). With respect to morphological evidence, they are generally characterised by the presence of trenches, gulls and uphill-facing scarps in the upper parts of the slopes and bulges in the lower parts (e.g. Tofane, Lastoni di Formin and Faloria groups). However, it has been observed that the presence of these phenomena favours or induces "collateral movements" (rock falls, slides and flows) in the surrounding areas.

17.3
Landslide Investigations in the Dolomites

Since 1991, landslide investigations have been carried out in the frame of European projects by the Geomorphology group of the Earth Science Department of Modena University, in collaboration with the National Research Council (CNR) of Padua, with other Italian universities and with the technical support of public administrations.

The research has involved geologists, geomorphologists and engineers with the aim of carrying out geomorphological and landslide hazard mapping of selected sites, defining the main conditions that cause mass movements and their temporal occurrence since the Late-glacial. In addition, the understanding of the processes involved in slope movements, achieved also by means of hydrological and stability models, has lead to the installation of monitoring systems on some landslides.

The research on a European basis started within the EC EPOCH Programme which financed a project concerning "The temporal occurrence of landslides in the European Community" (contract 90 0025) during the period 1991–1993 (Casale et al. 1994; Soldati 1996). The project involved nine research institutions under the co-ordination of the University of Strasbourg (France) (Fig. 17.1). The investigations in the Dolomites, directed by the University of Modena, have been carried out in the areas of Cortina d'Ampezzo and in Alpago (province of Belluno) together with the Universities of Milan and Padua, the CNR of Padua and in collaboration with the Veneto Region (Mantovani et al. 1994; Panizza et al. 1996b; Pellegrini and Surian 1996).

Particular attention was focused on the area of Cortina d'Ampezzo by most of the Italian and foreign institutions participating in another research project funded by the European Union in the frame of the ENVIRONMENT Programme (1994–1996). The research dealt with "The temporal stability and activity of landslides in Europe with respect to climatic change" (TESLEC, contract EV5V-CT94) (final report in progress). The University of Heidelberg (Germany) co-ordinated this project which involved six principle contractors (Fig. 17.1). The University of Modena and the CNR of Padova were in this case linked to the CNR of Cosenza.

The successful results and the experience achieved through the above mentioned projects enabled the participating laboratories, suitably integrated with new partners, to have a third project funded concerning "New technologies for landslide hazard assessment and management in Europe" (NEWTECH, contract ENV4-CT96-0248) for the period 1996–1998 (Fig. 17.1). The network of institutions involved is in this case co-ordinated by the University of Catalonia (Barcelona, Spain). The study areas in the Dolomites are Cortina d'Ampezzo and Corvara in Badia. The aim is to carry on the monitoring of the Alverà landslide (Gasparetto et al. 1996) and to test new technical devices with the aim of improving the knowledge of the dynamics of the movement, in the area of Cortina d'Ampezzo. In the Badia area instead, basic geological and geomorphological investigations will be carried out aiming at the recognition of the most suitable locations of drillings to be performed on the Corvara landslide and the selection of the best monitoring equipment.

Fig. 17.1. Sketch of the participants in the EPOCH, TESLEC and NEWTECH research projects

17.4
Landslide Hazard Investigations in the Area of Cortina d'Ampezzo

17.4.1
Introduction

The first detailed geomorphological investigations carried out in the area of Cortina d'Ampezzo (Panizza and Zardini 1986; Panizza 1990b; Soldati and Pasuto 1991) showed this as an ideal site for landslide hazard investigations in the frame of European research projects. In fact, the whole area has been affected by landslide phenomena of various types and sometimes of considerable dimensions, some of which are still active today. As a result of the favourable morphological conditions, the area has witnessed progressive urban development, which has also been tied to intensive tourist development. Because of this, intense urbanisation and the interest which this region holds for tourism, the presence of active landslides and dormant landslides makes this area particularly vulnerable and subject to high geomorphological risk (Panizza 1990a).

The investigations carried out in the frame of European projects have enabled the various types of landsliding to be recognised, mapped and characterised from a morphological and geotechnical point of view. The frequency of mass movements and the present degree of activity have been also outlined. This has lead to the achievement of a precise knowledge of landsliding phenomena, in terms of type, mechanism and evolution characteristics, as well as, the main factors which have conditioned or could condition their temporal and spatial development, including the rate of human activities in the area studied. This has enabled the researchers to define landslide hazard and risk and to plan and install monitoring systems on two active earth flows. The huge amount of data collected during the last five years by means of inclinometers, piezometers, deformometers and climatic stations allowed the definition of a rheological model capable of predicting displacement rates on the basis of recorded pore pressure changes.

After a brief geological introduction, the main results achieved in the area of Cortina d'Ampezzo will be described below.

17.4.2
Geological and Geomorphological Setting

From a geographical point of view the valley of Cortina d'Ampezzo is located in the Eastern Dolomites; it is crossed in a N-S direction by the Boite River, right tributary of the Piave River and surrounded by mountain groups reaching heights over 3 200 m, such as Tofane, Cristallo, Sorapis (Fig. 17.2).

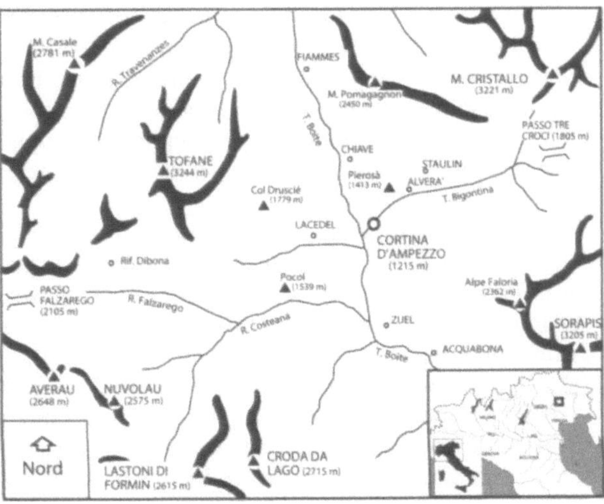

Fig. 17.2. Geographical setting of the area of Cortina d'Ampezzo (Eastern Dolomites)

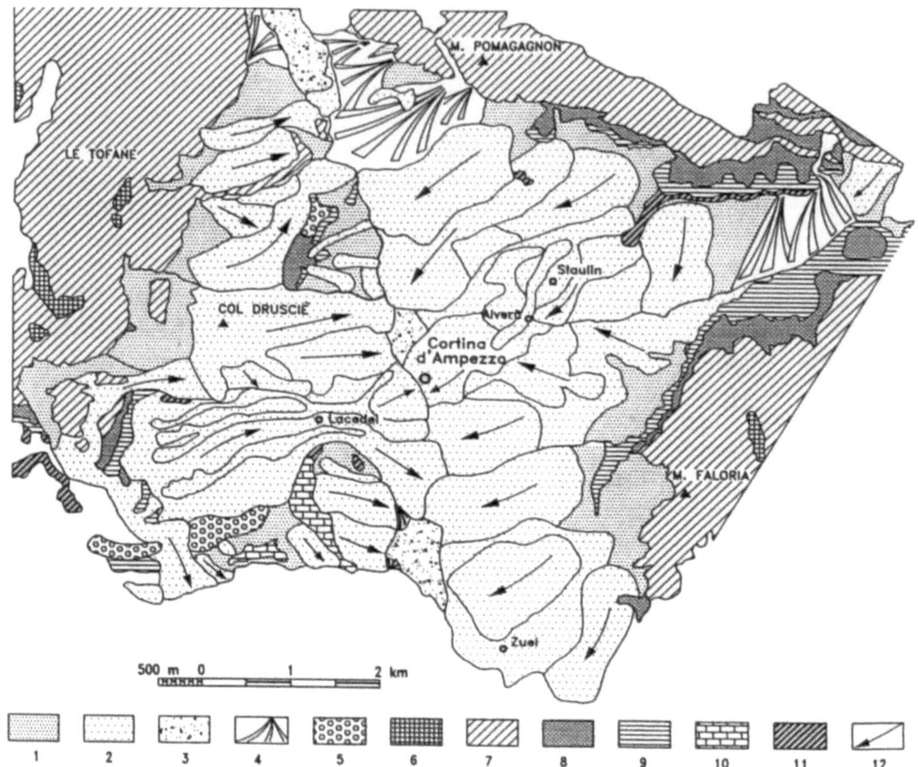

Fig. 17.3. Geological setting of the area of Cortina d'Ampezzo (Eastern Dolomites). Legend: *1* scree slopes; *2* landslides; *3* alluvial deposits; *4* debris flow fans; *5* glacial deposits; *6* Dachstein Formation; *7* Dolomia Principale; *8* Raibl Formation; *9* Dürrenstein Formation; *10* Dolomia Cassiana; *11* S. Cassiano Formation; *12* mass movement direction

The geological structure of the area, characterised by a repeated succession of dolomites and pelitic rocks, has remarkably influenced the morphological evolution of the slopes after the retreat of the Würmian glaciers. The resulting morphology is softly degrading in the medium and lower parts of the slopes where incompetent rocks outcrop, while steep dolomitic walls rise up in the peripheral parts of the basin, eventually interrupted by thick scree slopes, located in correspondence with intercalations of more erodible formations.

The rocks outcropping in the area belong to an Upper Triassic sedimentary sequence (De Zanche et al. 1993) and are largely covered by Quaternary deposits, especially in the central part of the valley, and consist of the following formations: S. Cassiano Formation (Lower Carnian), Dolomia Cassiana (Lower-Middle Carnian), Dürrenstein Formation (Upper Carnian), Raibl Formation (Upper Carnian), Dolomia Principale (Upper Carnian-?Rhaethian p.p.) and Dachstein Formation (Rhaetian) (Fig. 17.3).

From a tectonic standpoint, the study area has undergone an intense activity, attributable to different tectonic phases, whose effects are quite evident on the field and significant for slope evolution. Several faults have been recognised, as well as, a dense network of cracks affecting the dolomitic formations. In addition, overthrusts gave origin to local doublings of the stratigraphic sequence.

17.4.3
Landslide Causes

More than thirty landslides of different type, size, age and degree of activity have been identified in the area of Cortina d'Ampezzo.

Table 17.1 lists these landslides and, in particular, shows the most significant dates obtained through radiometric methods applied to wood samples found within landslide accumulations; these dates enabled a wider reconstruction of the geomorphological evolution of the study area, which appears to have been strictly associated with gravitational phenomena since the deglaciation.

The area of Cortina d'Ampezzo has always been prone to slope instability for different reasons.

First of all the structural conditions of the valley should be considered. The stratigraphic succession is, in fact, characterised by an alternation of dolomitic rocks showing a brittle mechanical behaviour (Dolomia Cassiana, Dürrenstein Formation and Dolomia Principale) and rocks with a ductile mechanical behaviour (S. Cassiano Formation and Raibl Formation). This situation has favoured the development of mass movements and deep-seated gravitational slope deformations; the latter, which were widely recognised in the area, may have favoured or induced the occurrence of landslides (Soldati and Pasuto 1991).

Furthermore, the incidence of tectonics is also significant; in fact, the dolomites are affected by an intense jointing in correspondence with the principal faults, thus creating discontinuities which became potential sliding surfaces and preferential seepage zones for water which could reach and moisten the underlying marly and clayey formations.

As regards the influence of earthquakes on landslides, the only confirmed case of correlation refers to an earth flow which occurred near Cinque Torri in concomitance with the seism which took place on the 15th September 1976 in the Friuli-Venezia Giulia Region. In general, no relationship was found, mainly because the time scale of landslide occurrence and of seismic series recorded are not comparable, since the latter refer only to historical time.

The effects of glaciers on the slopes must not be overlooked. It is likely that the pressure of ice on the valley sides, particularly intense at the confluence of more glacial tongues (Panizza 1973), determined rock deformations in correspondence with surfaces of structural discontinuity, favouring the formation of sliding surfaces; in fact, the occurrence of some ancient large-scale landslides has been linked to this cause (Fig. 17.4). As regards the relationships between landslides and rainfall, significant data have been collected for the specific mass movements which were monitored, but for most of the remaining landslides, no consideration can be made, since they are ancient movements and the rainfall data cover a time span of only seventy years. Particularly rainy periods occurred during the years 1965 and 1966, when many Italian regions suffered disastrous floods. On that occasion the Cortina d'Ampezzo area was affected by numerous mass movements which mostly consisted of reactivations of older landslides.

Some considerations on the relationships between landslides and climatic variations will be made below.

Table 17.1. List of landslides identified in the area of Cortina d'Ampezzo (Eastern Dolomites)

Name	Type	First known failure (years B.P.)	Reactivations (years B.P.)
Chiave landslide	Complex	4520 ±60	
Brite de Val landslide	Fall		
Crepe de Cianderou landslide	Complex		
Comate landslide	Complex		
Sponates landslide	Slide		
Sote Crepe landslide	Fall		
Cadin landslide	Slide	12150 ±435	
Cadin di sotto landslide	Flow		
Cadin di sopra landslide	Flow		
Cadelverzo di sopra landslide	Flow		
Col Drusciè landslide	Slide	9000 ±150	
Colfiere landslide	Fall		
Ronco landslide	Flow		
Pierosà landslide	Slide	10850 ±80	
Cortina d'Ampezzo landslide	Flow	8170 ±70	4350 ±60 1460 ±30
Alverà landslide	Flow	2560 ±80	
Staulin landslide	Flow	3315 ±140	2465 ±125
Chiamulera landslide	Complex	4700 ±60	
Malga Larieto landslide	Slide		
Col da Varda landslide	Slide		
Pecol landslide	Flow	155	
Albergo Cristallo landslide	Slide		
Rio delle Vergini landslide	Flow		
Pomedes landslide	Complex		
Lacedel landslide	Complex	10035 ±110	9270 ±105
Rio Roncatto landslide	Flow		
Rutorgo landslide	Flow		
Son dei Prade landslide	Flow		
Col landslide	Flow		
Pocol landslide	Complex		
Grotte di Volpera landslide	Fall		
Rio Costeana landslide	Slide		
La Riva landslide	Complex	8280 ±100	4220 ±60
Zuel landslide	Slide	9440 ±105	
Pezzié landslide	Flow	6570 ±70	6190 ±50 5170 ±100
Acquabona landslide	Flow		

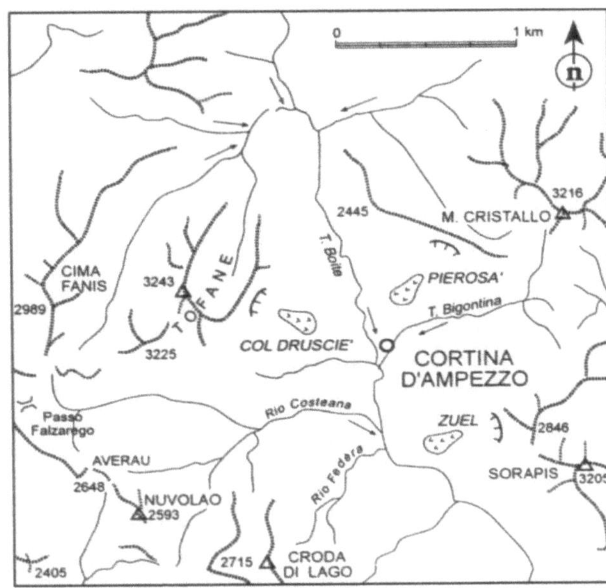

Fig. 17.4. Early post-glacial land-slides linked to "glacio-pressure" effects in correspondence with valley confluences

17.4.4
Landslide Types and Temporal Distribution

The geomorphological investigations carried out in the area of Cortina d'Ampezzo, together with archive research and ^{14}C age determinations enabled the reconstruction of the slope evolution, which is particularly important for landslide hazard assessment, to be outlined since the retreat of the Würmian glaciers.

On a large time-scale, the data gathered so far show a number of landslides distributed in two main periods; if the type of mass movements is considered, it appears that the events of these two periods are substantially different. A third group has been outlined including mass movements which are likely to have had a rather continuous activity during the Holocene due to the visco-plastic behaviour of the materials involved. Therefore, mass movements can be subdivided into three groups (Fig. 17.5).

The first group includes landslides which occurred in the Late-glacial and early Post-glacial (e.g. Pierosà, Zuel and Col Drusciè landslides) when, after the retreat of the glaciers, rock walls no longer sustained by the ice masses became prone to landsliding, giving rise to several rapid mass movement. They consist of rock slides and rock avalanches of considerable size detached from the steep rock walls surrounding Cortina d'Ampezzo at an altitude of 2 000–2 300 m. These events have left clear morphological evidence in the landscape (crowns and hill-shaped accumulations), because of their magnitude and the characteristics of the rock masses involved (dolomites). The period during which these landslides took place is considered the oldest of frequent post-glacial slope movements in Europe by various authors (e.g. Starkel 1985; Corominas et al. 1994).

The second group of landslides includes a number of events which occurred mainly between 5 000 and 4 000 B.P., that is at the end of the Upper Atlantic and during the Sub-Boreal. These mass movements seem to have evolved more slowly than the Late-

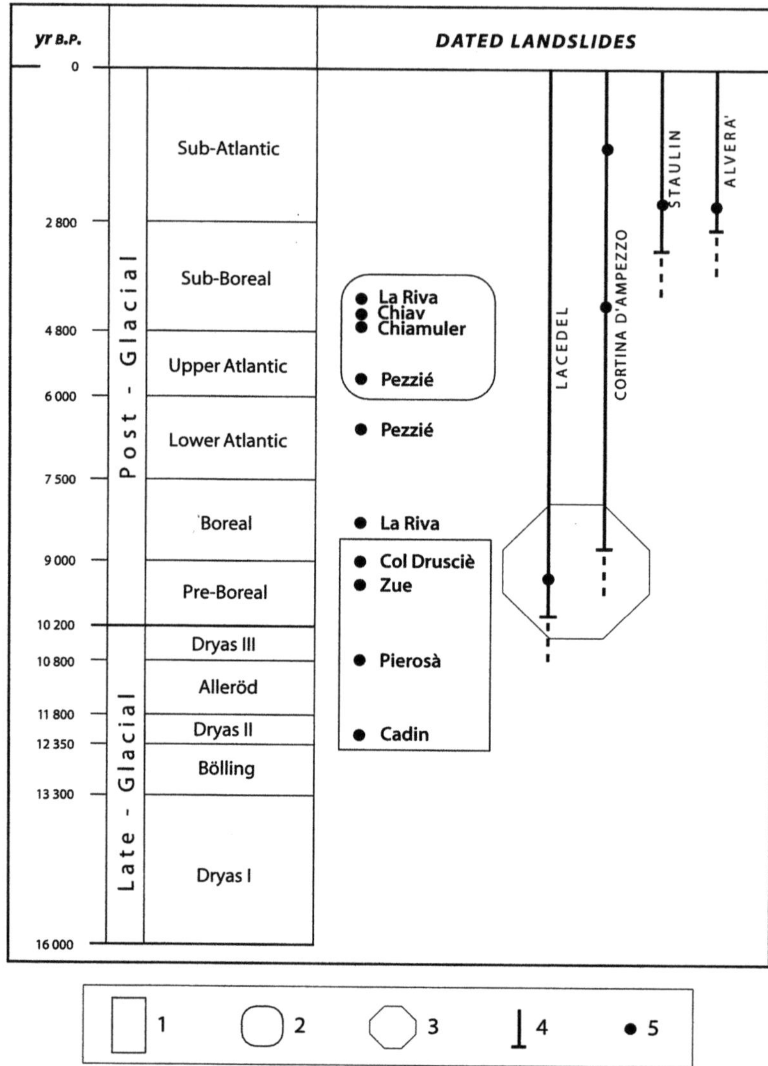

Fig. 17.5. Temporal distribution of landslides in the area of Cortina d'Ampezzo. Legend: *1* glacio-pressure-related landslides; *2* landslides connected with a wetter period; *3* landslides related to permafrost melting; *4* oldest known movement; *5* dated movement

and early Post-glacial ones. Moreover, they do not correspond to single events, since the slopes are likely to have been remobilised more than once. The movement types vary from flows and slides to complex movements, consisting of rock falls affecting the dolomites and rotational slides or flows affecting the clayey formations (e.g. Chia-mulera, Chiave and La Riva landslides). These mass movements consist of flows and slides prevalently affecting marly formations (S. Cassiano Formation). The period of occurrence comes just after the "climatic optimum" of the Upper Atlantic. This wet period probably made the slopes prone to landsliding. This hypothesis, partially confirmed by other European case studies, needs anyway to be further verified through the radiocarbon dating of wood samples recently found in the area of Cortina d'Ampezzo. Further confirmation will be searched by means of new physical dating techniques (e.g. optical stimulated luminescence) which enable non-organic material to be dated.

A third group of landslides, consisting of recurrent mass movements which took place since the retreat of the glaciers, has been identified. The onset of these movements is likely related to permafrost melting, though no direct evidence of this has been found so far. They have been rather active through the entire Holocene till the present time, even if they probably underwent periods of increasing activity due to climatic changes (long term) and seasonal variations (short term) which resulted in reactivations of various intensity (e.g. Lacedel and Cortina d'Ampezzo landslides). This recurrent activity has been made possible because the materials of the S. Cassiano Formation, which are largely involved in the movement, show a marked ductile behaviour.

As far as landslide hazard is concerned, it may be assumed that mass movements of the first group are almost inactive (except partial reactivations of the original landslide scarp). Mass movements of the second group are mostly dormant, but still susceptible of remobilisation in particular conditions. The most hazardous are however, those of the third group that are mostly active and show in some cases a high rate of movement. In addition, they endanger buildings, roads and ski-pistes.

Landslide hazard in the area of Cortina d'Ampezzo is also connected with debris flows which periodically affect the scree slopes and talus cones of Mts. Pomagagnon and Faloria, often obstructing the main road, respectively at Fiammes (north of Cortina d'Ampezzo) and Acquabona (south of Cortina d'Ampezzo).

17.4.5
The Study of the Active Landslides

The three largest active landslides of the Cortina d'Ampezzo area have been analysed in detail (Fig. 17.6).

These are the Alverà, Staulin and Rio Roncatto earth flows (mudslides according to British usage) which affect clayey material deriving from the weathering of the S. Cassiano Formation. The movements at Alverà and Staulin should be considered as partial remobilisations of the Cortina d'Ampezzo landslide, while the Rio Roncatto landslide is a partial reactivation of the Lacedel landslide. These phenomena show different rates of displacement, although they evolve with similar mechanisms. The Rio Roncatto landslide shows displacements of about 2 m per year in the vicinity of Lacedel; the Alverà landslide of almost 20 cm per year and the Staulin landslide, which is almost dormant at present, of a few centimetres per year.

Historical records refer above all to landslides which have affected the villages of Alverà and Staulin: major events occurred in 1879, 1882, 1924, 1927, 1935, 1942 and 1951; in 1935, both these localities were included in the list of towns to be relocated at the expense of the Italian government. As for Lacedel, which consists almost entirely of farming land, only a few records regarding the periodic maintenance works on the roads were found. Control works, consisting of superficial drainages, retaining walls and soil mass remodelling, have been carried out on these landslides since the beginning of this century with the aim of mitigating landslide effects on constructions.

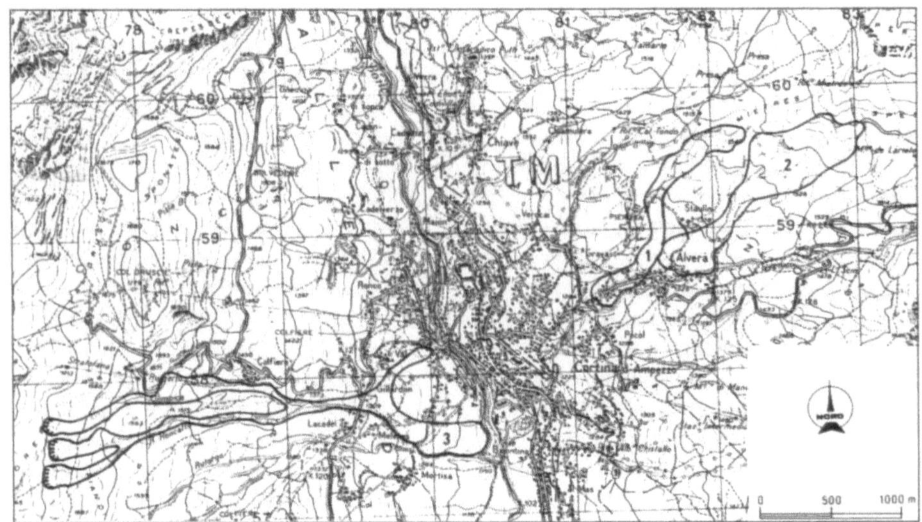

Fig. 17.6. Location of the active mass movements of *1* Alverà, *2* Staulin and *3* Lacedel

Owing to their potential risk for human activities, these slope movements have been monitored by means of automatic systems since 1989 (Deganutti and Gasparetto 1992; Angeli et al. 1996; Gasparetto et al. 1996). A number of boreholes has been equipped with inclinometric tubes, Casagrande piezometers and wire extensometers. A climatic station, consisting of a tipping bucket raingauge, an air thermometer and an ultrasonic snow gauge (giving the snow cover depth), has been also connected to the recording system. The sensors are linked to four peripheral units which acquire data by using instructions sent via radio from the central unit located in Cortina d'Ampezzo. The system has been also connected with a modem which enables the real-time transmission of the data acquired (Fig. 17.7).

As for the Staulin landslide, a well defined sliding surface was found at 19.5 m depth (Fig. 17.8). The movement rate recorded is of 15.6 mm per year.

As for the Alverà landslide, inclinometric measurements revealed two sliding surfaces at a depth of about 5 m and 23 m (Fig. 17.8). The movement rate recorded is of 182.5 mm per year.

As regards the Rio Roncatto landslide a principal sliding surface has been recognised at a depth of 10.5 m (Fig. 17.8). The average rate of movement recorded is 2.55 m per year for the recorded period. Another sliding surface has been found at a depth of 7.5 m. The life of the inclinometric tubes has been quite short due to the significant displacement occurred. Nevertheless, the data obtained afterwards from the wire extensometer has shown continuous deformation without significant variation in velocity. This has been confirmed by repeated topographic surveys (Angeli et al. 1996).

A good correlation between groundwater levels and displacements has been obtained, especially for the Alverà landslide to which a visco-plastic rheological model, capable of simulating the movement rate on the basis of the recorded groundwater levels, has been applied. The model, appropriately calibrated by means of the large

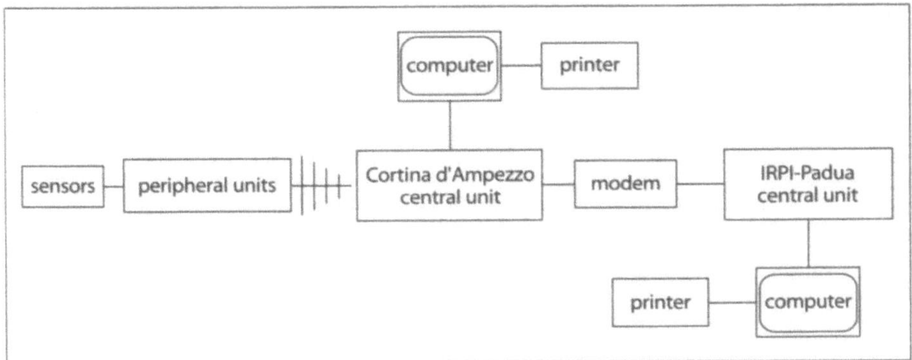

Fig. 17.7. Configuration of the automatic monitoring system installed at the Alverà and Staulin landslides

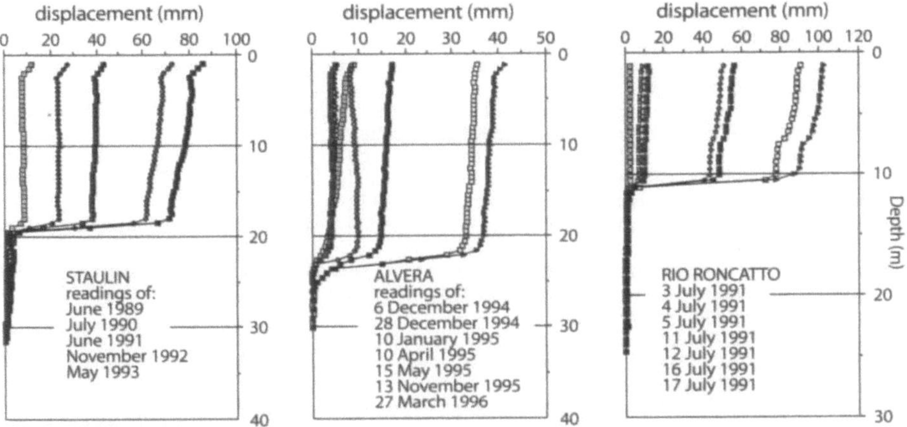

Fig. 17.8. Inclinometric curves of the Staulin, Alverà and Rio Roncatto landslides

number of data collected, gives an explanation of the relationship between groundwater level and the rate of movements of the landslide. The validity of the model has been verified on the basis of a four year record of displacements and groundwater levels.

17.5
Conclusions

Landslide investigations in the area of Cortina d'Ampezzo have been carried out on a multi-disciplinary basis, involving especially experts in geology, geomorphology and engineering. This enabled the researchers not only to achieve an outline of landslide causes, occurrence and evolution in the study area but also to assess landslide hazard and therefore, plan and install monitoring systems on the slopes which are more prone to landsliding with the aim of forecasting future movements.

References

Angeli MG, Gasparetto P, Menotti RM, Pasuto A, Silvano S (1996) Examples of mudslides on low-gradient clayey slopes. In: Senneset K (ed) Landslides. Proc. VII ISL, Trondheim, 17–21 June. Balkema Rotterdam 1:141–145

Brabb EE (1991) The world landslide problem. Episodes 14(1):52–61

Burton I, Kates RW, White GF (1978) The environment as hazard. Oxford University Press, New York

Casale R, Fantechi R, Flageollet JC (eds) (1994) Temporal occurrence and forecasting of landslides in the European Community – Final Report. European Commission, Bruxelles, 957 pp

Corominas J, Weiss EEJ, Van Steijn H, Moya J (1994) Use of dating techniques to assess landslide frequency, exemplified by case studies from european countries. In: Casale R, Fantechi R, Flageollet JC (eds) Temporal occurrence and forecasting of landslides in the European Community – Final Report. European Commission, Bruxelles, 1, pp 71–93

De Zanche V, Gianolla P, Mietto P, Siorpaes C, Vail PR (1993) Triassic sequence stratigraphy in the Dolomites (Italy). Mem Sci Geol 45:1–27

Deganutti A M, Gasparetto P (1992) Some aspects of mudslide in Cortina, Italy. In: Bell DH (ed) Landslides. Proc. 6th ISL, Christchurch (NZ). Balkema Rotterdam 1:373–379

Eisbacher GH, Clague JJ (1984) Destructive mass movements in high mountains: Hazard and Management. Geological Survey of Canada, Paper 84–16, 230 pp

Gares PA, Sherman DJ, Nordstrom KF (1994) Geomorphology and natural hazards. Geomorphology 10:1–18

Gasparetto P, Mosselman M, Van Asch TWJ (1996) The mobility of the Alverà landslide (Cortina d'Ampezzo, Italy). Geomorph 15(3–4):327–335

Hansen A (1984) Landslide hazard analysis. In: Brunsden D, Prior DB (eds) Slope instability. Wiley & Sons, New York, pp 523–602

Jones DKC (1992) Landslide hazard assessment in the context of development. In: McCall GJH, Laming DJC, Scott SC (eds) Geohazards. Chapman & Hall, London, pp 117–141

Mantovani F, Pasuto A, Silvano S (1994) Research in the Alpago area. In: Casale R, Fantechi R, Flageollet JC (eds) Temporal occurrence and forecasting of landslides in the European Community – Final Report. European Commission, Bruxelles 2:769–793

Panizza M (1973) Glacio pressure implications in the production of landslides in the dolomitic area. Geol. Appl. e Idrogeol. 8(1):289–297

Panizza M (1990a) Geomorfologia applicata al rischio e all'impatto ambientali. Un esempio nelle Dolomiti (Italia). In: Gutiérrez M, Peña JL, Lozano MV (eds) Actas 1 Reunión Nacional de Geomorfología, 17–20 Septiembre 1990, Teruel, pp 1–16

Panizza M (1990b) The landslides in Cortina d'Ampezzo (Dolomites, Italy). In: Cancelli A (ed) ALPS 90–6th ICFL, Switzerland-Austria-Italy, Aug. 31st–Sept. 12th, Conference Proceedings. Università degli Studi di Milano, pp 55–63

Panizza M, Zardini R (1986) La frana su cui sorge Cortina d'Ampezzo (Dolomiti, Italia). Mem. Sci. Geol. 38:415–426

Panizza M, Pasuto A, Silvano S, Soldati M (1996a) Landsliding during the Holocene in the Cortina d'Ampezzo Region, Italian Dolomites. In: Frenzel B, Matthews JA, Gläser B, Weiß MM (eds) Rapid mass movement as climatic evidence for Holocene times. Paläoklimaforschung/Palaeoclimate Research, 19

Panizza M, Pasuto A, Silvano S, Soldati M (1996b) Temporal occurrence and activity of landslides in the area of Cortina d'Ampezzo (Dolomites, Italy). Geomorphology 15(3–4):311–326

Pasuto A, Silvano S, Soldati M, Tecca PR (1994) Research on deep-seated gravitational deformations in North-eastern Italy. In: Crescenti U, Dramis F, Prestininzi A, Sorriso-Valvo M (eds) Deep-seated gravitational deformations and large-scale landslides in Italy. Special volume for the International Congress IAEG, Lisboa (sept. 1994), Dipartimento di Scienze, Storia dell'Architettura e Restauro – Univ. Pescara, pp 3–27

Pellegrini GB, Surian N (1996) Geomorphological study of the Fadalto landslide. Venetian Prealps, Italy. Geomorphology 15(3–4):337–350

Soldati M (1996) Landslides in the European Union. Geomorphology, Special Issue 15(3–4):183–364

Soldati M, Pasuto A (1991) Some cases of deep-seated gravitational deformations in the area of Cortina d'Ampezzo (Dolomites). Implications in environmental risk assessment. In: Panizza M, Soldati M, Coltellacci MM (eds) European experimental course on applied geomorphology. vol 2 – Proceedings. Istituto di Geologia, Università degli Studi di Modena, pp 91–104

Starkel L (1985) The reflection of the Holocene climatic variations in the slope and fluvial deposits and forms in the European mountains. Ecologia Mediterranea 11(1):91–97

Varnes DJ (1984) Landslide hazard zonation: a review of principles and practice. Unesco, Paris, 48 pp

Chapter 18

Monitoring and Warning Systems: Methodological Approach and Case Studies

M.-G. Angeli · A. Pasuto · S. Silvano

18.1
Introduction

Hydrogeological disarrangement is one of the most destructive natural events which every year strikes civil populations, urban settlements and infrastructures world-wide, causing thousands of casualties and remarkable damage.

The direct observation and monitoring of natural phenomena has acquired great importance for the scientific community in the past few years since human activities have started to affect wider and wider areas, often subject to hydrogeological risk.

Therefore, among the initiatives planned by the United Nations within the framework of the International Decade for the Mitigation of Natural Catastrophes, those concerning hydrogeological risk occupy a very important place.

Several research projects have been enterprised with the purpose of reducing the loss of human lives and assets. The identification of areas directly or indirectly involved in these events and the understanding of the mechanisms which govern them should help in defining adequate prevention measures for the mitigation of their effects.

18.2
Methodological Aspects

Mass movements show great variability not only from the typological but also from the kinematic and geometrical standpoints. Each landslide is therefore, characterised by its own evolution history and this is necessarily reflected also in the kind of sensors which should be set up, as well as in the number and location of measurement points, on the sampling frequency of the parameters etc. As a consequence, the standardisation of intervention methods and of the sensors which should be used is a rather difficult choice.

The instrumentation scheme should therefore be based on in-depth geological and morphological knowledge of the site to be investigated, besides an accurate back-analysis of all available data and historical and bibliographical information.

A monitoring system which is derived neither from a preliminary analysis of the phenomena nor from the result of logical deductions but is only the expression of a sum of technologies, will not lead to significant results. Similarly, a set of sensors haphazardly placed inside and along the boundary of a landslide body will produce a series of not easily comparable measurements. Hence the importance of correctly planning a monitoring system, in order to obtain sufficient and reliable data when it is necessary to alert the population, define the risk level synd modelling or stabilisation measures.

It is in any case desirable for data to be obtained both manually and automatically (Angeli et al. 1988). Indeed, operator-controlled data acquisition allows the main characteristics of a landslide to be identified and therefore, is the first step for those who wish to take advantage of automatic control and alert instruments. Moreover, it is advisable to choose easy instruments that could be operated also by non-specialised personnel during the first stages of an emergency.

Moreover, particular attention should be paid to the choice of adequate data sampling intervals, since this is of strategic importance in the delicate phase of data analysis and interpretation. If on the one hand a short sampling interval is desirable, on the other it implies the management of huge quantities of information. On the contrary, in many cases too long a sampling interval hinders the acquisition of important data (Angeli et al. 1989, 1992).

Apart from the methodological indications, monitoring should essentially represent an effective means for understanding mass movement mechanisms and, as such, it should lead to the correct analysis and interpretation of these events.

18.3
Technical Aspects

Before reviewing the various types of instruments, it would be useful to outline the goals which can be achieved by means of a monitoring system:

- First of all, the identification of the movements before the appearance of clear morphological signs at the surface;
- The possibility to define with precision the buried geometry of the rock mass in movement and therefore, identify its kinematic mechanisms (velocity, acceleration, etc.) and their possible correlations with hydrological parameters;
- The possibility to provide a continuous and possibly safe surveillance;
- Finally, the safeguard of public safety.

The three most important physical parameters which should be monitored in the study of a slope subject to landsliding are:

- Kinematic parameters;
- Hydrological parameters;
- Climatic parameters.

18.3.1
Kinematic Parameters

Within this category all those magnitudes which help considerably in assessing the degree of activity of a mass movement are included; basically they correspond to deformations, in both space and time, at the surface and in depth.

First of all the evaluation of these magnitudes will allow the geometry of the landslide to be defined, as well as, the amount and direction of the movement.

The following is a list of the most suitable sensors for this purpose:

- Surface deformation:
  - Crack gauges;
  - Tiltmeter;

 - Multi-point liquid level gauges;
 - Wire extensometers;
 - Automatic topography system.
- Subsurface deformation:
  - Inclinometers;
  - Shear plane indicators;
  - In-place inclinometers;
  - Multiple deflectometers;
  - Acoustic emission monitoring;
  - Wire extensometers;
  - Deflectometric rods (lowered);
  - Electrolytic bubble sensor;
  - Coaxial electrical cable;
  - Optic fiber cables.

18.3.2
Hydrological Parameters

The survey of these magnitudes becomes particularly important when correlated with kinematic parameters concerning a mass movement. The fundamental datum to be considered is given by pore-water pressure affecting the slope which should be measured by means of adequate sensors, such as the piezometers here listed:

- Open standpipe piezometer;
- Vibrating wire piezometer;
- Pneumatic piezometer;
- Multi-point piezometer.

Among hydrological parameters, also those characterising surface hydric circulation, such as rillwash, percolation etc. should be considered. For this purpose the following sensors may be used:

- Bucket and stopwatch;
- Water level behind weir;
- Pipe flowmeter.

18.3.3
Environmental Parameters

The environmental conditions, which are very important when compared with the other two types of parameters, may be surveyed by means of the following instruments:

- Pluviometer;
- Thermometer;
- Barometer;
- Anemometer;
- Snow depth gauge.

As regards the systems which function also as alarms, it will be necessary to make use of other types of sensors and instruments which can vary according to the specific problem. For this purpose, it is opportune to illustrate, although in a summarised way, the experiences arising in some case studies of monitored landslides, regarding both a possible alarm and also ordinary research activity.

18.4
Case Studies

The experiences carried out on three mass movements showing high risk for the population are presented. The cases considered vary both in the geological domain and in the typology. As a consequence, also the final aims are shown to be quite different.

18.4.1
The Ru Delle Roe Landslide

On the 31st January 1991 in the municipality of Zoldo Alto (Italian Eastern Alps), in an area already known for its precarious stability conditions a mass movement took place which, owing to its rapid evolution determined an alarm situation and well defined hazard conditions for the underlying village of Molin (Angeli et al. 1992; Govi et al. 1993).

The landslide affected the right flank of the Ru delle Roe stream, which stretches for a length of 2 km and is surrounded by the steep dolomitic slopes of Cime di San Sebastiano and Moiazza, whose toes are characterised by the exposure of the Raibl Formation units.

The rock types involved in the movement belong to the La Valle Formation which in this area is made up of fine-grained sandstones and thin levels of argillites and ash tuffs, sometimes completely argillified. The formation is deeply fractured with extremely variable attitudes.

From the structural viewpoint, the area is characterised by several NNE-SSW trending faults, linked to the Valsugana and Pieve di Cadore lineations, which are observable respectively near Agordo and in Val Zoldana.

The movement, ascribable to a rotational slump, affected a 80 000 m^2 area and occurred during a very rigid climatic phase (night temperature around $-15°$ C, $-20°$ C), coinciding also with abundant snowfall. The volume of material mobilised was reckoned at about 1.5×10^6 m^3.

During the first days the displacements were about 10 m/d. The movement continued consistently during the first two months and later decreased, showing a trend correlated to various meteorological parameters; in particular: snow melt and precipitation.

The cause of the failure, as corroborated also by direct observations, was attributed to the freezing of spring water, which by hindering the groundwater flow would cause a rise of the piezometric level within the landslide, sufficient to overcome residual strength.

On the basis of the displacements measured, the evolution hypotheses foresaw also the possibility of a sudden collapse of the whole fluidified rock body which would be channelised in the Ru delle Roe Riverbed as a debris flow (Fig. 18.1). For this reason it was necessary to plan a control system of the landslide area, in order to identify critical phases possibly well in advance.

Besides visual control of the landslide body, a monitoring system was set up in collaboration with regional boards, in order to record changes of activity or passage of

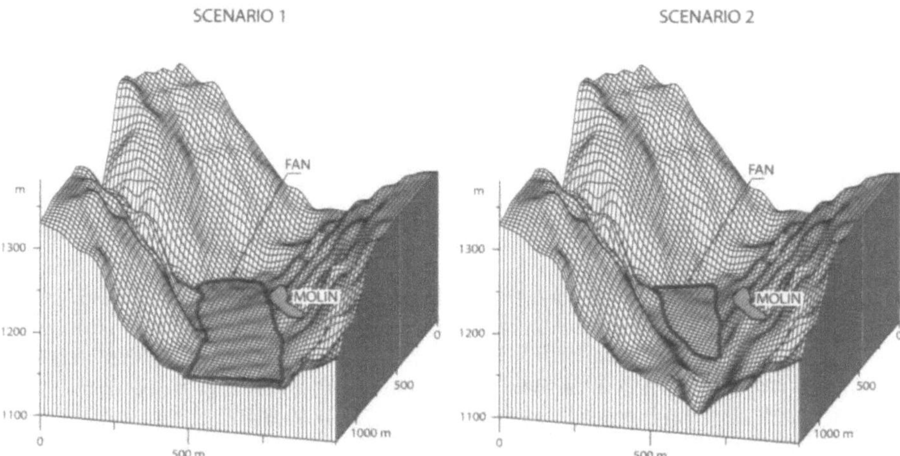

Fig. 18.1. Hypotesis on the evolution of the area in front of the Molin village in case of sudden flow of the landslide. *Scenario 1*: built up of a fan with slope inclination of 4% and a volume of 2.91 million of cubic meters, overflow level 1172 m a.s.l.. *Scenario 2*: built up of a 10% hanging fan with a volume of 0.341 million cubic meters, overflow level 1165 m a.s.l.

debris flow within the Ru delle Roe riverbed. The system consisted of:

- Three geophones connected to a recording system, in order to assess intensity and frequency of rock noises correlated to landslide activity;
- A climatic station for the measurement of rain and snow fall and temperature;
- Two rivergauges, installed along the Ru delle Roe stream, respectively upstream and downstream of the landslide, in order to measure flow rate differences correlated with a possible damming of the watercourse;
- Two inclinometric rods, with function of alarm in case of debris flow, situated along the watercourse just downstream of the landslide;
- A bench-mark system for measuring topographic features (measurements were carried out twice a day during the first phase) and air-photogrammetric filming.

At present the studies carried out and the data collected exclude the possibility of a sudden collapse of the whole unstable mass, but point to an evolution of the phenomenon by means of successive flows whose dimensions should not create problems for the inhabited centres affected.

These investigations are an example of monitoring carried out in order to define landslide risk.

18.4.2
The Tessina Landslide

This landslide was reactivated on the 17th April 1992 causing a situation of serious hazard for the inhabited centres of Funes and Lamosano in the municipality of Chies d'Alpago (Italian Eastern Alps). The landslide is a complex phenomenon made up of a vast rotational slump in the upper part which, in the mid-lower portion, evolves into a 2 km long flow. This disarrangement, known since 1960, has undergone several reactivations in the following years, with a progressive widening of the displaced area (Angeli et al. 1994).

From the geological viewpoint the event affected the Flysch formation (Middle Eocene), made up of alternances of low-permeability 1 000–1 200 m thick marly-clayey and calcarenitic layers. In some points this formation is buried by Quaternary talus fans and Würm moraine deposits of the river Piave glacier and other small local glaciers.

The collapsed sector occupies a 40 000 m² wide area, on the left hand-side of the Tessina stream, with an approximate volume of 1 million m³. The movement corresponds to a rotational slump with a 20–30 m deep failure surface, affecting also the flyschoid bedrock. Initially it caused the formation of a 15 m high scarp and a 100 m displacement downstream with consequent disarrangement of all the unstable mass and destruction of the drainage systems set up some years earlier.

The movements in this area continued with a certain intensity up to June, causing the mobilisation of some other 30 000 m², with a total volume displaced of about 2 million m³.

The material from this area, which is intensely fractured and dismembered, was channelised along the riverbed where, owing to its continuous remoulding and increase of water content it was more and more fluidified, giving rise to small earth flows converging into the main flow body.

As a remedial measure, the Civil Engineers Board of the Belluno Province constructed three embankments for the defence of the inhabited centres of Funes, Tarcogna and Lamosano, respectively. Moreover, in the proximity of Lamosano, an experimental plant, made up of a reinforced concrete raft, was built. This structure was provided with several nozzles capable of ejecting high-pressure water in order to fluidify the flow material and avoid its accumulation and consequent damming of the riverbed.

The planned and installed monitoring instruments (Fig. 18.2) consist of two multiple-based wire extensometric systems, 280 and 390 m long respectively, in order to continually measure the surface movements affecting the landslide's upper portion. They consist of a series of 12 measurement exchanges which, by means of an adequate demultiplication system can appreciate displacements as small as 1 mm.

In the landslide's upper portion, also a topographic system with automatic surveying of the benchmarks for the measurement of surface deformation was installed.

In the main flow body, upstream of the Funes and Lamosano hamlets, two control sections were set up and equipped respectively with three directional rods, one ultrasound echometer and two rods and one echometer, with exclusive alarm function.

Besides the above described systems, three televison cameras were also installed, for the recording and visual control of the movements in correspondence with the most critical areas, which are the upper accumulation and the two zones upstream of Funes and Lamosano, respectively.

This system is connected with a central station situated in the premises of the Town Hall whilst another one is placed in the headquarters of the Belluno Fire Brigade which, if alerted, must be prepared to intervene.

This intervention is an example of monitoring carried out with the aim of alerting the civil population.

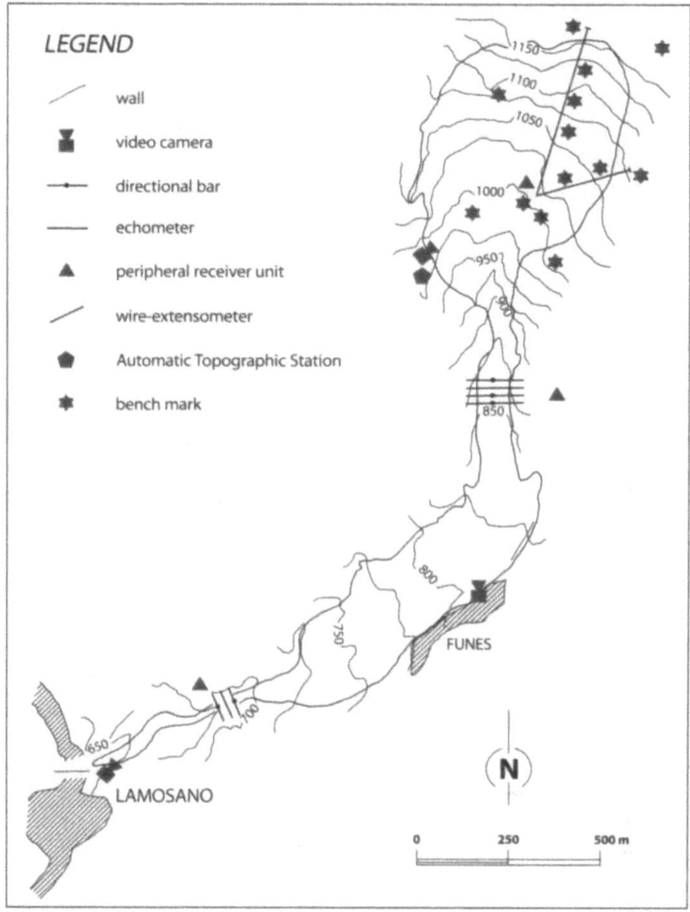

Fig. 18.2. The Tessina landslide alarm system (Chies d'Alpago – Belluno)

18.4.3
The Staulin-Alverà Landslide

This case study differs substantially from those so far examined. In fact, the investigations carried out in the Cortina d'Ampezzo area are not finalised to the safeguard of the population but only to the identification of the evolution mechanisms of landslides and the reliability of the instruments installed. Therefore, this is an important sample area equipped for testing different types of sensors and other instruments for automatic data acquisition.

In the Cortina d'Ampezzo valley the Middle-Upper Triassic series, stretching from the Cassiano Formation up to the Dolomia Principale crops out, whereas on top of the highest peaks also Jurassic formations are exposed. This alternance of ductile and brittle rock types has developed morphological conditions favourable to both the progressive urban development of the valley and the evolution of several mass movements of various kinds, some of which are particularly large and still active. With respect to these phenomena, radiometric dating has identified two main activity periods: one between 10 000 and 9 000 and another between 5 000 and 4 000 years B.P.

Among the numerous landslides still active, present investigations are concentrated on those of Staulin and Alverà, located on the hydrographic left hand-side of the Boite stream and covering a surface of 40 and 10 ha respectively.

The mass movements may be defined as mudslides and are made up of two different bodies moving with different velocities, ranging from a few millimetres to some tens of centimetres per year.

The material affected by landslides belong to the Cassiano Formation, which is prevalently made up of fossil-rich marls and argillites derived from basin deposits and intercalated with grey sandstones and calcarenites with yellowish alteration levels. In agreement with the values of the Atterberg limits, the material is classified as middle-high compressibility clay ($W_1 = 55\%$, $W_p = 32\%$). Laboratory permeability tests produced values of 10^{-10} to 10^{-7} m/s. Nevertheless, the presence of several cracks filled with calcite, inside the rock mass, point to a much higher original permeability. Shear strength tests performed on reconstructed samples gave ϕ_r values of 20° to 23°. Finally, seismic and electric soundings and repeated inclinometric measurements identified a 5 m deep main failure surface and a secondary one at a depth of about 23 m.

The monitoring system installed in 1989, consists at present of a set of inclinometric pipes and Casagrande piezometers equipped with electric sensors in order to measure the water table fluctuations. Moreover, each measurement hole is equipped with a wire extensometer for the measurement of displacements at depth whereas, a network of benchmarks for the topographic control allows surface movements to be recognised. Also a meteorological station, equipped for the collection of precipitation (both rain

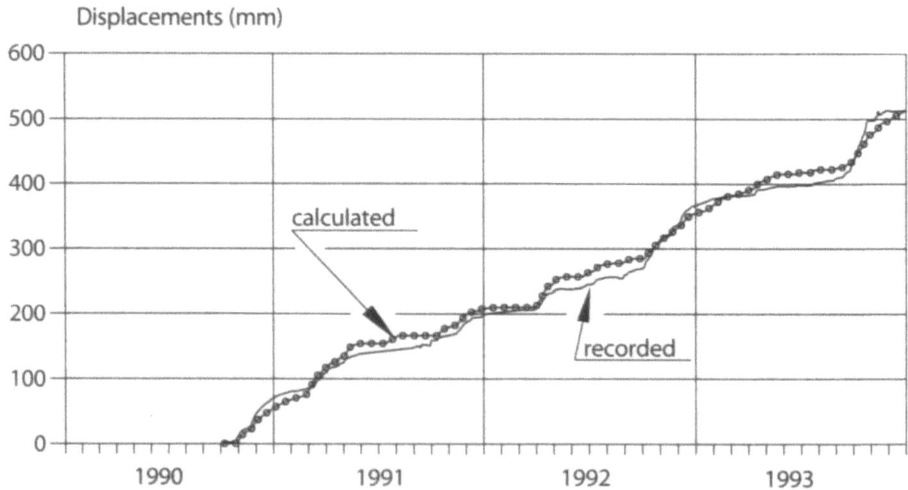

Fig. 18.3. Comparison between the measured and calculated displacements in the Alverà landslide (Cortina d'Ampezzo – Belluno)

and snow-equivalent) and air temperature data, as well as the thickness of the snow cover, is active on the site. The whole instrumentation is connected to peripheric acquisition exchanges, in turn linked to a main information centre which at regular intervals reads and stores the data collected. One of the most interesting characteristics of the whole system is the possibility of automatically changing the sampling frequency when the range of variation of the magnitudes measured reaches predetermined threshold values. This allows the peak values of the magnitudes measured during critical events to be recorded.

The great amount of data collected in the past few years with the instruments installed has led to the formulation of a mathematical model (Angeli et al. 1996) through which it is possible to obtain, starting from the piezometric level measurements, a theoretic curve of the displacements quite superimposable on the measured one (Fig. 18.3), thus helping in defining to a certain extent the evolution of the event.

This research is therefore, an example of monitoring carried out in order to develop a model of the phenomenon.

References

Angeli M-G, Gasparetto P, Silvano S, Tonnetti G (1988) An automatic recording system to detect the critical stability condition in slopes. In: Bonnard C (ed) Landslides. Proc. 5th ISL, Lausanne. Balkema Rotterdam 1:375–378
Angeli M-G, Gasparetto P, Pasuto A, Silvano S (1989) Examples of landslide instrumentation (Italy).. Proc. XII ICSMFE, Balkema Rotterdam 3:1531–1534
Angeli MG, Menotti RM, Pasuto A, Silvano S (1992) Landslides studies in the Eastern Dolomites Mountains (Italy). In: Bell DH (ed) Landslides. Proc. 6th ISL, Christchurch. Balkema Rotterdam 1:275–283
Angeli MG, Gasparetto P, Menotti RM, Pasuto A, Silvano S (1994) A system of monitoring and warning in a complex landslide in Northeastern Italy. Landslide News 8:12–15
Govi M, Pasuto A, Silvano S, Siorpaes C (1993) An example of low-temperature-triggered landslide. Eng. Geol. 36:53–65

Chapter 19

Mass Movements in Austria

B. Bauer

19.1
Introduction

Mass movements are part of the normal landscape evolution. The two antagonistic processes:

1. *Endogenic build-up* and
2. *exogenic denudation*

shape the surface forms of the earth.

Both are impermanent; their work is in stages of higher and lower energy distribution.

Exogenic processes are generally stochastic. But in the field of mass movement their orientation structure is strongly connected with tectonic features. These are expressed in the formation of joints and fissures which are the result of the neotectonic stress field. This joint orientation determines to a great extent then the larger landscape features like valley systems, surface mass movement etc. An inherent predesign exists that will be the guideline for exogenic forces then (Scheidegger 1991).

Austria is prone to have a lot of mass movements because of:

1. The high relative relief of most of the area which was glaciated in the Pleistocene;
2. A sequence of different rock types that are prone to movement especially when wetted;
3. Neotectonic movement of most of the area.

About 100 000 mass movements of different size, typology and age may be counted in Austria.

The distribution of landslides corresponds widely to the petrographic zones of the Alps. Most landslides lie in the calcareous Alps (northern and southern zone) due to the wider spacing of fractures and bedding planes. Fewer landslides and smaller features are to be found in the Old Massifs. In the crystalline Central Alps there are more landslides than in Old Massifs though the rocks are similar. Their size is less than in the calcareous Alps though. The big exception is the landslide of Köfels in the Crystalline Alps which is the 2nd largest in the Crystalline Alps and as big as all the others of this zone together (Abele 1974).

A statistical comparison demonstrates that the extent and morphology of landslides is mainly influenced by petrography, volume, vertical drop and pre-existing relief of the deposition area. The size mainly depends on petrographic quality. Usually there are more and bigger landslides in limestones than in crystalline rocks. The exception being the Köfels slide mentioned above.

I would like to take up three examples of Austrian landslides and demonstrate by them some of the mechanisms involved, damage done and possible reactions. This will be the Dobratsch limestone landslide, Bad Goisern a complex rockfall/topple- mudslide event and Köfels the big crystalline landslide/rockfall which gave rise to all sorts of discussion concerning the release incident.

19.2
Dobratsch (Villacher Alpe) Landslide

It is of interest because the release of the landslide is known to have been an earthquake, well documented by historical sources and we could differentiate two events an older postglacial/prehistoric and a younger at 1348 (see Fig. 19.1).

The Dobratsch is situated between Gail and Drau River close at their confluence in the Austrian province of Carinthia. It is a well defined mountain block bordered at all sides by tectonic fault lines that are visible in the valleys bordering it. The main lithological strata are Triassic limestones and dolomites dipping gently to the east and north underlain by sandstones and shists.

The strata show many joints and fractures that are more or less vertical. Some fissures and crevices are meters in width and may reach 10 m depth. They may be seen close to the edge of the plateau and are of course signs of tectonic stress of the mountain block.

The two events (there have been probably more but the glaciation has removed proofs!) are explained as follows. Due to deglaciation the vertically standing rock walls of Villacher Alpe suffered stress release fracturing parallel to the wall and weakening by weathering processes might have prepared the area for a landslide. Though we do not know the triggering factor for the "old" prehistoric landslide event we assume it was the same as for the historic one- an earthquake.

This region of Austria (Gail valley) is earthquake prone like the adjacent region of Italy. Many historic documents in monasteries and of course recent seismic measurements give proof of this. Such an earthquake shock will exert its energy on the different fissure planes that are nearly at right angle to the shock waves that come from the Gail valley. The first postglacial/prehistoric landslide triggered by the earthquake must have hit an area that was still covered with a lot of morainic debris because the landslide mass is mixed with striated morainic gravel. This lead former researchers to the conclusion that the first landslide fell on the retreating glacier. Only recently a piece of wood was found in the masses which could be dated and gave an age of 7 700 B.P.

The scars of the source area of this first slide at the top of the mountain are already rounded and weathered today (Till 1907).

The sedimentation area covers about 24 km² with a thickness of about 30–40 m. The length of the source area is about 10 km along the Gail valley and the deposits are still found upslope the southern Gail valley border to 640 m elevation.

The overall volume was estimated to be 0.535 km³ (535 million m³).

As a comparison Flims stands with 52 km² area and 15 km³ volume, 600 m thickness.

The angle of the slide (from top-end of the deposit) is 14–16°. (Flims-6-8).

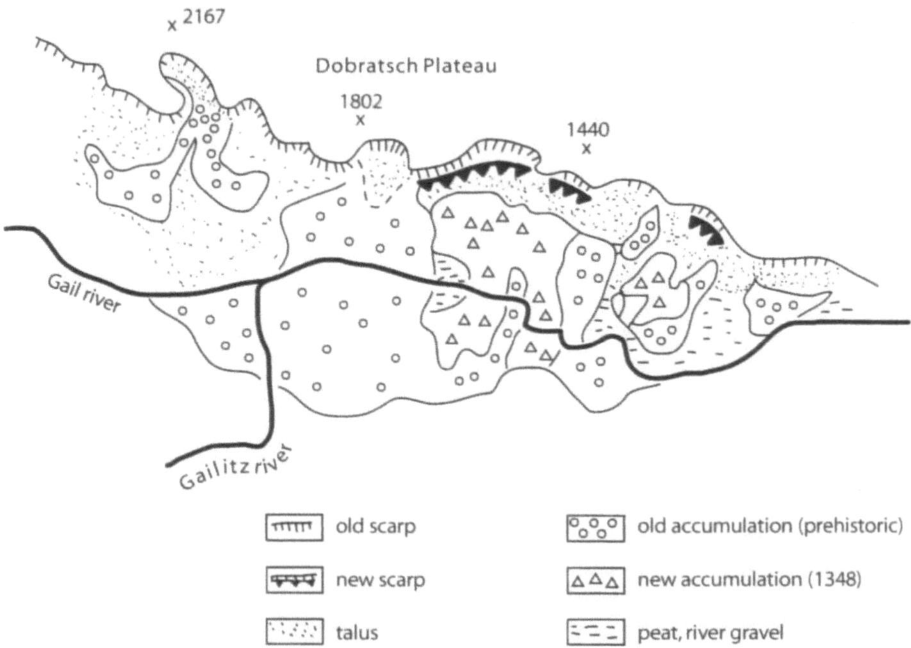

Fig. 19.1. The Dobratsch landslide from 1348 and an postglacial/prehistoric event

If we look at some specific localities along the 10 km length of the slide area we also find angles of 25° (Rote Wand e.g.).

The new historic landslide that happened 1348, was much smaller and covered some areas of the older one with its deposit. We have good records of that because of many written sources in monasteries and churches.

The deposit area is 7 km², the average thickness is only 5 m, 30 million m³ volume is estimated and the slope angle is between 18–22°.

The two events may be separated because the younger one has scars in the source area which are much fresher, not so weathered and has a sharper edge at the plateau. The sediments in the deposition area are rather unweathered blocks not yet covered by a thicker humus layer or soil development. Also the river terrace gravels do not cover it as they do the older masses.

The consequence of the blocking of the Gail valley with the mass deposits was of course a damming up of the river Gail which resulted later in flooding downstream and the creation of an erosion gully through the landslide masses of 20 m depth.

Most historical written sources mention water and fire as the main destructive forces and not the landslide masses. The reason may be that we do not have records from the landslide destroyed settlements and the ones that were farther away suffered from the earthquake (fire) and flooding (when the dammed up Gail River broke through). The interesting phenomenon of lineaments that run through the old and new landslides deposits (easily seen in air photos) prove that the Gail valley area is a very active tectonic zone still today and new landslides will endanger the area.

19.3
Bad Goisern – Zwerchwand Rockfall/Topple and Stambach Mud Slide

The whole landslide comprises an area of 32 ha and the mudslide alone destroyed a forest area of 6 ha valuable forest growth. Around 10 million m^3 of rock volume is involved while the mudslide itself that was in movement was about 300 000 m^3 (Schäfer 1983) (see Fig. 19.2).

Stambach valley is situated at the southern part of a giant anticlinal system consisting of rocks of the Hallstädter nappe (Triassic limestones, dolomites and halites-evaporites) with a thickness of about 2 000 m.

Along the southern wing of the anticline the strongly deformed and easily soluble, "Haselgebirge" reaches the surface. This characteristic formation locally named Haselgebirge (consisting of a series of lower Mesozoic halites, shales, anhydrites and other evaporites) when in contact with water becomes highly unstable and moves easily. The overlying carbonate blocks (upper Triassic) that kind of "float" on it undergo noticeable displacements. Many rockfalls and landslides are caused by these two factors:

1. The lithological situation described above and
2. neotectonic movements of the Hallstadt nappe.

Zwerchwand (Triassic limestone wall) rockfalls coincide with typical joint and fracture zones and is definitely tectonic enforced. The geomorphologic system in this region at least since deglaciation always worked like this: After a collapse of a part of a limestone wall the underlying, "Haselgebirge" became mobile leading to huge clay slides/flows. The initial movement is tectonically enforced.

The oldest proven mass movement must have happened after the deglaciation of the Traun valley (after 17 000 B.P.). This is proven by old halites and marls that are located about 1 300 m away from its origin and at their borders peat bogs have been dated.

The "Haselgebirge" and the old mudslide was mobilised again when in the 20th century new rockfalls forced it to move again. Most prominent movements took place from 1978 onwards. The mass transport was about 380 m in the upper part of the slide and about 180 m in the lower part. So the movement became slower downhill. It stopped at a rock barrier that was uplifted at a tectonic line. Looking at a map of lineaments we find a correspondence of tectonic lines with the rock wall above and the edge

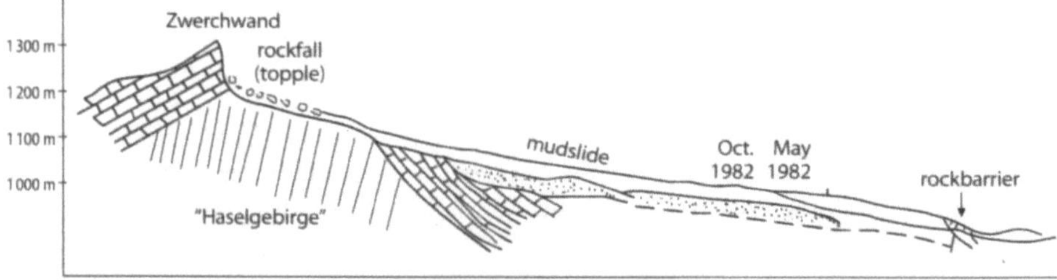

Fig. 19.2. Zwerchwand rockfalls with "Haselgebirge" and old mudslide

of the mass movement. By seismic sounding the depth of the slide was found to be 45 m (30–40 m average); the highest speed was up to 120 m/d.

Hydrology. The marls which are usually aquicludes show signs of doline growth therefore, one might conclude again that this area is in tectonic tensional stress and defragmentation.

Due to the rather loose consistency of the landslide material and the many new fissures caused by the movement a large amount of water is able to infiltrate this area. Some water intakes that were measured took between 5–20 l/s. Even the lower value of 5 l/s would amount to about 150 000 m³ per year. For the whole area (32 ha) this would be a water level of about 50 cm. But probably 2–3 times as much is the reality. The mechanical properties of the rocks will definitely be altered by that large amount of water drastically. Seismic refraction analyses has proven a dramatic worsening of the physical properties of these already stressed and loosened rocks. The especially soft and weak stratum was lowered to 10–15 m in two years (after the slides of 1981/82).

To cope with the mudslide that endangers a village downslope the main important task is to remove the water from the sliding area. This was done by a series of pipes and by biological countermeasures. Vegetation was planted on the slide area that uses a large amount of water. Of course one tried not to use plants like trees that would exert a heavy weight on the area when maturing but instead bushes like willow and alder which will be cut every few years and regrow from the sprouts.

The displacements and the slide are evidently gravitational and hydrologically induced. But the breakoff of the limestone wall on top is definitely due to tectonic young movements which was also proven by in-situ stress measurements in 1984.

19.4
Köfels Landslide

The Köfels landslide is known to be the largest landslide in the crystalline Alps. It covers an area about 13 km² and has the form of a semi-circular basin of about 4 km width. The original volume of displaced rocks is estimated to have been 2.1–2.5 km³, sliding a distance of several hundred metres in less than a minute reaching velocities of 50 m/s (Erismann 1977). The event happened when a large part of the Fundus crest (the western, left side mountain ridge of the Ötz valley) collapsed (Fig. 19.3 and Fig. 19.4).

This is easily recognisable geomorphologically but also petrographically because the landslide deposits consist mainly of augengneiss which builds up the Fundus crest

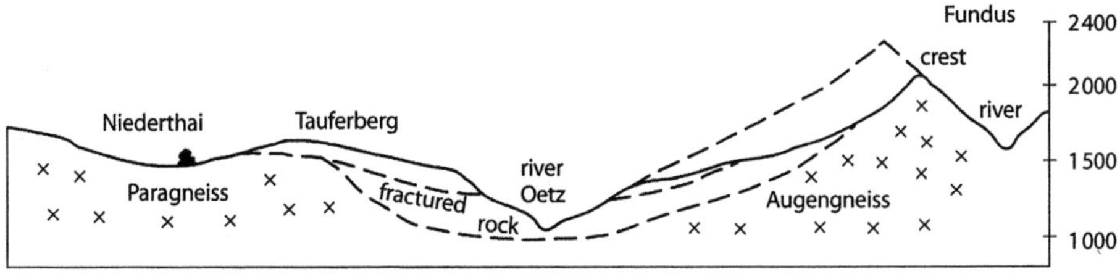

Fig. 19.3. The Köfels landslide. At this event a large part of the Fundus crest collapsed

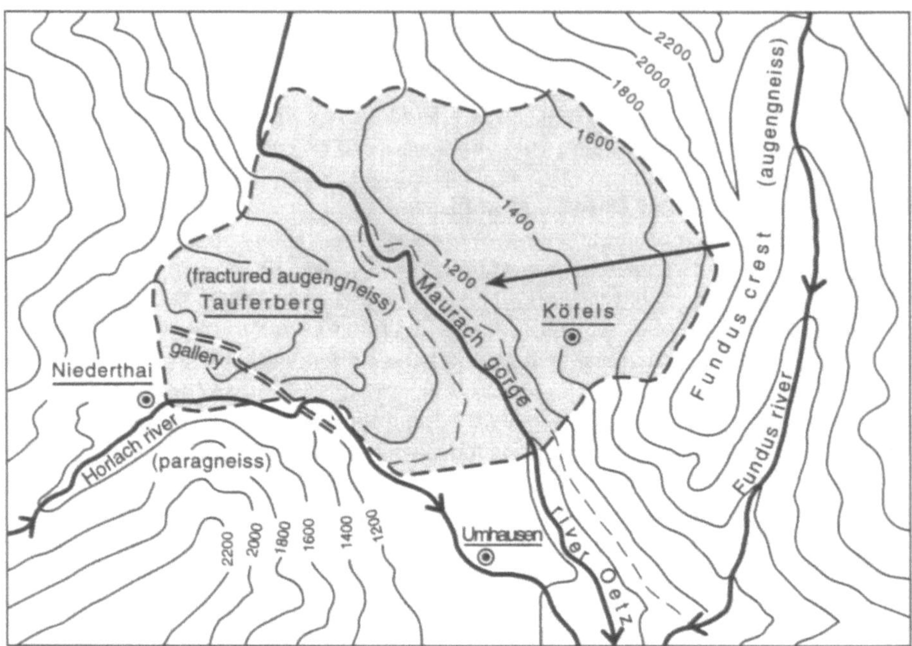

Fig. 19.4. Map of the area of the Köfels landslide

and vanishes to the east and west. The broken crest must have been about 400 m higher than today and situated farther to the east (closer to the Ötz valley). The primary (basal) sliding surface is easily recognisable at various locations along the headwall of the Köfels niche (Preuss 1971).

There are slabs with slope angles of 25–30° to be seen along trails from Köfels to Fundus valley. At the foot of the headwall the slide surface disappears below the landslide deposits. Pumice outcrops have not been found at the primary sliding surface but close to the secondary sliding surface north of Köfels. The landslide moved down like a huge sled crashing against the opposite slope of the Ötz valley, splitting into two parts: The lower part remained in the main valley and was heavily shattered and pulverised, as is easily seen along the main road into the Ötz valley.

The upper part continued to move above it and filled the Horlach valley. As a result a secondary slip surface occurred and it is this surface which is visible around Köfels. Close to this the famous pumice have been found. The upper part rests on a high rock terrace. A blockade of the Ötz and Horlach valleys dammed up both rivers (see Ötz valley). The resulting lake deposits are known by coring to be more than 90 m in the Langenfeld area.

Due to the rather low inclination of the sliding angle, large parts of the landslide deposits remained more or less coherent and look like bedrock. Subsequently they have been mapped as bedrock in the geological literature of the past. A tunnel was built in 1951 in 1 200 m a.s.l., on the western slope of the Ötz valley for hydrogeological purposes which revealed the contrast between, "real" bedrock (paragneiss) and the in part fractured landslide deposits (augengneiss). In addition, this tunnel showed the pre-existing now buried gorge of the Horlach valley. The Horlach River was deflected to the north by the landslide and plunges now 150 m over a rock cliff. This waterfall is called Stuiben fall and has eroded 6–7 m into bedrock which is a good mark for erosion when one knows the time span.

In the tunnel, also a strongly deformed wood was found in the shattered landslide sediment. Dating by ^{14}C was 8 710 ±150 B.P.

By this the landslide could be dated and also the erosive force in bedrock which seems to be in this case (Horlach River at Stuiben fall) about 1 m in 1 000 years.

19.4.1
Chronology

Before the tree was found in the tunnel the dating of the slide was complicated by the fact that glacial deposits were lacking on the western side of the Maurach gorge (Köfels area) but are ample on the eastern side (Niederthai-Tauferberg with glacial moraines, erratics, tilted and turned around roches moutonnees with crescentic gouges on the surface pointing in the "wrong" for the Ötz valley glacier-direction!). If one assumes a landslide before the last glacial advance where are the glacial features on the other side?

Heuberger (1966, 1975, 1994) and Aulitzky et al. (1994) explain this:

The glacial features of Tauferberg do not continue beyond the landslide deposits. The idea is that all the glaciated surface from the other side (west) was carried with the landslide to the east and remained at the surface to form the area of Tauferberg today. A comparable event was the landslide Vaiont 1963; also there trees etc. remained on the surface throughout the slide. There are also no signs of periglacial action (solifluction etc.) subsequent to the sliding. The knick point between slope and basin of Niederthai is pretty sharp, This would have been impossible had the slide occurred in late glacial time.

So whereas many other slides in the Alps took place in late glacial time, this slide as well as the Tschirgant slide occurred in the Holocene.

19.4.2
Köfels Pumice

The Köfels event became well known when in 1863 pumice or fused rock was found at the site (Pichler 1863). This raised the question of the source of the energy that would be required to produce such a rock structure (Heuberger et al. 1984; Leroux and Doukhan 1993; Pichler 1863; Preuss 1971). At first, the prevalent hypothesis attributed these energies to a volcanic event. The volcano hypothesis lost support when it was seen by drillings that neither the shattered Maurach barrier rock nor the pumice carried on downwards.

A second idea was a meteorite impact because a rather rapid cooling, necessary for the pumice production, made a volcano hypothesis untenable. In both theories the formation of the pumice relates to the process which triggered the landslide.

Preuss (1971) one of the investigators of the meteorite impact event at the Nördlinger Ries, Bavaria was the first to refute both above theories by comparing the mineralogical data of both events. He could rule out that the two situations resulted from the same kind of event. His idea was that the required energy to produce rock fusion had been generated by the action of the landslide itself.

Studies of deformation of quartz grains (microstructure density and type of dislocation) and the total lack of glassy planar deformation features clearly indicate that deformation at Köfels occurred in a regime characterised by a lower stress than an impact. No evidence of shock metamorphism can be detected. These defect microstructures in the quartz grains of the gneiss wall rock are typical of plastic deformation at high temperature with very efficient recovery. At relatively short distances from this surface (dimensions of cm) the dislocation microstructure is markedly different. It is typical of a deformation regime controlled by lattic friction with no efficient climb processes. These dislocations microstructures demonstrate the very strong temperature gradient which must have occurred during the landsliding. The glassy veins detected in the crystal relicts have nothing to do with the planar deformation features found in naturally shocked quartz. The silica glass particles found in the pumice is mostly irregular and is not identical at all with the straight narrow amorphous lamellae with controlled orientation found in shocked quartz. Landslide processes release their energy much more slowly than impacts in such a way that this energy is dissipated in the form of heat without radiation or shock wave effects. The temperature along the glide surface could have been in excess of 1700° C during about half a minute. Under these conditions the base of the rock slide must melt and most of the mechanical energy generated along the friction zone must be transferred into heat in the surrounding wall rock. Other processes for dissipating the energy (cracking, wave radiation etc. must have been negligible. The fractured gneiss rock involved in the landslide was degassed and amphiboles and biotite where preferentially consumed to form the melt which still contain about 30% glassy quartz and feldspar. The gneiss nearby the shear zone is introduced by the melt along many fractures. Erismann (1977) did experimental work to confirm the calculations and the term "frictionite" was coined for rock modified by landslides for all stages of deformation ranging from fraction to fusion.

19.5
Conclusions

Three different types of landslides have been presented. They differ in size, age, lithology, type of movement and triggering of the event.

The Dobratsch landslide in calcareous rocks in the south of Austria was definitely triggered by an earthquake and we can differentiate two incidences, a postglacial and a historic event. Destructional was not only the massmovement itself but the resulting fires (historic event) and flood that happened after river damming. The Bad Goisern landslide is complex in that a rock topple in the limestone top part triggered a mudslide further downslope. This happened a few times in historic time, the last one took place in the 1980's. Villages below the mudslide are in constant danger. The Köfels landslide is the largest in the whole crystalline Alps and a prolonged discussion in the scientific literature took place about the triggering incidence (volcanic, meteor impact etc.). Also the dating could only recently be clarified when a buried tree was found, This huge landslide blocked a main tributary valley in the Alps and created a lake that lasted many centuries.

References

Abele G (1974) Bergstürze in den Alpen. Wiss. Alpenvereinshefte 25, München

Aulitzky H, Heuberger H Patzelt G (1994) Mountain hazard geomorphology of Tyrol and Vorarlberg, Mountain Research and Development, vol 14, N° 4

Bauer B (1995) Köfels landslide. 4. Intensive Course on Geomorphology, High Alpine Environment, Tirol

Erismann T (1977) Der Bimstein von Köfels, Impactit oder Friktionit? Material und Technik 77(4) Dübendorf, Zürich

Heuberger H (1966) Gletschergeschichtliche Untersuchungen in den Zentralalpen zwischen Sellrain und Ötztal. Wissenschaftliche Alpenvereinshefte, Innsbruck 1966, Heft 20

Heuberger H (1975) Das Ötztal. Bergstürze und alte Gletscherstände. Innsbrucker Geographische Studien, 2

Heuberger H (1994) The Great Landslide of Köfels, Ötztal, Tyrol. Mountain Research and Development, 14, p 290-294.

Heuberger H, Masch L, Preuss E, Schocker A (1984) Quaternary landslides and rock fusion in Central Nepal and in the Tyrolean Alps. Mountain Research and Development, 4

Leroux H, Doukhan H (1993) Dynamic deformation of quartz in the landslide of Köfels, Austria. European Journal of Mineralogy, 5

Pichler A (1863) Zur Geognosie Tirols. Die vulkanischen Reste von Köfels. Jahrbuch der Geologischen Reichstanstalt, 13

Preuss E (1971) Über den Bimstein von Köfels/Tirol. Fortschritte der Mineralogie, 49 Beiheft 1:70

Schafer G (1983) Massenbewegung Stambach-Zwerchwand. In: Arbeitstagung d. Geologischen Bundesanstalt, Wien

Scheidegger A (1991) Mass movements in Austria. In: Bell (ed) Landslides Balkema Rotterdam

Till A (1907) Das große Naturereignis von 1348 und die Bergstürze des Dobratsch, Mitt.d.Geogr.Ges. Wien, Bd. SO, Wien

Landslides and Precipitation: the Event of 4–6th November 1994 in the Piemonte Region, North Italy

S.C. Bandis · G. Delmonaco · F. Dutto · C. Margottini · G. Mortara · S. Serafini · A. Trocciola

20.1
Introduction

The large number of landslides released in the Langhe Hills (Piemonte region – Northern Italy) during the intensive rainfalls of 4–6th November 1994 manifests the degree to which water may affect the stability of natural and man-excavated slopes.

Water may affect the stability of soil and rock slopes in several ways, namely by:

a Altering the stress state through transient or permanent pore pressure generation;
b Adding a driving force through flow (seepage) force development;
c Changing the bulk density of the materials forming the slope;
d Affecting the mechanical properties of the soil or rock masses;
e Altering the slope profile by erosion.

All the above influences are believed to have been responsible (albeit to variable degrees) for the rainfall induced extensive land instability phenomena in the Piemonte region. The types of landslides were clearly related with local geology and previous slope instability. Mudflows, with generally shallow depth (<3 m), were observed along steeper slopes; rock-block slides, which involved volumes up to millions of m³, in sliding prone areas and gently sloping ground (≤15°).

The meteorological event of November 1994 triggered thousands of landslides, mostly mudflows and over 500 rock-block slides.

The present paper has been carried out in the framework of the CEC Project "Meteorological Factors Influencing Slope Stability and slope movement types: evaluation of hazard prone areas" (MeFISSt), contract EV5V-CT920180.

20.2
Geology and Topography of the Region

The territory of Langhe Hills, affected by landslides, belongs to the Piedmontese Tertiary Basin comprised of late-Miocene; marly-silty and arenaceous-sandy alternations. Layers show a general isoclinal bedding with a NE direction and a NW dip with an inclination usually lower with respect to slope angle and not higher than 15°.

The general geomorphological feature is characterised by strongly asymmetrical valleys with very steep SE slopes, with opposite dipping bedding, and more developed NW gentle slopes, with bedding with the same dip direction of slopes. The steepest slopes experienced phenomena definable as mudflows (Varnes 1978) or as "soil slips" (Campbell 1975; Govi et al. 1985) which involved the upper portion of soils for a thickness not higher than 3 m and volumes generally lower than 1 000 m³. On slopes with bedding plane parallel to the slopes themselves, rock-block slides (Varnes 1978) developed involving the bedrock at depths up to 20 m.

The problem of evaluation of slope instability in the Piemonte region has been faced since 1908 when the Law 445 of 9th July by the Ministry of Public Works established the possibility to consolidate or move the towns threatened by landslides at Italian State total cost (Luino et al. 1993). In the context of historical analysis, it is significant that all the towns of the study territory are placed over the top of the narrow ridge-lines or sometime over the scarp slopes, due to the behaviour of the cropping flysch sedimentary rocks; on the contrary, the largest and less sheer dipping slopes are mostly devoted to agricultural activities with rare farmhouses placed along slopes.

From 1801–1990 the Piemonte suffered several meteorological events which caused 87 disasters with an average of one large displacement every 26 months (Luino et al. 1993). Likewise with past experiences, the exceptional intensity and persistence of precipitation triggered thousands of landslides in the Tertiary hills of Langhe.

Fig. 20.1. Localisation of high concentration of mudflows (*asterisks*) and rock-block slides (*dots*) in the area of Langhe. The outlines show actual or historical unstable areas (From Ayala et al. 1996)

20.3
Precipitation Data

The meteorological event of 4–6th November 1994 caused catastrophically consequences: 70 victims, 86 people wounded, 2 226 homeless, 496 municipalities affected, damages for over US $16 000 million.

Large floods along many of the main water-courses affected densely populated areas while precipitation triggered numerous landslides, particularly concentrated in the hilly territory of Langhe (Fig. 20.1).

Precipitation commenced on 3rd November and terminated on 7th November. The highest intensities were recorded on the 5th and 6th: in 19 pluviometric stations, on 100 long-term pluviometric series (>60 years), the previous annual maxima were exceeded, for 1 and 2 days consecutive precipitation. Intensities higher than 25 mm/h were reached in large areas of the Piemonte while, the zero degree isotherm, was around 2 000 m so that also the upper portions of slopes contributed to feed the water-courses.

All the hydrographic network of Piemonte has been interested by a very high volume of precipitation, in large sectors of its territory, so that, where rains reached extraordinary values, storage values have been saturated and the drainage capacities of slopes exhausted, changing, in this way, the whole rainfall into effective precipitation. In addition, the duration of the phenomenon exceeded the concentration time of the hydrographic basins so that the river-beds have been interested by long-term runoff. The Tanaro River at Farigliano recorded a height of 7.80 m, 2.50 m higher than the previous maximum (November 1962) and at Montecastello it reached the height of 8.50 m, exceeding the previous gauged maximum since 1905 (7.74 m in November 1951). The Po River, at the hydrometric station of Becca, reached the height of 7.60 m, a value that was exceeded only three times from 1801.

20.4
Landslides

Mudflows occurred mostly on the steepest slopes (25–40°). Even if very diffused (thousands cases), they are not uniformly distributed: from some units up to 170–180 per km^2 (Cerreto Langhe, Lequio Berria).

This irregular distribution, as a first analysis, seems to depend on local variability of precipitation rather than by land use, since both cultivated slopes (vineyards, grasslands) and wooded ones, have been indistinctly involved.

The areal extent of a single mudflow is, as a rule, modest, but where situations of continuity and coalescence occurred, the whole involved areal extension increase considerably. The same considerations can apply for the mobilised volumes.

They have been triggered by rains of strong intensities for saturation and subsequent fluidification of loose materials, involving, sometime, even fragments of the weathered bedrock.

Mudflows affected seriously the hilly road system which, in many cases, promoted slide triggering with a rapid afflux of wild waters along the slopes. In fewer cases, but sometimes with very severe consequences (Feisoglio, 2 victims), buildings have been involved, especially if sited at the slope foot. The rule of soil slips has been to feed the solid transport in the hydrographic network and to cause temporary dams to water defluxion.

Rock-block slides are gravitational movements generally involving slopes in accordance of bedding and joints which separate large terrain blocks, previously interested by systems of discontinuities. These movements involve Tertiary flyschoid complexes represented by alternations of marly-silty and arenaceous-sandy lithotypes. The sliding surface coincides with a thin marly-clayey level and the movement was triggered by the action of waters, penetrating in depth along systems of discontinuities and circulating on the contact between layer joints. Rock-bock slides involved volumes varying from thousands m³ up to millions m³. Most of them occurred on slopes which have showed evidences of previous movements (Govi et al. 1985).

Continuous observations inferred from geological and morphological features of slopes between Tanaro and Bormida Rivers, from literature and technical reports, have emphasized that instability along bedding partings presents the same typology of evolution. In general, two different development phases are recognized where sliding is characterised by a progressive displacement.

The first phase is characterised by sudden opening of ground cracks, tension gashes, particularly in the upper parts of the slope and well-developed in depth. Frequently these gashes may coincide with pre-existing fractures from previous slope movement in the middle and lower part of the slopes where ground bulging may be observed locally with moderate vertical movement.

The second phase of development involves a pronounced increase of the density of fractures, whose spatial distribution generally indicates incipient destabilization of the entire slope. At this phase, the overall extension of the unstable rock mass body is delineated by predominant structural features of pre-existing fractures.

Fig. 20.2. Mudflows in the Langhe hills (Piemonte)

Planar slides involve generally very large ground movements altering completely the previous morphological shape. Such failures in the Piemonte region were comparatively limited in number and varied from thousands m³ up to millions m³.

In the Langhe (Fig. 20.2), according to Govi et al. (1985) common features can be distinguished as follows:

- All planar failures developed in slopes which had experienced similar movements in the past.
- In large mass movements the geometry of unstable masses was determined by pre-existing discontinuities from ancient slide scarps and opened fractures.
- The average depth of all planar slides was ca. 10 m. Translational movements ranged from some tens up to hundreds meters.
- Slide movement velocities were variable between a few m h⁻¹ to some hundred m h⁻¹.
- The sliding (failure) surfaces, more gently inclined at 8–10° correspond to the interface contact between sandy-arenaceous and marly-silty levels.
- The dominant direction of translation was generally coincident with the dip of the strata, even when the slope face dip direction was different.

Triggering of the rock-block slides (Fig. 20.3) generally occurred in the toe zones, where materials generally show high degrees of structural disorder. Zones of accumulation of the toe of the slopes created temporary dams.

Rock-block slides, in the area of Langhe, were the most distinctive of the November 1994 event, both in terms of volumes involved and patterns of development. They

Fig. 20.3. Rock-block slide at Murazzano

are certainly the most representative phenomena for the endemic morphological instability of Langhe and the principal morpho-genetic factors of slopes.

The meteorological event of November 1994 presents some analogies with the February 1972 and February 1974 ones, which were responsible of the activation of numerous landslides where the phenomenon of Somano emerges (1 km² of surface, 10 million m³ of volume), as the biggest occurrence in the present century (Govi et al. 1985).

20.5
Threshold Rainfall Heights

The following rainfall data were recorded during 4–6th November 1994 in the high basin of the Tanaro River:

- Maximum rainfall intensity : 54 mm/h;
- Total rainfall height (3 days): 350 mm.

Most of the mudflows occurred 24–36 h after the onset of the rainstorm, indicating a potential threshold of 200–250 mm/d.

Similar rainfall data are indicated as critical for the release of so-called first-time landslides as deduced from review of world-wide sources (Gostelow 1991). In addition, the time elapsed from the onset of the rainstorm until mobilisation of the majority of events was within the range of 9–72 h deduced from relevant case studies world-wide (Gostelow 1991).

For the landslides in Piemonte and more specifically in Langhe, the relationship between precipitation and landslide release in November 1994 followed trends known to have occurred in the past.

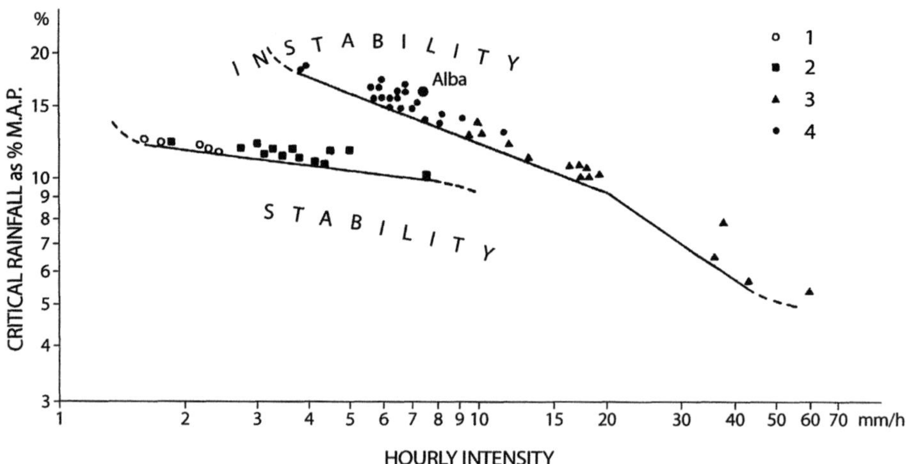

Fig. 20.4. Mudflows: correlation between initial stage of landsliding and rainfall parameters: hourly intensity and cumulated precipitation of the event expressed as mean annual precipitation (%) (events in: *1* winter; *2* spring; *3* summer; *4* autumn). The dotted area indicates mudflows of November 1994 (by Govi et al. 1985, modified)

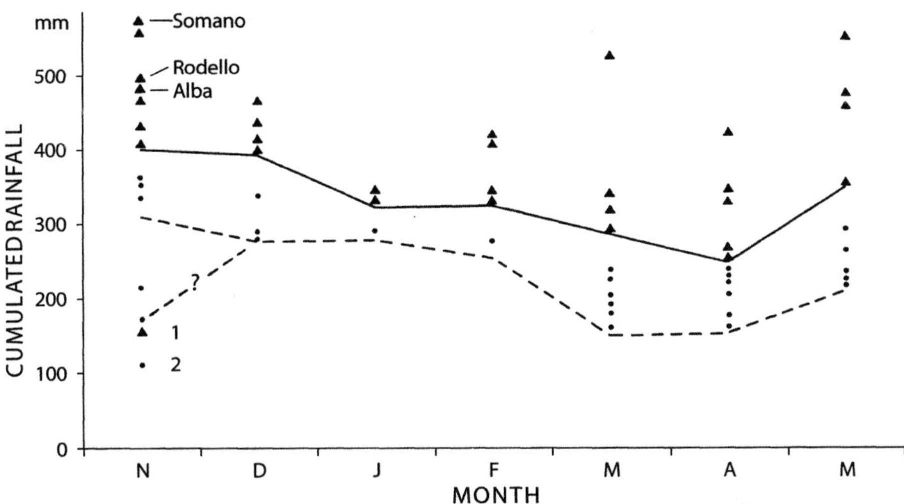

Fig. 20.5. Rock-block slides: diagram of minimum amount of precipitation triggering planar rock slides in the Piedmontese Tertiary Basin. The dotted line defines the field of the minimum value of precipitation of the 60 d before landsliding. The continuous line corresponds to the threshold global values (previous precipitation + event precipitation). *1* cumulated precipitation induced landsliding; *2* cumulated precipitation not induced landsliding (by Govi et al. 1985, modified)

Mudflows are triggered by very intense rainfalls (Fig. 20.4), while for rock-block slides the previous 60 d total rainfall has to be taken into account (Fig. 20.5) (Govi and Sorzana 1982).

20.6
Mudflows – A Preliminary Geotechnical Evaluation

Numerous mudflows in the Piemonte region were observed on 20–35° slopes. The depth of the flow tracks were generally ≤3.0 m and usually 1–2 m.

It can be postulated that during the rainstorm, water percolation and seapage occurred at certain depth in the subsoil resulting in the development of a shallow saturated zone. The apparently consistent thickness of 1–3 m of unstable ground was controlled by the mechanical characteristics of the soil types and the geometry of the slopes in the region. Surficial zones are obviously the most vulnerable to flow during heavy rainfalls, since desiccation cracks or tension gashes from ongoing creeping facilitate water percolation even under rapid runoff conditions.

An approximate analysis of a mudflow can be based on the conceptual model of shallow slides on infinite slopes (Skempton and DeLory 1957). It can be shown that for cohesive soils and inferring full seepage parallel to the surface, there is a critical thickness (h) corresponding to limiting equilibrium:

$$h = c' \sec^2 \beta \, / \, \gamma [\tan \beta - (\gamma_b \, / \, \gamma) \tan \phi'] \, , \tag{20.1}$$

where γ_b is the buoyant unit weight of the soil. Application of the Eq. (20.1) may provide an estimate of the depth of zones susceptible to flow sliding as a function of the parameters c', ϕ' and the slope inclination β. For the following inferred set of parameters:

$$c' = 5 \text{ kN/m}^2, \quad \phi = 25\text{-}30°, \quad \beta = 20° ,$$

the predicted maximum thickness of soil at limit equilibrium would be 2.2–3.8 m which is reasonably close to the observed situation.

20.7
Block Slides – A Preliminary Geotechnical Evaluation

The block slides developed in alternating weak sandstone/marl formations are known to have been involved in past landslide events.

The mechanisms involved translational shearing along persistent, clay coated, low angled (10–15°) bedding partings. Major subvertical cracks (open tension cracks from past movements) and a generally orthogonal subvertical structure comprising two joint systems facilitated the rapid infiltration of the water through the fracture network.

The following observations are worthwhile noting:

a Most of the block slides were shallow with the failure planes at depths generally <10 m;
b The apparently considerable difference between the static and the dynamic frictional resistance, which resulted in very significant translational movements (few tens and up to 200 m) of the sliding blocks.

The sliding planes were all pervasive (length 100–200 m, width 50–150 m), planar and covered by a layer of silty clayey material of 5–20 mm thickness.

Diffractometric analyses performed by ENEA of the soft materials showed considerable inclusions of moscovite, while the grain-size distribution was 65–75% silt, 20–25% clay and 5–10% sand.

It can be shown that for given shear strength and inclination (a) of the sliding plane, the critical thickness (h) of a block ($b \times h$) under self-weight loading $W = b\,z_W\,\gamma$ and hydraulic loading $v = 1/2\;\gamma_W h_W z_W$ and $u = \gamma_W h_W b$ can be found from:

$$h = 2b[\tan\phi(\gamma\cos\alpha - \gamma_W) - \gamma\sin\alpha]/\gamma_W \tag{20.2}$$

For the set of parameters considered representative of the Piemonte block slides:

$$\phi = 22°, \quad \gamma = 24 \text{ kN/m}^2, \quad \gamma_W = 10 \text{ kN/m}^2, \quad \alpha = 10°$$

the following values of z_W can be back-calculated from Eq. 20.2:

b (m)	10	20	50
h (m)	2.7	5.4	13.4

The back-calculated estimates of $h = 5$–10 m correspond to individual block lengths of ca. 20–40 m.

This is probably a physically realistic estimate in relation to the rock mass structure geometry and the presence of vertical tension cracks from previous movements.

20.8
UDEC Modelling of a Typical Block Slide

A further investigation of the stability phenomena was undertaken with the aid of a numerical modelling. The Distinct Element Method (DEM) was adopted to simulate the discontinuous rock mass system and the Universal Distinct Element Code (UDEC) was adopted as a state-of-art numerical implementation of DEM (Cundall 1987; Hart 1991; Bandis et al. 1990).

20.8.1
Model Geometry – Input Parameters

The simulated slope section comprised alternating layers of weak sandstone and marl with low friction interfaces and the following geometry: slope length: 500 m, slope inclination: ≈1 : 7, angle of bedding: 10–12°, bedding spacing: 2 m. A subvertical jointing structure was also modelled discretizing the rock mass into rectangular blocks of variable aspect ratios ranging from elongate to square with increasing sizes.

The input parameters for the sandstone and marls materials and for the joints are listed under Table 20.1.

The intact materials were modelled as linear elastic, isotropic. The shear behaviour of the discontinuities was simulated according to the linear Coulomb model.

The modelling steps included:

a Model consolidation under self-weight loading;
b Model saturation (static pore pressure condition);
c Simulation of ground water flow;
d Gradual decrease of frictional resistance along the bedding partings.

Table 20.1. UDEC model input parameters

| Rock materials | | Rock discontinuities | |
Sandstone	Marl	Bedding planes	Subv. joints
$E = 5\,000$ MPa	$E = 2\,000$ MPa	$c_j\ =\ 0$	0
$v =\ \ 0.2$	$v =\ \ 0.25$	$\phi_j =\ 15°\,(z \leq 6\ \text{m})$	30°
$K = 2\,800$ MPa	$K = 1\,300$ MPa	$22°\,(6\ \text{m} < z \leq 15\ \text{m})$	30°
$G = 2\,100$ MPa	$G =\ \ 800$ MPa	$27°\,(>15\ \text{m})$	30°
$c =\ \ 0.5$ MPa	$c =\ \ 0.3$ MPa		
$\phi =\ \ 40°$	$\phi =\ \ 35°$	$K_{ss} =\ 10\ \text{MPa}\,\text{m}^{-1}\,(z \leq 6\ \text{m})$	
		$50\ \text{MPa}\,\text{m}^{-1}\,(z \leq 15\ \text{m})$	
		$100\ \text{MPa}\,\text{m}^{-1}\,(>15\ \text{m})$	

E: Young's Modulus; v: Poisson's ratio; K: Bulk Modulus; G: Shear Modulus; c: cohesion of intact material; ϕ: angle of internal friction; c_j: cohesion of interface; ϕ_j: frictional resistance along interface; K_{ss}: shear stiffness along interfaces.

20.8.2
Principal Results

The key analytical findings can be summarized as follows:

a Significant shear overstressing was induced along the sandstone-marl interfaces as a result of the significant decrease in the effective normal stresses associated with the simulated conditions of full saturation, with the maximum displacements occurring in the lower parts of the slope, as shown in Fig. 20.6a.

b Translational shear movements of the order of 0.05–0.10 m were predicted over a length of >200 m of the slope, involving a 5–10 m thick zone and penetrating down to 15 m in the toe region.

c It was evident from the analyses, that the major instability mechanism was initiated in the toe zones and developed upslope in a retrogressive fashion. The open/at slip joints plot in Fig. 20.6b illustrates the significant loosening of the mass structure in the toe region (many subvertical joints opened). The ongoing shearing evident in the middle parts of the slope is a demonstration of the retrogressive development of the slide associated with the reducing passive toe reaction.

d It is considered likely that the block sliding phenomena in the Piemonte region was associated both with a dramatic decrease in the effective normal stresses and a further loss of shear strength along the mobilised sliding planes. Measurements of the dynamic (true) residual friction of similar materials involved in slope problems comparable to Piemonte in terms of geology, mechanism and recurrence (Hart 1991) yielded in ring shear $\phi_r = 9$–$13°$.

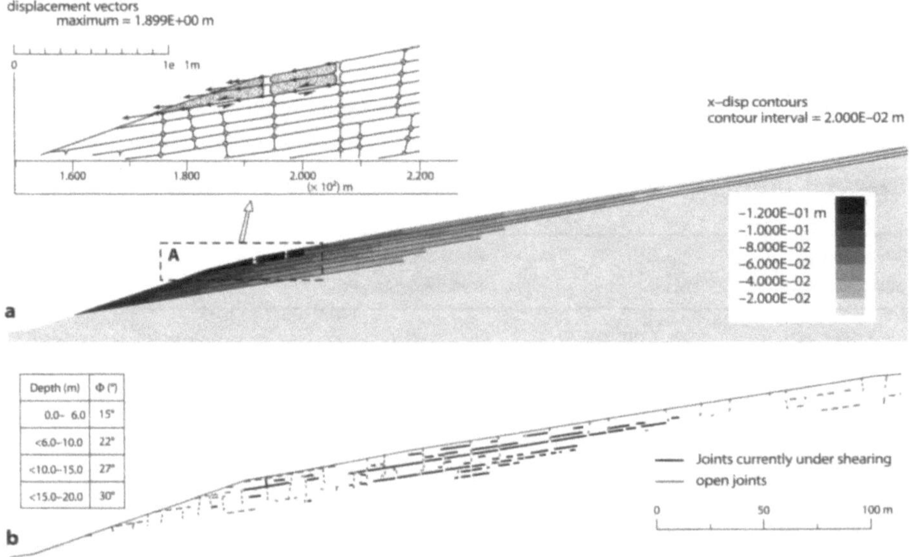

Fig. 20.6. Overview of induced displacement field and shear overstressing along the sandstone/marl bedding partings as predicted by the UDEC model

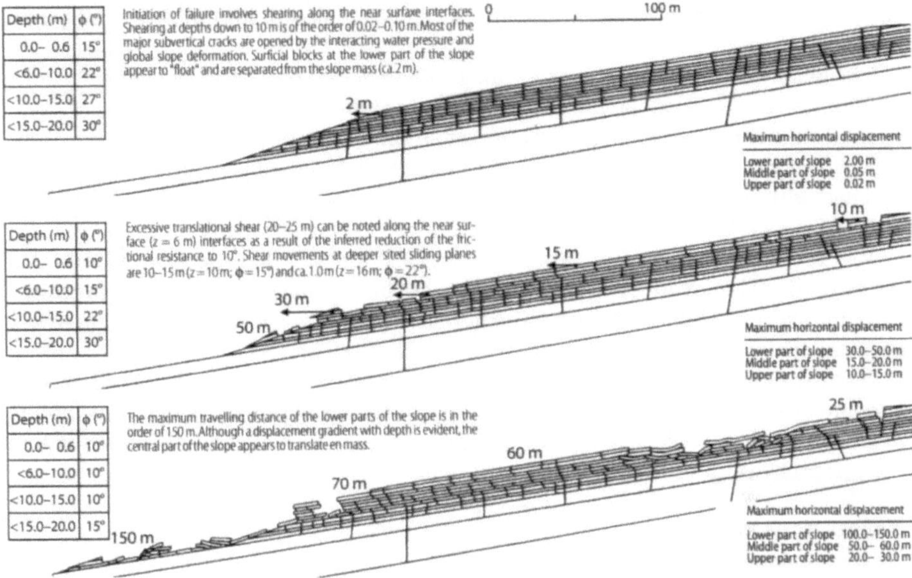

Fig. 20.7. UDEC simulation of successive stages of block slide development. An envisaged strain-softening behaviour was simulated by gradual reduction of the frictional resistance and shear stiffness along the sliding planes at depths down to 15–20 m

By assigning friction angles $\phi_r = 10\text{–}15°$ to the UDEC model interfaces the forms and extent of instability illustrated in Fig. 20.7 were obtained, which generally conform with the actual effects.

20.9
Summary – Conclusions

The intensive rainfalls of 4–6th November 1994 in the Piemonte region resulted in two distinctly different forms of land instability:

- First-time slides of mudflow type which were of relatively limited extent (typically $l < 50$ m, $w = 5\text{–}10$ m, $d = 1\text{–}2$ m) but very frequent;
- Reactivation of old slides involving shear translation of large masses along low strength failure planes, representing bedding partings in the typical sandstone-marl regional stratigraphic sequence. Typical dimensions were $l = 100\text{–}300$ m, $w = 50\text{–}100$ m, $d = 5\text{–}10$ m.

The rainfall records of maximum intensity circa 50 mm/h and 200–250 mm/d are generally within the critical threshold ranges according to the regional scenario (Ayala et al. 1996).

The mudflows occurred due to excessive pore pressure generation. An explanation of the consistently shallow affected zones can be obtained from back-analysis inferring an infinite slope model. Assuming slope angle 20°, cohesive soil $c = 5$ kN/m², $\phi = 25\text{–}30°$

and seapage parallel to the surface, the back-calculated depths for limiting equilibrium range from 2–4 m.

The block slides were re-activated along sliding planes probably involved in historic instability events; hence, the planes were already at the residual state in terms of shear strength, although a certain recuperation of strength may have taken place since then. Such a strength "gain" may be envisaged through an "over-consolidation" process associated with pore pressure dissipation during the post failure period and could explain the generally stable state of the slopes prior to November 1994.

Rigorous numerical investigations of a typical block slide generally indicate that the exceptional kinematic energy of the slides can be explained on the basis of the static/dynamic shear resistance concept. Critical values of dynamic frictional resistance were back-calculated at 10–15°.

Finally, as main result, the meteorological event of November 1994 confirms an agreement between a semi-quantitative approach, through the correlations among precipitation features and landsliding occurrence, and a mathematical modelling, in the comprehension of the environmental dynamics of the territory.

References

Ayala FJ, Bandis S, Delmonaco G, De Lotto P, D'Epifanio A, Dutto F, Ferrer M, Lied K, Margottini C, Mortara G, Palandri M, Rybar J, Sandersen F, Serafini S, Stemberk J, Trocciola A, Anselmi B, Crovato C (1996) Meteorological events and natural disasters: An appraisal of the Piedmont (North-Italy) case history of 4–6 November 1994 by a CEC field mission. Casale R and Margottini C (eds) Roma, 96 pp

Bandis SC, Demiris CA, Barton N, Makurat A (1990) Discrete modelling of jointed rock structures. Proc. 2nd European Specialty Conference on Numerical Methods in Geotechnical Engineering, ISSMFE, Santander Spain

Campbell RH (1975) Soil slips, debris flows, and rainstorm in the Santa Monica Mountains and vicinity, Southern California. US Geol. Survey, Prof. Paper 851, 51 pp

Cundall PA (1987) Distinct element models of rock and soil structure. In: Brown ET (ed) Analytical and computational methods in engineering rock mechanics. Allen & Unwin (London) Publ., pp 129–163

Gostelow TP (1991) Rainfall and landslides. Proceedings "Natural Hazard and Engineering Geology. Prevention and Control of Landslides and Other Mass Movements". European School of Climatology and Natural Hazards, E.E.C. Course, Lisbon

Govi M, Sorzana PF (1982) Frane di scivolamento nelle Langhe cuneesi, febbraio-marzo 1972, febbraio 1974. Boll. Ass. Min. Sub. Anno XIX, N° 1–2, pp 231–264

Govi M, Mortara G, Sorzana PF (1985) Eventi idrologici e frane. Geol. Appl. e Idrogeologia 20(2):359–375

Hart RD (1991) An introduction to distinct element modelling for rock engineering. Proc. 7th Int. Congr. ISRM, Aachen Germany

Luino F, Ramasco M, Susella G (1993) Atlante dei centri abitati instabili piemontesi. CNR-IRPI Torino-REGIONE PIEMONTE S.P.R.G.M.S. GNDCI, Pubbl. N° 964, pp 16–30

Skempton AV, DeLory FA (1957) Stability of natural slopes in London Clay. Proc. 4th Int. Conf. on Soil Mech. Found. Eng. London, vol 2, London, pp 378–381

Tsotsos S, Anagnostopoulos C (1992) Investigation of landslides in the Pieria region. Greece. Proc. 6th Int. Symp. on Landslides, Bell DH (ed), New Zealand, pp 237–242

Varnes DJ (1978) Slope movements type and processes. In: Schuster RL, Krizek RJ (eds) Landslides analysis and control. Transp. Research Nat. Sc., Special Report 176:11–33

Evaluation of Radar and Panchromatic Imagery for the Study of Flood and Landslide Events in Piemonte, Italy, November 1994

P.J. Mason · A.F. Palladino · J. MCM. Moore

21.1
Introduction

The intense storm of 5–6th November 1994 produced up to 300 mm rainfall in 36 h and produced slope instabilities and flooding across the region of Piemonte. Records and literature show that widespread storms, floods and landslides occur frequently and cyclically in this region every year.

The aims of the research include the delineation of terrain effected by the 1994 storm, the targeting and discrimination of landslides (which are often at incipient stages of development) the estimation of flood extents and the automatic discrimination of floods waters. This paper describes methodologies for enhancement and extraction of flood and landslide evidence from multi-temporal, satellite imagery and digital elevation data, illustrated by examples from test sites in the Belbo and Tanaro valleys, illustrated in Fig. 21.1.

21.2
Methodology

The data used provided by geo-referenced SPOT scenes dated 29th June 1994, 11th November 1994 and 9th April 1994; and a Digital Elevation Model (gridded at 50 m) created from 1992 stereo air photography. The simultaneous processing of these image data using simple techniques enables efficient identification of hazard features in the two test sites.

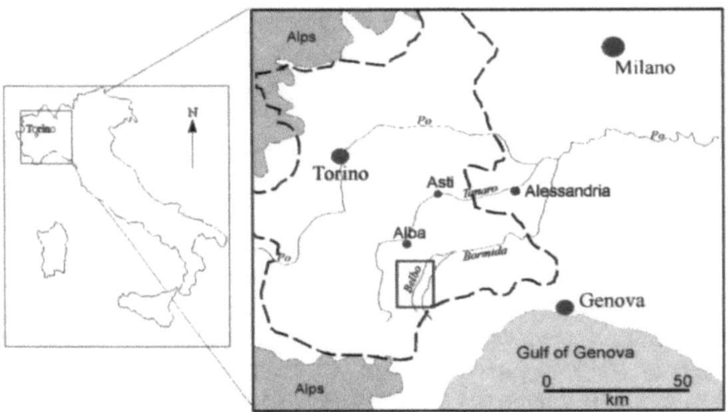

Fig. 21.1. Maps indicating the location of test sites A and B in Piemonte, Northern Italy

21.2.1
Preliminary Evaluation of Image Suitability

ERS-1 SAR. This image is of limited value for landslide study but ideal for flood analysis due to the effects of experiencing different types of slope movements. SAR's southeasterly illumination produces very high returns on the scarp slopes and extremely low returns on the shallow dip slopes. The overall effect of this geometry, wavelength and saturated ground conditions, is a lack of textural information which makes landslide discrimination impossible. Test site B comprises a mature flood plain, where SAR's sensitivity to moisture can be used to good effect in the detection of currently and recently inundate areas. The low relief of the flood-plain means that the terrain effects on the radar returns are negligible.

SPOT-Panchromatic. The SPOT scenes were acquired in different years and different seasons, and represent very different terrain and weather conditions. Solar illumination at acquisition time is constant, from the south-east, but at different elevation angles (high in June producing almost no topographic shadow, very low in November, producing a great deal of shadow and then moderate shadowing in the April image). The images are highly correlated so their illumination conditions can be relatively simply normalised via the linear regression of DN values from constant features in the two images.

21.2.2
Test Site A: Landslide Identification in the Langhe Hills

The three types of slope instability which have been identified are summarised in Table 21.1. They are; i) debris slides on scarp slopes, ii) planar rock-block slides, as described by Dikau et al. (1996) and Govi and Sorzana (1982) and iii) compound slides, as described by Cronin (1992) and they pose different problems for remote sensing. The debris slides are spectrally and texturally distinct except where their dimension approach the resolution limit of the imagery. The planar slides though larger, are distinguishable from the undisturbed ground only at the exposed crown area. The compound slides are sizeable but display complex textures which are related to NE-SW structural lineaments characteristic of the region.

To identify the slope movements, the processing is performed in the following stages:

i   *Change detection and the identification of slope movement target pixels.* The SPOT images were first stretched to the same mean and value ranges using the BCET method (Liu 1991) prior to further processing. The April scene contains all the landslide information but image complexity is high due to patterns of intense cultivation, woodland and scattered settlements. A change-detection image is produced by the subtraction of the June image DN from the April DN. This image shows a reduction in both textural complexity and topographic shadowing, and shows many bright pixels where bedrock and fresh soils are exposed. In many cases these were caused by slope movements during the winter of 1994/95.

ii *Separation of slope movements types using the DEM.* Thresholding of the change-detection image isolates these bright target pixels representing all slope movement

Table 21.1. Characteristic of the slope movements in test site A

Type	Slope angle	Slope aspect	Area (m^2)	Exposed area (% of total)	Textural complexity	Relation to drainage	Incipient indicators
Superficial (flows)	>15°	0 – 240°	800 – 20 000	100	Low	Follow gullies	–
Planar (slides)	<18°	240 – 360°	15 000 – 75 000	25 – 30	Moderate	Unknown	Sub-pixel fractures
Compound (slides)	<18°	240 – 360°	20 000 – 100 000	30 – 50	High	Bounded by streams	Sub-pixel fractures

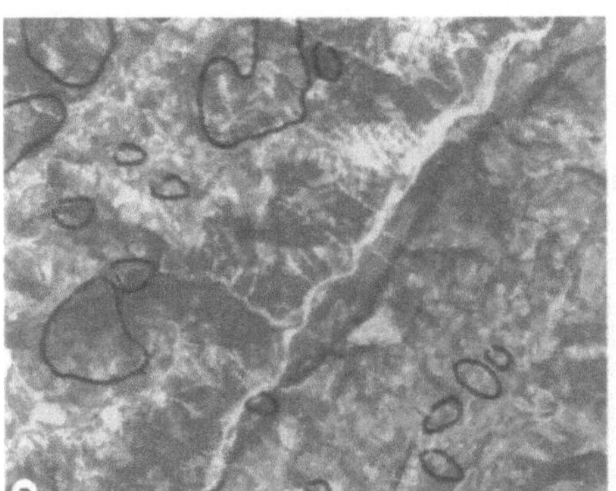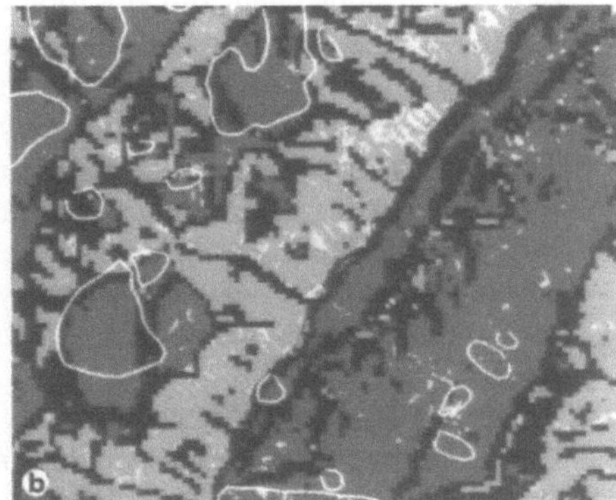

Fig. 21.2. a April SPOT image; **b** thematic image: *white:* superficial flows and landslide targets; *dark grey:* slopes susceptible to block and compound sliding; *pale grey:* slopes susceptible to superficial slope movements, polygons-landslides recorded since 1970 (scale approx. 1 : 68 000)

types. Susceptible slopes were identified on the basis knowledge and kinematic instability analysis (Maurenbrecher 1995) using recorded fracture measurements. Slope angle and aspect images are created from the DEM and these are classified into slopes susceptible to the different types of slope movement based on parameters listed in Table 21.1.

The target pixels are separated using the classified slope angle and aspect image, into debris flows on scarp slopes and slide targets on the dip slopes. This is performed using a multiple logic expression applied to the digital data layers simultaneously.

iii *Convolution filtering* of the slide target image using a diagonal gradient kernel, to enhance NE-SW edges, helps to identify linear features which may be related to compound slides textures and the scars of incipient slides.

iv *Display of all slope movement types.* Simultaneous processing of the four stages enables the rapid generation of a thematic image in which scarp slopes susceptible to debris slides and the debris slide pixels are separated from slopes susceptible to landsliding and the identified planar and compound slides (see Fig. 21.2).

21.2.3
Test Site B: Flood Mapping in the Tanaro Valley

Test site B comprises a section of the easterly flowing Tanaro River. It forms a broad flood plain dominated by intensive agriculture and many small towns which is particularly prone to flooding because of its large upland catchment and low lying nature. Processing for flood identification is performed in 3 stages:

i   *Identification of wet areas in the November SAR image (direct method).* The SAR image directly records the position of flood waters and of flood saturated soils, as very dark areas, a few days after the height of the flood. Image contrast is enhanced using a logarithmic function and speckle is reduced using a standard median filter to improve clarity. A multiple logical expression involving simultaneous thresholding of the SAR, slope angle and elevation data enables restriction of the enhancement to the flood-plain region and therefore, the elimination of other dark areas in the radar image which may confuse the interpretation.

ii *Delineation of flood debris covered areas in the April SPOT image (indirect method).* The SPOT image shows clearly the distribution of spectrally distinct flood debris (a thin covering of clayey, sandy silt) after the flood waters have subsided. This image is median filtered, to improve clarity, and thresholded to separate the flood debris covered areas. These relatively spectrally uniform areas are then processed separately, using DEM information, so that only the data within the flood-plain region are enhanced.

iii Combined display of all flooded areas.

Colour-draping of the thresholded SAR image over the selectively enhanced SPOT image allows the display of flood debris coverage (indicating maximum flood water extents) with areas of standing water and saturated soils (see Fig. 21.3). The ponded water in the SAR covers a smaller area than the flood-debris area in the SPOT image and may represent the areas of lowest ground level where flood waters pond naturally and therefore, take longest to drain from.

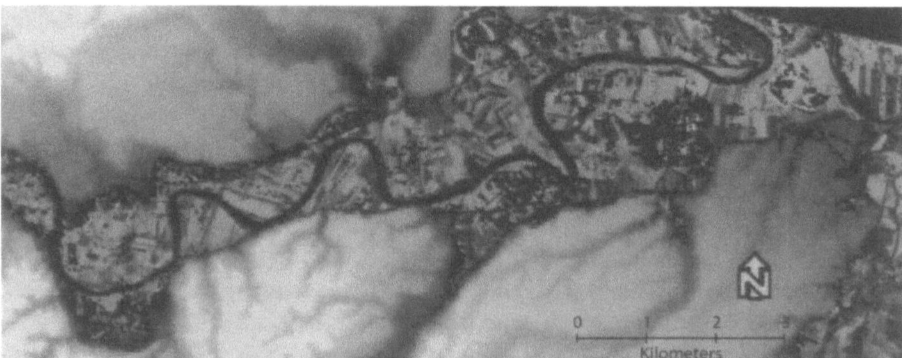

Fig. 21.3. Test site B: areas under water (direct from SAR image) dark grey; areas of bright flood debris (indirect from SPOT image); light greytones of the DEM show surrounding hilly terrain north and south of the river (Scale approx. 1 : 111 000)

21.3
Results and Discussion

21.3.1
Results of Processing

SPOT optical imagery provides necessary resolution for detailed studies of this nature and good temporal coverage. It is limited only by its restriction to cloud-free conditions so analysis of images at the time of events is not practical. A better radar image geometry for landslide studies would involve illumination at a similar elevation to that of ERS-1 but from the west i.e. facing the dip slopes, and acquisition during a drier period. Detection of moisture is relevant to landslide studies but spatial resolution and the effect of water on sensitivity to surface roughness and texture must be considered. Data acquisition some months after the event would of course be inconvenient for the rapid monitoring of catastrophic events so compromises must be made. Sideways Looking Airborne Radar (SLAR) X-band imagery has commonly been used for terrain analysis in tropical regions where ground saturation is constant. The choice of radar wavelength for the monitoring of slope movements may therefore be case dependent. Further potential lies in the development of SAR interferometric techniques where minute changes in phase may be used to monitor small scale and subtle changes in ground elevation which are far beyond the resolution limits of optical data.

From this work it is considered that slope angle, slope aspect and temporal texture changes are key parameters for slope instability studies and should be variably weighted according to the nature of the case. Further research involving spectral studies (satellite and spectrometer), is required to refine the segmentation of high risk areas on the basis of rock and soil type changes.

Threshold filtering effectively highlights the silty flood deposits in the SPOT imagery but some confusion arises from flood debris which is obscured by post-flood vegetation growth. Optimum delineation of the floodplain and identification of flood waters from radar shadow is made by direct visual interpretation of the SAR data. The two independently derived maps of the floodplain delineate are highly correlated and this substantiates their validity. The floodplain boundaries are also consistent with the maximum N-S extent of the flooded region (3–4 km) discussed in Tropeano (1995).

Modern image processing/GIS software combined with powerful computing capabilities enable efficient means of processing and analysing large datasets. Here the use of highly correlated optical images provides little/no problem for data registration. In addition, the simultaneous processing of multi-form datasets, which is routine functionality of most image processing packages, enables an efficient and flexible method of extracting spatially, spectrally and temporally dependent information.

21.3.2
Discussion

Landslide and flood research in the region have traditionally involved detailed studies concentrated on specific localities (Boni 1941; Cortemiglia and Terranova 1969; Dikau et al. 1996) and the use of aerial photograph archives. Until this storm event, no attempt had been made to systematically map the widespread slope instabilities or to develop a methodology for the assessment of risk to future failures (Boccardo et al. 1995). Satellite remote sensing provides the regional perspective previously lacking in such studies, in addition, to enabling detailed digital analysis. Recent research also includes studies of the meteorological characteristics, geotechnical evaluation and modelling of the block sliding (Bandis et al. 1996) and on the significance of clay minerals in the failure mechanism (Forlati et al. 1996) which provide vital information at different scales of study.

A difficulty in identifying planar block-slides and compound slides arises because only 25–30% of the total slide area (the area of exposed bedrock) is detectable from the satellite sensor and occupies only a few pixels. In addition, in the compound case, significant spectral complexity exists due to large, relatively undisturbed areas of rock, soil and vegetation making the slide area as a whole rather poorly defined and difficult to map confidently without multi-spectral data. Confident identification of these slides can only be made when in advanced stages of development. The identification of incipient slide targets is still important because of the recurrent nature of the landslide problem in this region, and will always be hindered by the sensor constraints.

Flooding in the area occurs almost every year but it has been estimated that the flood events of November 1994 were more widespread than had been experienced in perhaps the last 100 years (Maurenbrecher 1995). Until satellite radar imagery became readily available a cost-effective and routine method of monitoring the extents and dynamics of such a sudden event was not possible. The flexibility of computerised processing methods and the wealth of data available make such studies feasible.

21.4
Conclusions

21.4.1
Landslide Studies

Panchromatic imagery enables direct visual detection of landslide providing they are of sufficient size or are spectrally distinct. Multi-temporal data coverage enables change detection and thereby, a more reliable detection of recent landslides. High spatial resolution and a moderately high sun angle are key factors for efficient processing and interpretation. The DEM and its derivatives are crucial to the delineation of different types of slope movement and to the identification of high risk regions but the accuracy of this relies heavily on the quality of elevation data.

21.4.2
Flood Analysis

Radar is the ideal choice of imagery for stormy weather conditions and to detect water but may not be available at the time of maximum flood. Post-flood SPOT imagery indirectly captures the maximum flood extents via the flood debris which is spectrally and texturally distinct from the surrounding natural and cultivated ground. The DEM enables segmentation of the terrain so that confusion with spectral anomalies in areas of higher relief can be eliminated. Combined use of these data enables a semi-automated discrimination of flood extents and knowledge of flood dynamics.

Acknowledgements

We wish to thank MURST/British Council for pump-priming funding at the start of this project; Dr Piero Boccardo at the Politecnico di Torino (Dipartimento di Georisorse e Territorio) and authorities of Regione Piemonte for their assistance and the provision of data; and ULIRS for the use of computing facilities at Imperial College.

References

Bandis S, Delmonaco G, Margottini C, Serafini S, Trocciola A, Dutto F, Mortara G (1996) Landslide phenomena during the extreme metereological event of 4–6th of November 1994 in Piemonte Region, N. Italy. Landslides: Proceedings of 7th International Sympsium on Landslides, 17–21 June 1996, Trondheim Norway, vol 2, p 623

Boccardo P, Lingua A, Rinaudo (1995) Integrazione di dati planoaltimetrici con immagini acquisite da piattaforma satellitare: Il caso delle aree esondate durante l'evento alluvione del novembre '94 in Piemonte. Telerilevamento, GIS e Cartografia al servizio dell'informazione Territoriale, VII Convegno Nazionale AIT (Associazione Italiana di Telerilevamento) Chieri (TO), 17–20 Ottobre 1995, CSEA-Bonafus, p 223

Boni A (1941) Distacco e scivolamento di masse a Cissone, frazione di Serravalle delle Langhe. Geofisica pura e applicata, 3, 3, p 1

Cortemiglia GC, Terranova G (1969) La frana di Ciglié nelle Langhe. Memorie Società Geologica Italiana, 8(2):145–153

Cronin VS (1992) Compound landslides: Nature and hazard potential of secondary landslides within host landslides. Geological Society of America, Reviews in Engineering Geology, IX, pp 1–8

Dikau R, Brunsden D, Schrott L, Ibsen M (eds) (1996) Landslide recognition: identification, movement and courses. International Association of Geomorphologist, Publication N° 5, Willey and Sons

Forlati F, Lancellotta Osella A, Scavia C, Veniale F (1996) The role of swelling marl in planar slides in the Langhe region. Landslides: Proceeding of the 7th International Symposium on Landslides, 17–21 June 1996, Trondheim Norway, vol 2, p 721

Govi M, Sorzana PF (1982) Frana di scivolamento nelle Langhe Cuneesi Febbraio-Marzo 1972, Febbraio 1974. Bollettino della Associazione Mineraria Subalpina, Anno XIX (1–2):231–263

Liu JG (1991) Balanced contrast enhancement technique and its application in image colour composition. International Journal of Remote Sensing, 12(10):2133–2151

Maurenbrecher PM (1995) Stereographic projection wedge stability analysis of rock slopes using joint data. Mechanism of Jointed and Fracture Rock-II. In: Rossmanith HP (ed) Balkema Rotterdam

Tropeano D (1995) Evento Alluvionale del Novembre 1994 in Piemonte. Interventi di Studio Effettuati dall'IRPI/CNR di Torino: Sintesi delle osservazioni, in press on GEAM (Bollettino Associazione Georisorse e Territorio, Torino)

Chapter 22

Longitudinal Evaluation of the Bed Load Size and of its Mobilisation in a Gravel Bed River

C. Deroanne · F. Petit

22.1
Introduction

As water levels rise, the bed load of a gravel bed river is partly mobilised by drag forces. When the stream can no longer entrain these sediments, the pebbles are deposited on the bed and create locally depositional forms. These bars then perturb the flow and play an important role during subsequent floods.

That is why we set ourselves a dual goal target with regard to the bed load. First, to determine the sedimentological characteristics of the riverbed and more precisely the size of the material. Second, to analyse the processes of mobilization of these particles. But in both cases, we have focused on the longitudinal nature of the results, that is from the channel head to its mouth.

This paper is an abstract of a study (Deroanne 1995) carried out in the Department of Physical Geography at the University of Liège. Completed within the framework of a final year project, it was supervised by Professor Dr. F. Petit.

22.2
Presentation of the Drainage Basin of the River Hoëgne

The gravel bed river that was chosen to support this study is the river Hoëgne situated in the Belgian Ardennes. Table 22.1 summarizes its main drainage basin characteristics.

22.3
Longitudinal Grain-Size Distribution

Downstream fining of riverbed material has been an unanimously accepted principle for many years (Knighton 1980; Hoey and Ferguson 1994; Kodama 1994). But the explanation of this reduction caused by processes such as abrasion and selective transport is not quite clear.

22.3.1
Methodology

To modelize this reduction tendency, 17 study-sites have been selected along the river from its head to its mouth. For each of these study-sites we took the appropriate measurements to determine the mean grain-size of a representative sample which was estimated by the D_{50} which is the median diameter.

Table 22.1. Main drainage basin characteristics of the Hoëgne River

Drainage area (km^2)	190
Length (km)	31
Difference in height (m)	520
General slope gradient (mm^{-1})	0.017
Discharges near its mouth (1979–1987) (m^3 s^{-1})	
Mean discharge	3.5
Minimum discharge	0.05
Maximum discharge	70 (recurrence of 16 years)
Full bank discharge	35 (recurrence of 0.7 year)

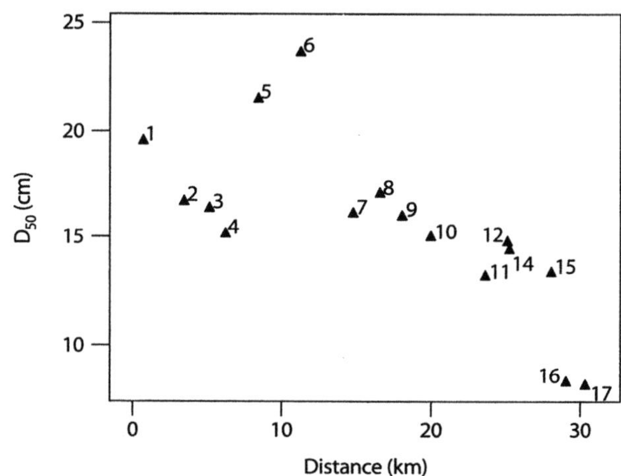

Fig. 22.1. Longitudinal variation of the D$_{50}$ along the 17 study-sites

Two measurement techniques were adopted. A *weighting technique* was used at those sites where the pebble-size and weight could be measured by hand and a *linear technique* where boulders could not be held (Grant et al. 1990). The data supplied by these two techniques are plotted on a grading curve for each study-site to measure the D$_{50}$.

The longitudinal variation in particles-size along the 17 study-sites is then analysed on a graph representing the D$_{50}$ as a function of the distance of each sites from the river head (Fig. 22.1). A variety of scales can be adopted to fit a specific model to the measurements.

22.3.2
Results

Figure 22.1 shows the longitudinal variations in particle-size (weighting technique only) along the 17 study-sites. In keeping with the theory, the size of the bed material is seen to decrease progressively downstream. But this size reduction is perturbed by a sudden increase after the 4th site.

This increase can be explained by the long profile of the river (Fig. 22.2) where the usually observed concave appearance is locally perturbed. The perturbing convex section pro-

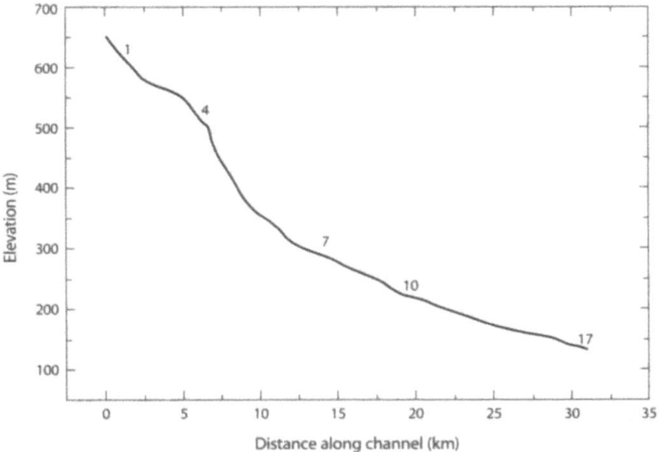

Fig. 22.2. Long profile of the Hoëgne River

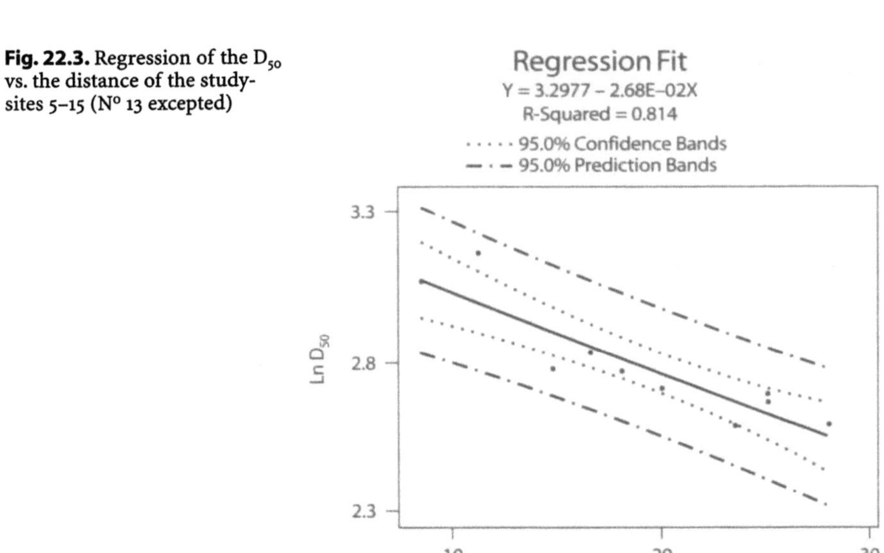

Fig. 22.3. Regression of the D_{50} vs. the distance of the study-sites 5–15 (N° 13 excepted)

vides evidence of a new erosion point which either has not yet reached the plateau or is held back by stony accumulations at its extremity (Pissart 1953). In fact, this sector is a significant source of material which occasions an increase of the D_{50} in study-sites 5 and 6.

Then we tried to fit an exponential model (Werrity 1992) to the measurements situated downstream of this sector. We also decided not to take into account the sites numbered 16 and 17 where the size of the material is affected by geological and human influences. The data fit the model reasonably well (Fig. 22.3) with a coefficient of determination R^2 above 80%.

22.4
Mobilisation of the Bed Load

Having analysed the longitudinal distribution of the riverbed grain-size, we then tried to describe the processes that cause the downstream fining in a qualitative and quantitative way. Therefore, we estimated longitudinally the drag forces and also the critical parameters of mobilisation.

22.4.1
Longitudinal Estimation of the Drag Forces

22.4.1.1
Methodology

Our longitudinal estimation of drag forces was predominantly based on the measurement of the most frequently used criterion namely *shear stress*. This represents the drag force operating parallel to the bed which act on a particle to entrain it. It has been defined by the Eq. 22.1 of Du Boys (1879). The shear stress τ can be divided (Petit 1989a) (Eq. 22.2) into:

- the *grain shear stress* τ' due to the resistance of particles,
- the *bed form shear stress* τ'' due to the additional resistance from bed forms and stream banks.

$$\tau = \rho \, g \, Rh \, S_e \; , \tag{22.1}$$

with the water density ρ, the gravity g, the hydraulic radius Rh estimated by the height of the water and the energy slope S_e estimated by the water surface slope.

$$\tau = \tau' + \tau'' \; . \tag{22.2}$$

Consequently, on the 17 study-sites we measured the height of the water and the water surface slope to estimate respectively the hydraulic radius and the energy slope.

22.4.1.2
Results

The results of shear stress for each study-site are plotted vs. their drainage area in Fig. 22.4. These measurements were, of course, taken in similar hydrological conditions ($0.17 \, Q_b$).

This distribution recalls the one obtained for the D_{50}. The downstream decrease in shear stress is again disturbed in the sites where large boulders were found. So, as for the D_{50}, the shear stress is highly influenced by this channel unit. Further analyses of the measurements allow us to conclude that it is more precisely the bed-form shear stress component that is influenced by this section and therefore, is partly responsible for the entrainment of the bed load.

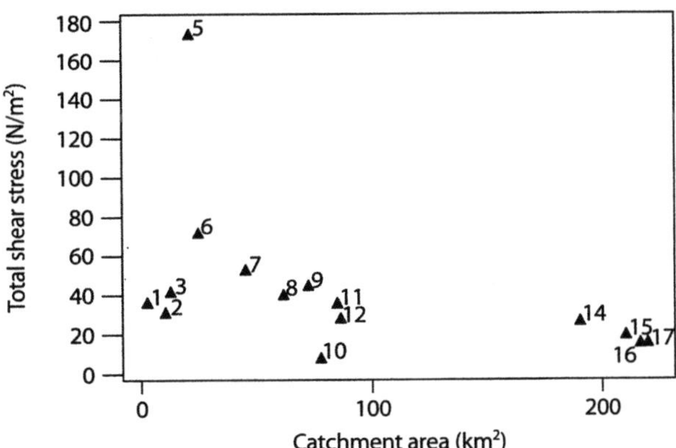

Fig. 22.4. Longitudinal estimation of the shear stress along the 17 study-sites

22.4.2
Critical Parameters of Entrainment

22.4.2.1
Methodology

To estimate the critical parameters of mobilisation, we observed the entrainment of tagged particles in seven study-sites for two rates of mobilising discharges.

After the floods, the size, weight and travel of each tagged particle were measured to ascertain the type of mobilisation which occurred. This objective was also achieved by calculation of the *critical shear stress* (Eq. 22.3) which requires measurement of the height of the water and of the water surface slope during the two mobilising discharges considered.

$$\tau_c = \rho\, g\, Rh\, Se \;. \tag{22.3}$$

Finally, we calculated the *critical dimensionless shear stress*, known as the criterion of Shields (Eq. 22.4), which is the ratio of fluid forces keeping a particle in motion and gravity forces tending to keep the particle at rest.

$$\theta_{ci} = \frac{\tau_{ci}}{(\rho_s - \rho_f)gD_i} \;, \tag{22.4}$$

where θ_{ci} is the criterion of Shields, τ_{ci} the critical shear stress responsible for the mobilisation of a particle of size D_i, ρ_s the density of the material ($\pm 2\,650$), ρ_f the water density ($\pm 1\,000$) and g the gravity.

Interested in the variation of that criterion of Shields in relation to the sedimental structures along the river, we also measured the D_{50} of the tagged pebbles before the rise of the water and the D_{50} of the entrained pebbles and of those not entrained. These measurements were carried out on a photograph taken before the floods to avoid disturbing the overlapping of pebbles, and by means of a weighting technique after the floods respectively.

22.4.2.2
Results

The two discharges considered as responsible for the mobilisation of more than 50% of the tagged particles are lower than the full bank discharge. Also, the recurrence of the observed discharges in the upstream sites (0.32 years) is higher than that for the downstream sites (<0.3 years).

The two regressions shown in Fig. 22.5 express the relationships between the travel distance of tagged pebbles and either the size, or the weight, of particles for one study-site; type of mobilisation that occurred can thus be describe.

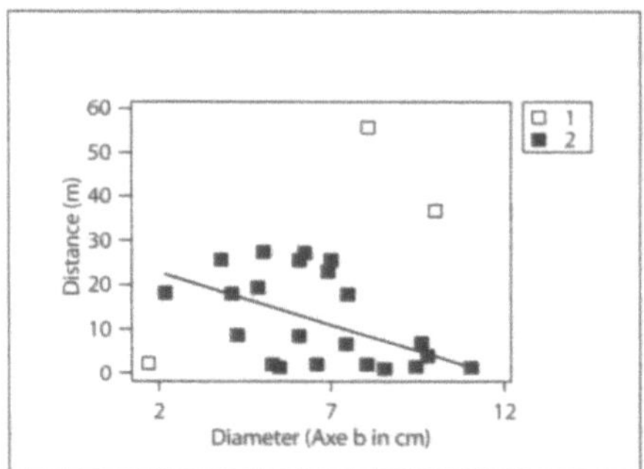
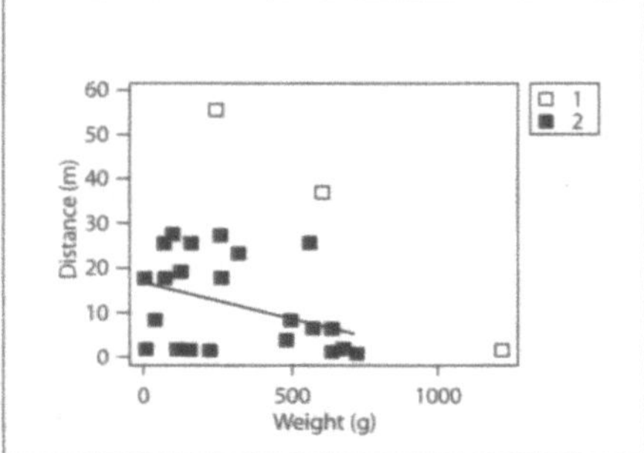

Fig. 22.5. Regression of travel distances vs. either the particle-size (left) or weight (right) at the Mousseux study-site for the mobilized pebbles (*1* points representing aberrant values, *2* points from which regression was plotted)

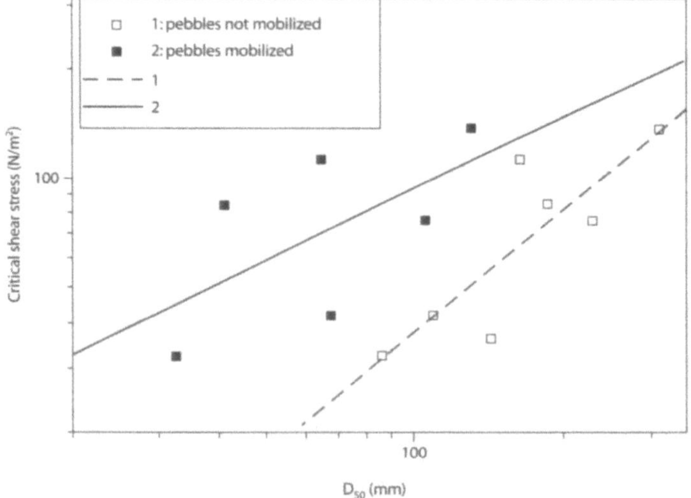

Fig. 22.6. Regression of the critical shear stress vs. the D_{50} of the pebbles (*1* not mobilized, *2* mobilized) for each study-site

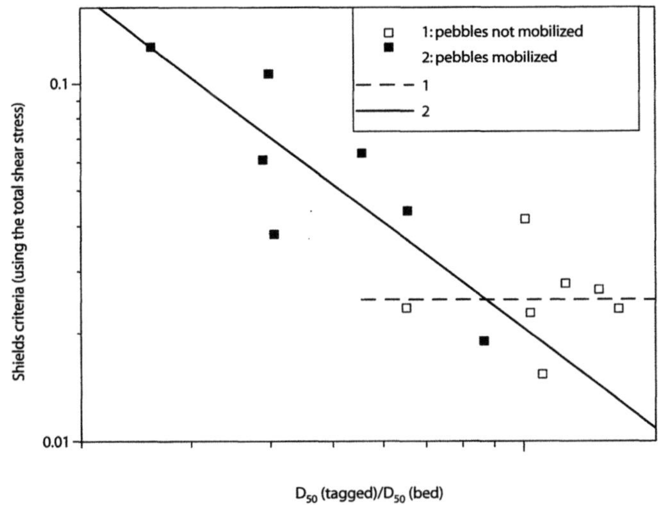

Fig. 22.7. Regression of the criterion of Shields vs. the ratio D_{50} of tagged pebbles (*1* not mobilized, *2* mobilized) and D_{50} of bed load for each study-site

In fact, in most study-sites, the slope of these regressions indicate selective transport of the particles (Ashworth and Ferguson 1989). This conclusion is reinforced by the calculation of the critical shear stress plotted vs. the D_{50} of each study-site (Fig. 22.6). Again, the slope of the regression defines selective transport.

Although the ratio D_{50} of tagged pebbles and D_{50} of bed load represents the processes of hiding and protrusion of the riverbed material (Andrews 1983; Richards 1991; Petit 1994), the regression slope of the criterion of Shields vs. that ratio (Fig. 22.7) shows the importance of these sedimental structures.

22.5
Conclusions

The hydrological processes responsible for the entrainment of the material of a gravel bed river are extremely complex. If this study can confirm that there is downstream fining of particles, it also shows that this tendency can be perturbed locally by sources of material.

Similar to downstream fining, the longitudinal distribution of the shear stress is also perturbed by bed forms. We can therefore, conclude that grain shear stress is not the only factor responsible for the mobilisation of the material.

Although the discharges considered were inferior to full bank discharge, they were sufficient to set up partial entrainment of the bed load. It will also be noticed that these partial mobilisations of particles were generated by a discharge of lower recurrence in the downstream sites than in the upstream sites.

Finally, we observed that the influence of the sedimental structure of bed load on the critical parameters of entrainment is important for those discharges inferior to full bank discharge.

References

Andrews ED (1983) Entrainment of gravel from naturally sorted riverbed materiel. Geological Society of American Bulletin 94:1225–1231

Ashworth PJ, Ferguson RI (1989) Size-selective entrainment of bed load in gravel bed streams. Water Resources Research 25:627–634

Deroanne C (1995) Dynamique fluviale de la Hoëgne. Evaluation longitudinale des caractéristiques sédimentologiques du lit et des paramètres de mobilisation de la charge de fond., Département de Géographie physique et du Quaternaire, Hydrologie et Géomorphologie fluviale, Université de Liège, p 155

Du Boys P (1879) Etudes du régime et de l'action exercée par les eaux sur un lit à fond de graviers indéfiniment affouiable. Annales des Ponts et Chaussées, Série 5, 18:141–195

Grant GE, Swanson FJ, Wolman MG (1990) Pattern and origin of stepped-bed morphology in high-gradient streams, Western Cascades, Oregon. Geological Society of American Bulletin 102:340–352

Hoey TB, Ferguson R (1994) Numerical simulation of downstream fining by selective transport in gravel bed rivers: Model development and illustration. Water Resources Reasearch 30(7):2251–2260

Knighton AD (1980) Longitudinal changes in size and sorting of stream-bed material in four English rivers. Geological Society of America Bulletin, Part I, 91: 55–62

Kodama Y (1994) Downstream changes in the lithology and grain size of fluvial gravels, the Watarase River, Japan: Evidence of the role of abrasion in downstream fining. Journal of Sediment Research A64(1):68–75

Petit F (1989) Evaluation des critères de mise en mouvement et de transport de la charge de fond en milieu naturel. Bulletin de la Société Géographique de Liège 25:91–111

Petit F (1994) Dimensionless critical shear stress evaluation from flume experiments using different gravel beds. Earth Surface Processes and Landforms 19: 565–576

Petit F, Pauquet A, Pissart A (1995) Fréquence et importance du charriage dans des rivières à charge de fond caillouteuse. Actes du colloque Crues, versants et lits fluviaux Paris, mars 1995, 11 pp

Pissart A (1953) Les coulées pierreuses du plateau des Hautes Fagnes. Extrait des Annales de la Société Géologique de Belgique 76: 203–219

Richards KS (1991) Fluvial geomorphology: Initial motion of bed material in gravel-bed rivers. Progress in Physical Geography 14 (3):395–415

Werrity A (1992) Downstream fining in a gravel-bed river in southern Poland: Lithologic controls and the role of abrasion. Dynamics of Gravel-bed Rivers. Billi P, Hey RD, Thorne CR, Tacconi P, John Wiley and Sons, pp 333–350

Chapter 23

Morphological Changes in Mountain Rivers During a Flood Event

C.V. Bellos

23.1
Introduction

This work deals with the development of a morphological model, suitable for computing the morphological changes in mountain rivers during a flood event. In most of the morphological river bed models developed earlier, the granulometric compo-sition of the sediment transport assumed to be uniform or represented by a material of some characteristic diameter, i.e. D_{50}. The granulometric composition of the river bed and, particularly in the mountain rivers, is characterised by a variety of dimension of particles, which vary from fine sand to boulders. It is obvious that the uniform material assumption is far from reality and more detailed approaches must be made.

In the last decade many authors examined the influence of the granulometric composition on the sediment transport process. Among others we refer the studies of Ribberink (1987), Di Silvio and Peviani (1989), Parker (1991a,b), Seal et al. (1995), Fergyson et al. (1995), etc. All the above mentioned studies assume the existence of an "active layer" in the river bed. The thickness of this layer depends on granulometric composition of the river bed considered "well mixed" and is potentially in motion. In this layer, the finer material is transported at a lower rate, because it is "protected" by the coarser particles. This effect is quantified in the corresponding transport equations by a coefficient called "exposure-correction coefficient".

During a flood event in a river, eventually the supply input of sediment will be either larger or smaller than the equilibrium transport capacity. In those cases, aggradation or degradation may occur in a reach. In the first case it is obvious that the active layer thickness changes, because there exists the influence of the input sediment composition. In the second case, the layer thickness depends on the granulometric composition of the material underneath.

The differential equations, describing the complex interaction of hydraulic motion and sediment transport phenomena, do not have analytical solutions except for a few simplified cases generally. Therefore, numerical solutions are common. In the present work, an attempt is made to develop a numerical model for the simulation of morphological changes of riverbeds and the corresponding granulometric composition changes. The model consists of a hydraulic part, either a full hydrodynamic or a kinematic model, that is combined with a sediment transport model. These models are based on the well known Mac Cormack numerical scheme. This is an explicit, second order of accuracy, two steps predictor corrector numerical scheme, suitable for hyperbolic types of differential equations. The hydraulic model is updated every time step for changes in the river bed (e.g. bed slopes and roughness) and then the entire model can be considered as "quasi coupled".

The model are applied to an extreme event, that took place in the Mallero River. This river is located in North Italy, in the Alpine zone. The comparison of results with a previously developed model and measurements in-situ are satisfactory.

23.2
Mathematical Formulation

23.2.1
Hydraulics of Floods

23.2.1.1
Hydrodynamic Equations

The unsteady flow equations in conservation law form for irregular cross sections are (Bellos 1995):

$$\frac{\partial A}{\partial t} + \frac{\partial Q}{\partial x} = q_l \ , \tag{23.1}$$

$$\frac{\partial Q}{\partial t} + \frac{\partial}{\partial x}\left[\frac{Q^2}{A} + \frac{F_h}{\rho}\right] = gA(S_0 - S_f) + q_l u_l + \frac{F_{lx}}{\rho} \ , \tag{23.2}$$

in which
- A = cross section area;
- Q = flow discharge;
- u = mean flow velocity;
- q_l = lateral inflow for unit length;
- F_h = hydrostatic pressure force in the cross section;
- F_{lx} = longitudinal pressure force;
- u_l = velocity component of the lateral inflow in the mean stream direction;
- g = gravity acceleration;
- ρ = water density;
- S_o, S_f = bed and friction slopes respectively.

All units are in the SI system.

23.2.1.2
Kinematic Wave Model

If the inertial and the pressure differential terms can be neglected, then, the Eq. 23.2 assume the form:

$$S_f = S_0 - \frac{F_{lx}/\rho - q_l(u_l - u)}{gA} \ . \tag{23.3}$$

Eqations 23. 3 and 23.1 constitutes the "Kinematic wave model".

23.2.2
Sediment Transport

The sediment transport continuity equation is:

$$\frac{\partial z_b}{\partial t} + \frac{\partial q_s}{\partial x} = q_{ls} \; , \tag{23.4}$$

where
- z_b = the bed level;
- q_s = the volumetric sediment discharge for the unit width;
- q_{ls} = the lateral sediment input for the unit length and unit width.

The sediment motion, per size fraction j $(j = 1, \ldots, N)$ (Seal et al. 1995) is:

$$\frac{\partial F_j \delta}{\partial t} + \frac{\partial q_{sj}}{\partial x} + F_j^* \left(\frac{\partial z_b}{\partial t} - \frac{\partial \delta}{\partial t} \right) = q_{lsj} \; , \tag{23.5}$$

where
- δ = the active layer;
- F_j = the percentage of the j fraction;
- F_j^* = the percentage of the j fraction of the undisturbed material below the mixing layer if;
- $dz_b < 0$ (erosion), otherwise $F_j^* = F_j$;
- q_{sj} = the volumetric discharge for each fraction for unit width;
- q_{lsj} = the lateral sediment input for each fraction for unit length and unit width.

The general form of the active layer thickness is (Ribberink 1987):

$$\delta = f(H, \, y, \, q, \, n, \, D_{m, \, \ldots}) \; , \tag{23.6}$$

where
- H = the dune-height;
- y = water depth;
- h = Manning's coefficient;
- D_m = mean diameter of a size fraction in a mixture of different sizes.

For gravel bed $H = 0.0$. In this study the active layer thickness assumed to be:

$$\delta = 2d_{90} \; . \tag{23.7}$$

A general form of a sediment transport equation per size fraction is:

$$q_{sj} = f_j(Q, \, F_j, \, D_m, \, D_j, \, n, \, \ldots) \; . \tag{23.8}$$

In the present study the transport equation of Di Silvio (Di Silvio and Peviani 1989) was used. For a rectangular cross section this equation reads:

$$q_{sj} = \alpha \frac{S_0^n Q^m}{B^{p+1} d_j^q} F_j r_j \ ,$$

(23.9)

where r_j is the "exposure-correction coefficient":

$$r_j = \left[d_j \Big/ \sum_1^N F_j d_j \right]^s \ ;$$

(23.10)

- B = stream width;
- F_j = fraction of the class j;
- D_j = grain diameter of the fraction;
- α, n, m, p, q, s are coefficients evaluated empirically.

23.3
Numerical Formulation

23.3.1
Hydrodynamic Model

The numerical method used is based on the Mac Cormack's numerical scheme. In order to smearing the oscillations in the discontinuities caused from the supercritical flow conditions and the variation of cross sections, a diffusive term was added. The numerical scheme used is an explicit, two steps, second order of accuracy.

Predictor

$$\hat{A}_i = A_i^k - \lambda(Q_{i+1}^k - Q_i^k) + \Delta t q_{li}^k \ ,$$

(23.11)

$$\hat{Q}_i = Q_i^k - \lambda(G_{i+1}^k - G_i^k) + \Delta t H_i^k \ ,$$

(23.12)

$$\delta\hat{F}_{j,i} = F_{j,i}^k - \lambda(q_{sj,i+1}^k - q_{sj,i}^k) + F_j^{*k}\lambda(q_{s,i+1}^k - q_{s,i}^k) + \Delta t(q_{ls,i}^k - F_j^{*k}q_{ls,i}^k) \ ,$$

(23.13)

$$\hat{z}_i = z_i^k - \lambda(q_{s,i+1}^k - q_{s,i}^k) + \Delta t q_{l,i}^k \ ,$$

(23.14)

$$\hat{q}_{sj} = \alpha \frac{S_0^n \hat{Q}^m}{B^{p+1} d_j^q} \hat{F}_j \hat{r}_j \ ,$$

(23.15)

where $\quad \lambda = \dfrac{\Delta t}{\Delta x} \ , \qquad G = \dfrac{Q}{A} + \dfrac{F_h}{\rho} \ , \qquad H = gA(S_0 - S_f) + u_l q_l + \dfrac{F_{lx}}{\rho} \ .$

Corrector

$$A_i^{k+1} = \frac{1}{2}\left[\omega_1 A_i^k + \frac{1}{2}(1 - \omega_1)(A_{i-1}^k + A_{i+1}^k) + \widehat{A}_i - \lambda(\widehat{Q}_i - \widehat{Q}_{i-1}) + \Delta t q_{li} \right] , \qquad (23.16)$$

$$Q_i^{k+1} = \frac{1}{2}\left[\omega_1 Q_i^k + \frac{1}{2}(1 - \omega_1)(Q_{i-1}^k + Q_{i+1}^k) + \widehat{Q}_i - \lambda(\widehat{F}_i - \widehat{F}_{i-1}) + \Delta t \widehat{D}_i \right] , \qquad (23.17)$$

$$\delta F_{j,i}^{k+1} = F_{\omega_2} - \lambda(\widehat{q}_{sj,i} - \widehat{q}_{sj,i-1}) + \widehat{F}_j^* \lambda(\widehat{q}_{sj,i} - \widehat{q}_{sj,i-1}) + \Delta t \widehat{q}_{li} , \qquad (23.18)$$

$$z_i^{k+1} = z_{\omega_3} - \lambda(\widehat{q}_{s,i} - \widehat{q}_{s,i-1}) + \Delta t \widehat{q}_{li} , \qquad (23.19)$$

where

$$F_{\omega_2} = \omega_2 F_i^k + \frac{1}{2}(1 - \omega_2)(F_{i-1}^k + F_{i+1}^k) + \widehat{F}_i ,$$

$$z_{\omega_3} = \omega_3 z_i^k + \frac{1}{2}(1 - \omega_3)(z_{i-1}^k + z_{i+1}^k) + \widehat{z}_i .$$

The ω_1, ω_2, ω_3 coefficients needs for a "controlling diffusion" of the scheme. For $\omega_1 = \omega_2 = \omega_3 = 1$ there is not a diffusion.

Initial conditions

- $Q(x, 0)$ = constant;
- $A(x, 0)$ = calculated with Marning's equation;
- $q_{sj}(x, 0)$ = calculated with transport equation;
- $z_b(x, 0)$ = geometrical characteristics of the river;
- $F_j(x, 0)$ = from measurements.   (23.20)

Boundary conditions

Upstream:
- $Q(0, t)$ = hydrograph data;
- $A(0, t)$ = calculated with Manning's equation;
- $F_j(0, t)$ = assumed to be constant;
- $q_{sj}(0, t)$ = calculated with transport equation.   (23.21)

Downstream:
- $Q(L, t)$ = $Q(L - Dx, t)$;
- $A(L, t)$ = $A(L - Dx, t)$;
- $F_j(L, t)$ = $F_j(L - Dx, t)$;
- $q_{js}(L, t)$ = calculated with transport equation.   (23.22)

Remark. These downstream conditions in the fully hydrodynamic model are valid only in supercritical conditions.

In sub-critical cases, a relationship between water depth and flow discharge is needed.

Landslides:

Two types of boundary conditions concerning landslides are considered (Di Silvio and Peviani 1989):

a When landslides occur in relatively flat tributaries, the sediment input in the main stream depends from the granulometric composition of landslide and the flow discharge of the tributary.
b When the tributary is steep, the landslide entries in the main stream like debris flow and the further transport depends from the discharge in the main stream.

Another assumption is that the river bed alluvial material is of two meters thick and over it exists a fixed bed.

23.3.2
Kinematic Model

In this case, as in the above mentioned case, the same two steps technique was used without the momentum equation (Eq. 23.2). The flow discharge predictor and corrector step are calculated by the Eq. 23.3 where the friction gradient is:

$$S_f = \frac{n^2 q^2}{h^{10/3}} \; .$$

(23.23)

Initial and boundary conditions as the above cases.

The advantage of the kinematic approach is that no boundary conditions downstream are needed. This is valid also in the full hydrodynamic approach but only in the case of supercritical conditions of the flow downstream.

23.4
Resistance to Flow

The Manning's coefficient n, related to grain roughness, is calculated by the following expression:

$$n = d_{90}^{1/6} / 24 \; .$$

(23.24)

However, total flow resistance factors must include free surface instabilities, secondary flows, non-uniform shear stress distribution, cross section irregularities, channel shape, obstructions, vegetation, channel meandering, suspended and bed load (Jarret 1984). Due to large bed roughness and subsequently to great energy dissipation, the flow resistance at steeper slopes is much greater than for flatter slopes. In fact

Jarret, based on measurements performed on 21 mountain rivers in Colorado (USA), proposes the following expression for Manning coefficient:

$$n = 0.32 S_o^{0.3} R^{-0.16} \; , \tag{23.25}$$

with $0.002 < S_o < 0.052$ and 0.15 m $< R < 2.2$ m, where R is the hydraulic radius. Equation 23.25, when applied to some reaches of the Mallero, gives values of n approximately 3–4 times greater than those calculated with Eq. 23.24. Then, in the applications, the values of n obtained by applying Eq. 23.24 were multiplied by a coefficient equal to 4 (Bellos et al. 1995).

23.5
Stability Conditions

23.5.1
Stability of the Scheme

In the models the choice of the time step size Dt is subject to Courant-Friedrichs-Lewy (CFL) criterion of stability:

$$\Delta t = \sigma \frac{\Delta x}{c_{\max}} \; , \tag{23.26}$$

where $0 < \sigma < 1$ and $c_{\max}$ is the maximum wave celerity that is for the hydraulic part:

$$c_y = Q/A + \sqrt{gA/B} \; . \tag{23.27}$$

For the grain size F_j:

$$c_{F_j} = q_{sj}\left(\frac{1}{F_j} - sr_j^{1/s}\right)/\delta \; . \tag{23.28}$$

For the z_b:

$$c_{z_b} = \sum c_{F_j}\left(\frac{\partial q_{sj}}{\partial q_s} - F_j\right) . \tag{23.29}$$

The constant σ is generally set to values close to unity. In addition to CFL criterion, the time step size in the Dynamic model must satisfy the following condition (Terzidis and Strelkoff 1970; Becker and Yeh 1972):

$$\Delta t \le \frac{R^{4/3}}{gn^2|u|} \; . \tag{23.30}$$

In the applications of the Kinematic version of the model the value of σ was assumed equal to 0.2, while in the Dynamic model it was equal to 0.1.

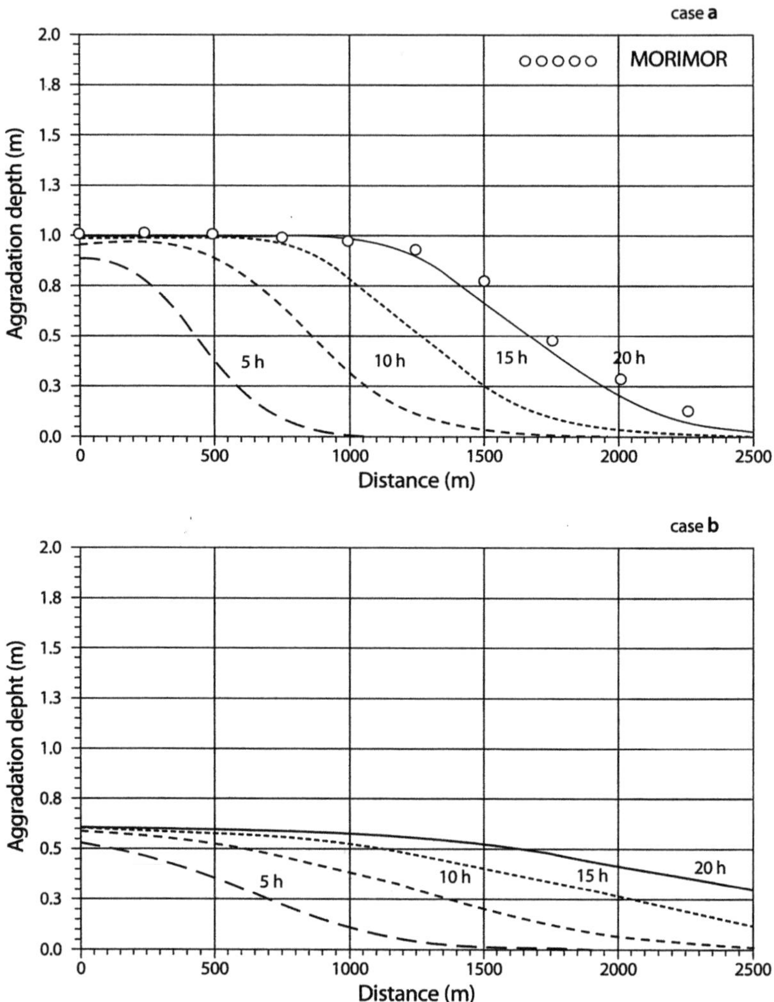

Fig. 23.1. Effects of initial bed composition

23.6
Applications

23.6.1
Verification Tests

Before applying the model in real cases some tests are performed. These concern a uniform flow where: $Q = 300$ m^3/s, $B = 67$ m, $S_0 = 0.02$ (supercritical flow conditions). The coefficients were $a = 0.011, m = 1.8, n = 2.1, p = 0.8, q = 1.2, s = 0.6$ and $\delta = 2\,d_{90}$. The granulometry of initial bed was:

- In the case a:

  $F_{01} = 0.04$, $F_{02} = 0.20$, $F_{03} = 0.50$, $F_{04} = 0.26$;
- In the case b:

  $F_{01} = 0.12$, $F_{02} = 0.28$, $F_{03} = 0.42$, $F_{04} = 0.18$;
- At the upstream boundary in both cases the granulometric composition was:

  $F_{01} = 0.25$, $F_{02} = 0.52$, $F_{03} = 0.20$, $F_{04} = 0.03$.

In Fig. 23.1 it can be seen that in the case of finer initial bottom composition, the disturbance propagates faster like the corresponding tests of Di Silvio and Peviani (1989). Also the height of the wave is smaller in the case of finer material.

23.6.2
Real Case Simulation

The above developed numerical model was calibrated and applied to simulate the flood event in the torrent Mallero (Valtellina-Northern Italy) during the catastrophic event of July 1987.

23.6.2.1
Description of the Mallero Basin

The Torrent Mallero is a tributary of the Adda River that is the main stream of Valtellina, in the central Alpine region of Northern Italy. The Torrent Mallero is 24 km long, start-

Fig. 23.2. Basin of torrent
Mallero

ing at the elevation of 1 636 m (m.s.l.) from the confluence of the Vazzeda and the Ventina torrents, and ending at the elevation of 282 m on its confluence with the Adda River (Fig. 23.2). The surface of its basin is approximately 319 km². The Valtellina region has always been, as has been known since the Middle Age as a place where severe storms accompanied by landslides and overaggradation have occurred. In July 1987 an exceptional event produced enormous disasters in Valtellina.

23.6.2.2
Implementation of the Model

The necessary data for the implementation of the models are: longitudinal profile and cross sections of the main stream, granulometric composition of the bottom material, hydrographs corresponding to each tributary as well as hydrograph at the upstream boundary.

The basin of the torrent Mallero and the schematization of the models are shown in Fig. 23.2. The available morphological and geotechnical data are reported in the previous report (Bellos et al. 1995).

Tributaries are not included in the model, except as a lateral input of water and sediment. In Fig. 23.3 there is a comparison of the results of the kinematic model with the hydrodynamic version. Essential differences cannot be seen as appears in the case examined. That can be justified because the flow motions in this case are dominated by a strong friction effect and the inertia terms are not important.

Figure 23.4 shows the comparisons with the MONMOR model (Di Silvio and Peviani 1989). Some differences may be due to the probable different positions of tributaries in the two models, different assumptions about the alluvional thickness and the diffusion coefficient used ($\omega_{zb} = 0.999$).

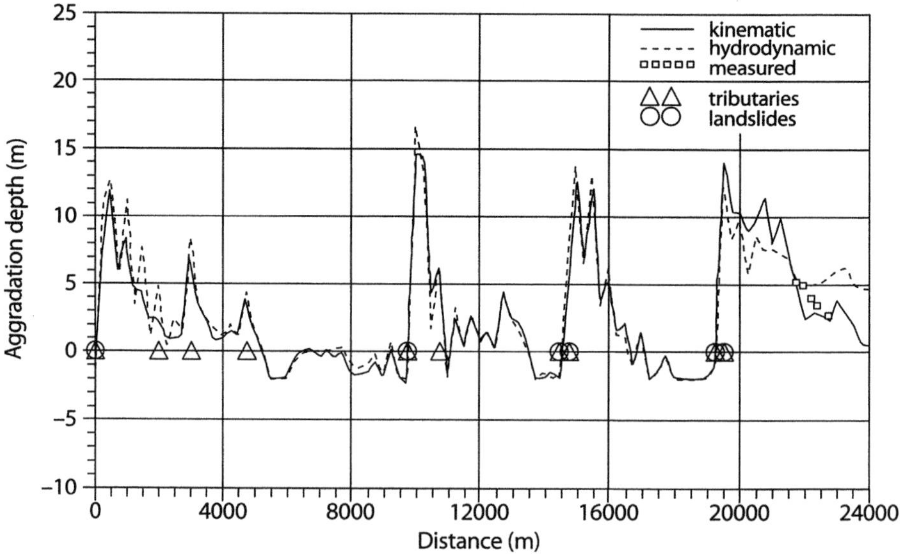

Fig. 23.3. Final bed level configuration after the flood event (kinematic – hydrodynamic)

In Fig. 23.5 a sensitivity analysis respect ad coefficient of diffusion ω_{zb}. The result from this analysis shows that the model is very sensitive in this coefficient and further studies on this must be made. The opinion of the writer is that the exponential form of the transport equation is the cause of this inconvenience.

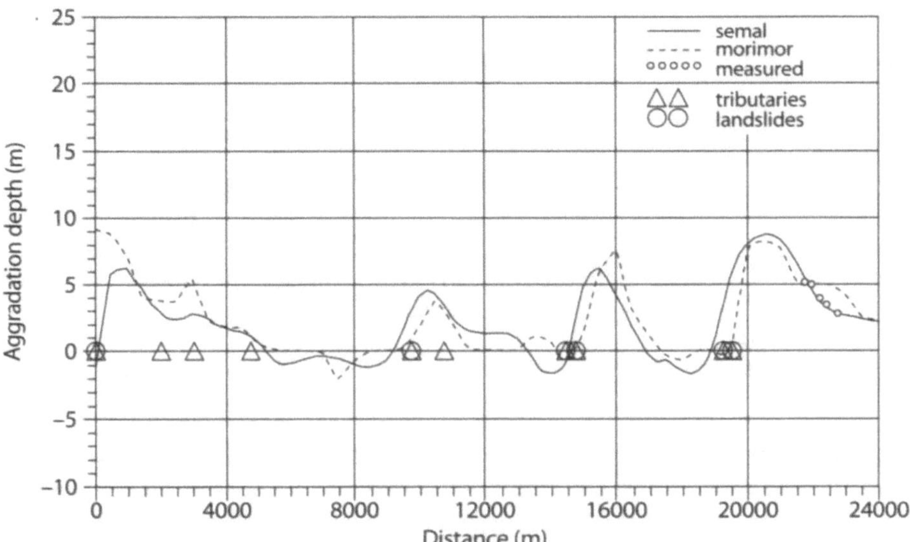

Fig. 23.4. Final bed level configuration after the flood event (kinematic SEMAL – MORIMOR)

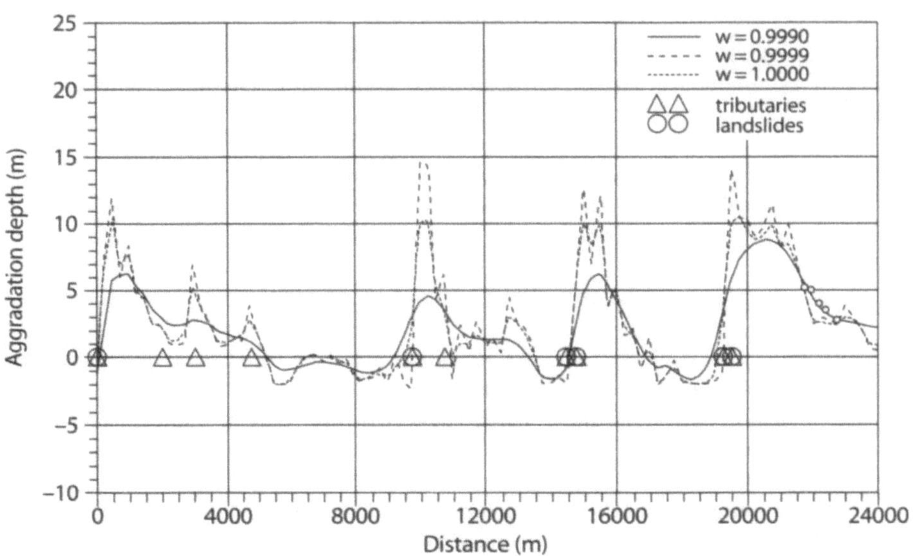

Fig. 23.5. Sensitivity analysis of Δz_b

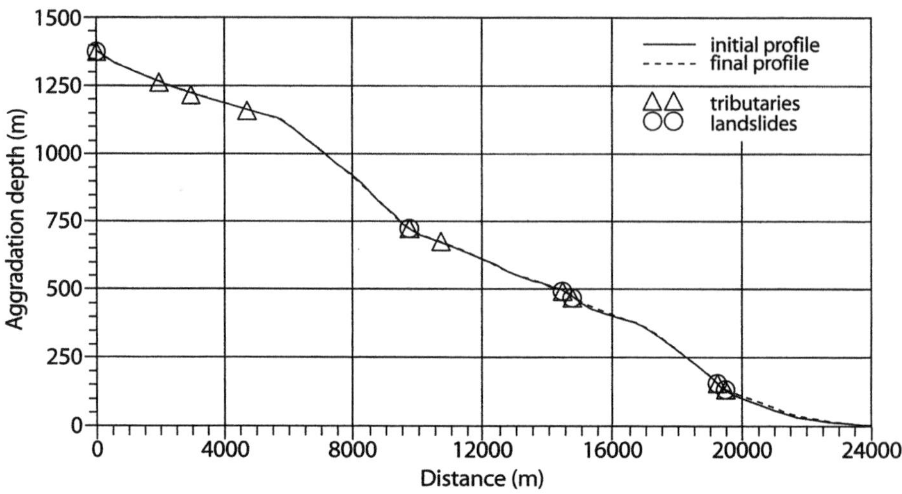

Fig. 23.6. Final bed level configuration after the flood event in real scale

Finally in Fig. 23.6 a configuration of the longitudinal profile of the riverbed before and after the flood event is reported. In this figure one can see that the final situation of the river bed is expectable e.g. deposition on the flat reaches and erosion in the steepest.

23.7
Conclusions

The above developed morphological model can describe the morphological changes in a mountain river stream during a flood event. These are the advantages of the model:

- The robustness of the model under various circumstances.
- The simplicity of the model-structure that allows to control the procedure.
- The relatively speed computation.

As disadvantages can be considered:

- The very small time step that results sometimes from the CFL criterion creating inconvenients like numerical diffusion, round off errors etc.
- The sensibility in the diffusion coefficients. A further analysis must be carried out on this aspect.

Certainly, more verifications with laboratory experiments and measurements in-situ must be made. Also another transport equation that contains more physical factors of the sediment transport phenomena may be proved.

Acknowledgements

This paper is part of a series of reports prepared under the FRIMAR (Flooding Risks in Mountain Areas) project sponsored by European Union. The project is funded under the "Environment and Climate Programme" under contract EV5V-CT94-0462. This support is gratefully acknowledged.

Furthermore, the author would like to acknowledge to Delfi Laboratory administration for the technical support during this work.

References

Becker L, Yeh WG (1972) Identification of parameters in unsteady open channel flows. Water resources Research 6(4):956–965

Bellos C (1995) Computation of flood propagation in natural channels. Project FRIMAR, Internal Report presented in Norwich meeting, May 1995

Bellos C, Basile P, Peviani M (1995) Numerical models for flood wave propagation in mountain rivers: An application to the torrent Mallero (ITALY). Project FRIMAR Internal Report presented in DELFI meeting, December 1995

Di Silvio J, Peviani M (1989) Modelling short-arid long-term evolution of mountain rivers.: An application to the torrent Mallero (Italy). International Workshop on Fluvial Hydraulics of Mountain regions. Trent Italy October 3–6, B145–B167

Di Silvio J, Peviani M (1991) Sediment exchange between stream arid bottom: a four layer model. Grain sorting Seminar October 21–26, Switzerland

Fergyson RI, Hoey TB, Wathen SJ, Werritty A, Hardwick RI, Sambrook Smith GH (1995) Downstream fining of river gravels: an integrated field, lab and modelling study. 4th Intern. Workshop on Gravel-Bed Rivers and the Environment, Gold Bar, Wa USA, August 1995

Jarret R (1984) Hydraulics of high gradient streams. Journal of Hydraulic Engineering, ASCE 110(6): 1519–1539

Parker G (1991a) Selective sorting arid abrasion of river gravel. I: Theory. Journal of Hydraulic Engineering 117(2) 131–149

Parker G (1991b) Selective sorting arid abrasion of river gravel. II: Applications. Journal of Hydraulic Engineering 117(2):150–171

Peviani M (1989) Long term evolution of mountain streams: Influence of hydrological and sediment feeding fluctuations. International Conference on Hydroscience & Engineering Washington DC, June 1993

Ribberink JS (1987) On mathematical modelling of one-dimensional morphological changes in rivers with non-uniform sediment. PhD Thesis Delft University of Technology, Report No. 87-2, TU Delft

Seal R, Toro-Escobar C, Cui Y, Paola C, Parker G, Soythhard JP, Wilcock PR (1995) Downstream fining by selective deposition: theory, laboratory, and field observations. 4th Inter. Workshop on Gravel-Bed Rivers and the Environment, Gold Bar, Wa USA, August 1995

Terzidis G, Strelkoff TH (1970) Computation of open-channel surges arid shocks. Journal of hydraulics Division, ASCE, vol 96, N° HY12, pp 2581–2610

Methodological Approach in the Analysis of Two Landslides in a Geologically Complex Area: The Case of Varenna Valley (Ligury)

P. Brandolini · S. Nosengo · F. Pittaluga · A. Ramella · S. Razzore

24.1
Introduction

This paper provides a brief study of Varenna basin (Ligury) through the analysis of its instability features and mass movement proneness; this area, during the five years from 1991–1995, has gone through frequent episodes of flood damages and landslides. The quick slide of topsoil debris and its displacement into the stream has dramatically increased the over-alluvion along the riverbed reaches with a strong dissipation of energy, and has made the river more liable to an esondation; besides, in correspondence of a restricted hydraulic section it reaches, several new bank erosions, being themselves responsible for the triggering mechanism in the slopes. For mountain basins, as well as the present one, it is necessary to define geomorphologic and geotechnical characteristics of the landslides, rainfall triggering threshold and thickness critical rate of the materials involved in the mass movements, in order to set out new activities for the reduction of the flood and landslides integrated hazards. These criteria have been used in the analysis of two mass-movements which occurred in Varenna valley during the autumnal rains in 1993–1994; they are still representative of the geological characteristics of most of the examined basin: the figures derived from the study show that these typical landslides always occur in presence of serpentine-schists strongly influenced by tectonic movements and widely spread along the whole basin. The geomorphologic and geotechnical data have been compared to those derived from the survey for the assessment of the slope stability, using the "back analysis" method suggested by Jambu. The results allow to define several degrees of proneness to the instability and, as well as, the degree of geomorphologic hazards in all the above-mentioned area.

24.2
A Geological and Geomorphological Outline of Varenna Valley

Varenna valley, along which the homonymous stream runs, lies in the western part of Genoa, near Pegli, it extends for 23 km² and it stretches for 9 km perpendicular to the coast line, with a maximum width of 4.5 km (Fig. 24.1). The highest points are those belonging to the Po valley-tirrenic watershed between the M. Pennello and Mt. Fasciallo ridge-line, slightly inferior to 1 000 m a.s.l. This area is geologically influenced by the overlapping of the two units belonging to the "Voltri Group" (metaophiolites) and "Sestri-Voltaggio group" (serpentinites, dolomite rocks, marly limestones, phyllite, slates). Both the structural order, formed by two main lines with approximate N-W orientation, and the local lithological groups have strongly modified the slope morphology and the hydrographical network; the two factors are responsible for an evi-

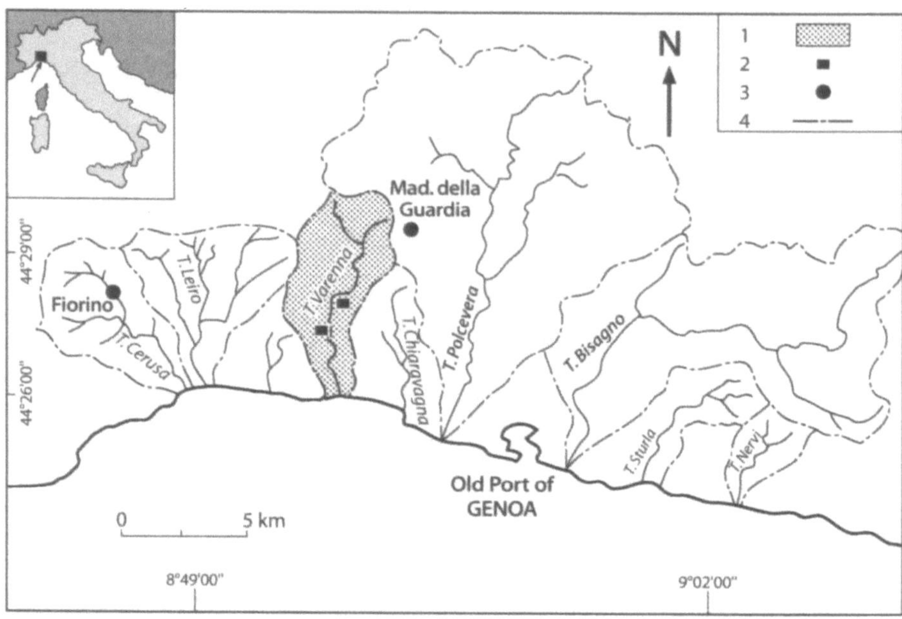

Fig. 24.1. Location of studied area: *1* Varenna valley; *2* analysed landslides; *3* pluviometric stations; *4* watersheds

dent valley asymmetry (right slope being more developed than left one). The most frequent types of mass movements are "mud and debris flows", of little thickness (up to 2 m) and generally composed by calc-schists and schists; landslides of higher complexity and volume have been found by tectonised serpentine-schists and calc-schists contact layers or inside the serpentinites. From their first triggering on they are characterised by a great regressing and widening activity. In autumn, owing to unfavourable meteorological and geomorphologic conditions, approximately 50% of the examined area was exposed to the integrated flood and landslides hazard.

24.3
Rainfall Analysis and Return Periods Estimation

The valley climate is strongly influenced by its orography. A peculiar slope exposure, as well as a brief distance between the Po valley-tirrenic watershed and the coast-line provoke strong changes in the climate, also influenced by the altitude; the rainfall has therefore, a discontinuous behaviour, with higher monthly mean values recorded on tops, with an absolute maximum value in winter and one in springtime.

The rainfall analysis has been carried out using data furnished by the weather stations in Fiorino, (slightly western to the Val Varenna, 236 m a.s.l.) and at Madonna della Guardia, eastern to the watershed next to the Val Polcevera, 809 m a.s.l.

The monthly rainfall mean values in 1991, 1992 and 1993, measured near the station in Fiorino, show that they have dramatically increased in comparison with the long set

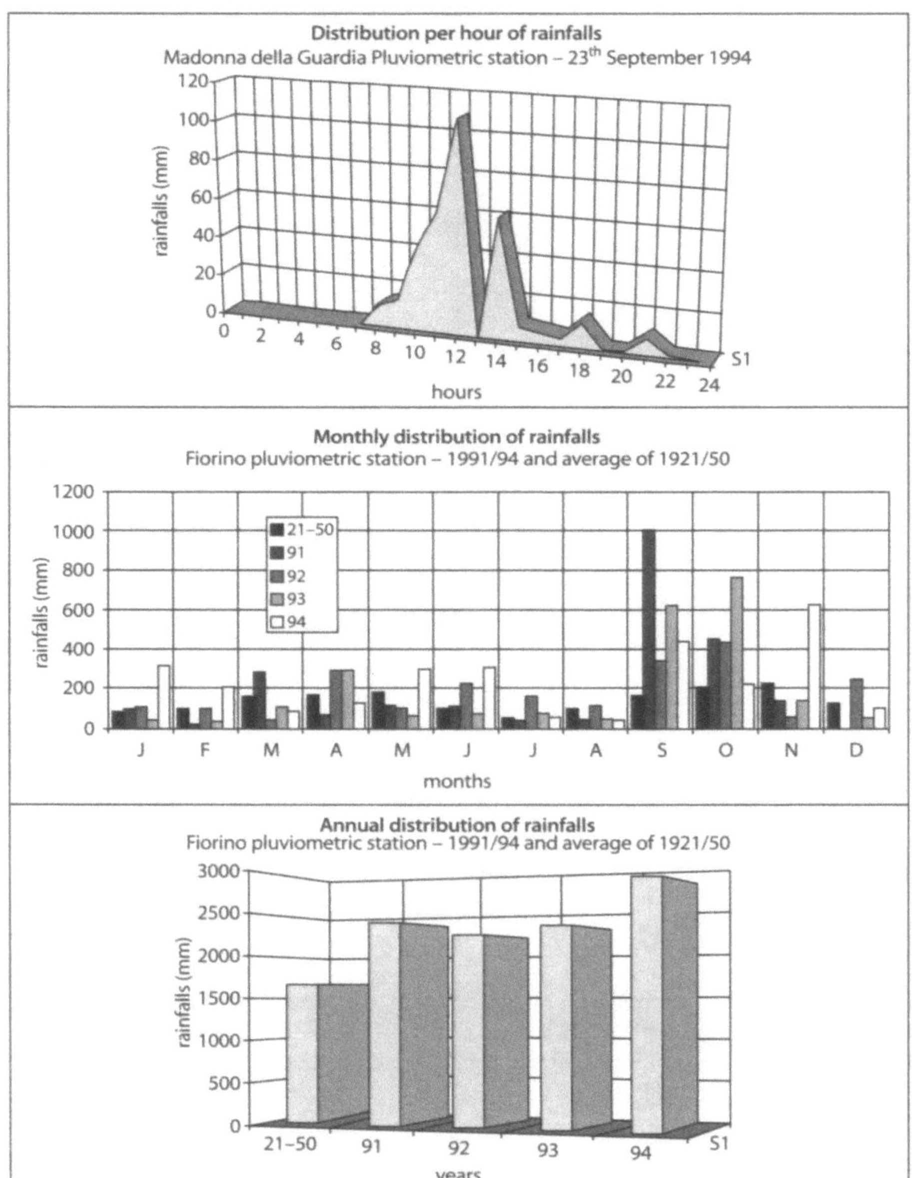

Fig. 24.2. Distribution of rainfalls

of rainfall mean values recorded during the historical thirty years from 1921–1950 (Fig. 24.2): in particular, it should be remarked the high value recorded in September and October 1993 (623 mm and 754 mm compared to the thirty years rainfall average, i.e. 164/210 mm).

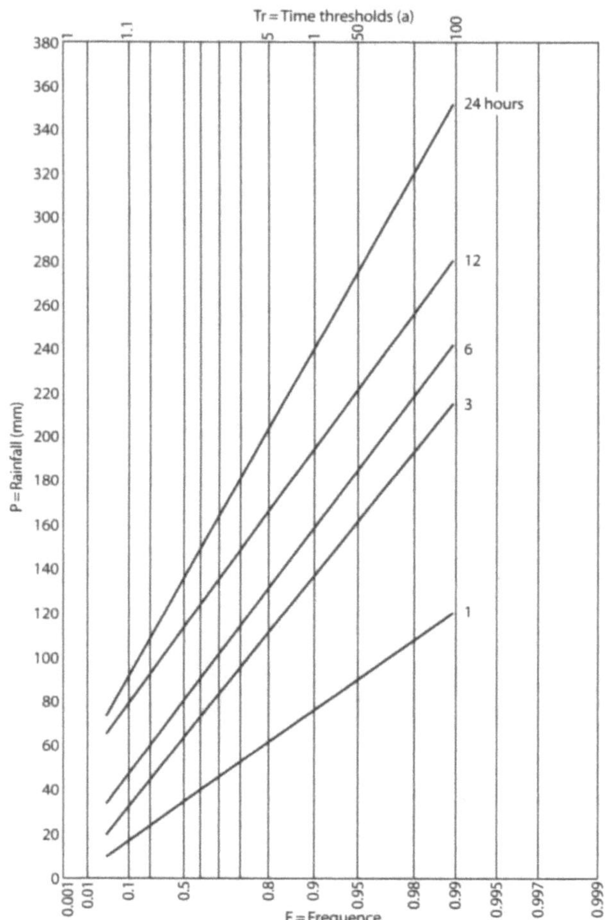

Fig. 24.3. Diagram "Frequence-Rainfall" (Gumbel Pluviogram)

Annual rainfall peaks related to the 24 h, recorded by the raingauge at Madonna della Guardia, underwent a statistic elaboration with the aid of the Gumbel Method. Return periods (T_r) derived from the diagram "Frequence-Rainfall" (Fig. 24.3) using maximum rainfall (P) and the corresponding duration (D) data in terms of hourly rainfall intensity (I).

24.4
Landslides Analysis

The meteorological events of high intensity have the strongest effects on bedrock topsoils, in which first-time landslides are triggered; the high amount of quick-filtering water brings to saturation the most superficial layers, being also the reason for their collapse.

First-time landslides made from "soil slip", and "mud and debris flow" have been detected on serpentine slopes; they generally become active along altered talc-schist terraced lands, on calc-schists and prasinites slopes, on limestone rocks.

A correlation between superficial landslides and terraced slopes has also been found.

Translational and roto-translational ground and debris displacements are found in great number and localised all along the valley, even in proximities of ridges; they are brought about either by bank erosions and the consequent widespread landsliding and rilling, and an increased load on slopes, owing to local groves (mainly chestnut trees).

24.4.1
Chiesino Landslide

This landslide is next to Chiesino settlement, on the right bank of Varenna Torrent, (Fig. 24.4). The mass movement event occurred suddenly, on the 4th November 1994, following intense rainfall and without any previous evidence of dangerous instability.

The raingauge data show that rainfall values referring to the days immediately antecedent the landslide event were about 290 mm, 72% of which refer to the 4th November (more or less 210 mm).

The landslide reactivated an old debris accumulation, which underwent a secular stabilisation in a terrace leaning on a serpentine-schist layer belonging to the "Voltri Group" (see Table 24.1).

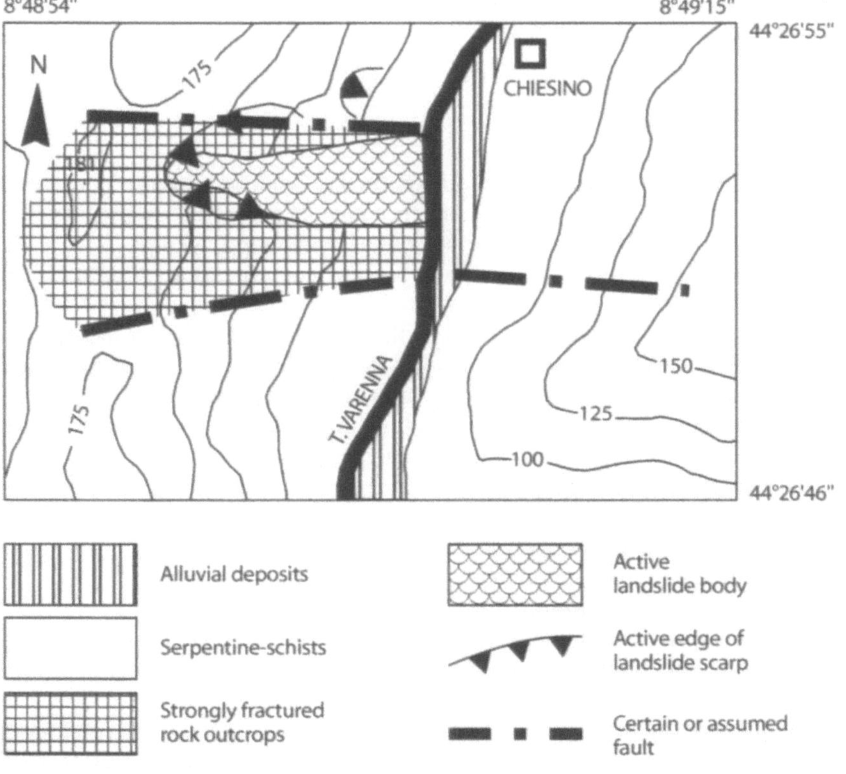

Fig. 24.4. Chiesino landslide

Table 24.1. Typical parameters of material in the landslide body

γ	γsat	Wn (%)	Sr (%)	Wl	Wp	Ip	n	$neff$	k (m s^{-1})
1.97	2.02	22.8	0.54	38	28	10	0.61	0.2	5.8×10^{-7}

These are the most relevant morphometric features of this mass movement:

- Longitudinal mass development: 140 m;
- Mass movement body medium width: 60–70 m;
- Mass thickness: 5–8 m;
- Landslide global surface area, measured on plan, of about 8 000 m²;
- Landslide body estimated volume: 40 000 m³.

The landslide can be classified as a rock and debris flow, for which both an undermined slope toe and a saturated topsoil are equally responsible. The material collapsed is mostly made from incoherent terrain (slime gravels) deriving from an alteration in the serpentinite-schist rocks. The granulometric curves result well graded, with a high amount of fine material, varying from 12–30%.

On the above mentioned materials several shear tests have worked out a friction angle $\varphi' \approx 35°$ and a cohesion $c' = 0$

24.4.2
Carpenara Landslide

The landslide occurred near Carpenara, on the left bank of Torrente Varenna, at the confluence with a rill (Fig. 24.5) and affects a part of serpentine-schists altered soils belonging to the "Voltri Group", of about 7 m of maximum thickness.

The slope shows a very wide alteration band in the serpentinte-schists, covered by a coarse debris topsoil.

The mass movement event occurred on the 24th October 1993, one month after the intense showers of the last days of September, followed by a period of downpours (about 25 mm per day) for a total amount of 1 350 mm. It should be noted that 350 mm of rain fallen on the 24th September, more or less one month before the landslide event, did cause a severe flood without any landslide trigger sign.

These are the relevant morphometric data referred to the landslide:

- Maximum longitudinal development: 100 m;
- Mass movement body approximate width: 60–80 m;
- Maximum thickness: 5–7 m;
- Maximum areal extension of the landslides, measured on plan: about 4 500 m;
- Mass movement body estimated volume: 10 000 m.

The landslide can be classified as a rock and debris translational and rotational flow, while the entire slope shows some evidence of instability under the appearance of little landslide scarps and small traces of debris flow.

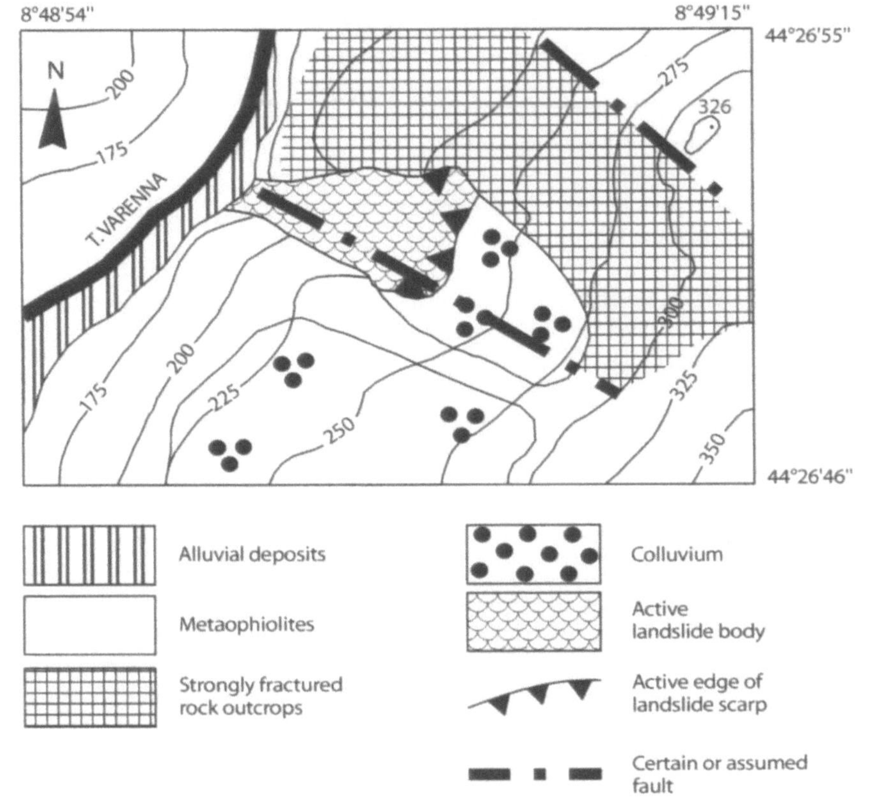

Fig. 24.5. Carpenara landslide

Table 24.2. Typical parameters of lime material and in the landslide body

γ	γsat	Wn (%)	Sr (%)	Wl	Wp	Ip	n	neff	k (m s^{-1})
2.04	2.17	18.1	0.84	29	20	9	0.4	3	2.0×10^{-7}

Materials mostly consist of sands and lime gravels, well graded and showing great affinity with the Chiesino samples (see Table 24.2).

Also, in the landslide body, it has been found the presence of thin levels (10–30 cm) mostly made from lime material and with a very low degree of stress resistance. The recorded parameters are resumed in the following table.

The results of dimensional analysis using the "thin section" method show that the material is very altered, it ranges uninterruptedly from "very fine" (1/1 000) to "coarse" and features a very low stress resistance.

The fragments are mostly made from highly oxidised serpentinites, sometimes in lamella and sometimes in fibroid shape. A certain amount of amphibole is also present.

On the serpentinites, particles of some dimensions also show an oxidation rim more compact and more resistant than the original material, hint of an advanced pedogenesis.

24.5
Soils Stability and "Back Analysis"

The available data enabled the back analysis on stability, exploiting the "equilibrium limit" method proposed by Jambu, with imposed slide-surfaces.

Chiesino landslide tests ($y = 1.97$ t/m³, $\varphi^\prime = 35°$, $c^\prime = 0$) show that in absence of the water table the slope parameters maintain themselves within the stability threshold ($Fs = 1.4$); on the contrary, the presence of the water table is responsible for the steep decrease of the security factor below unity (Fig. 24.6).

With reference to the circumstances preparing the landslide event (after four days of intense rains) and to the pre-existing natural conditions (steep acclivity of 33°, untended terraced lands no longer tilled) the landslide event may be ascribed to both an increase in the water table level and the lack of intervention for the surface-waters drainage.

The analysis carried out on the Carpenara landslide have been more difficult for the presence, in the serpentinites topsoil, of strongly altered fine material, (mostly lime) with very poor mechanic qualities, ($\varphi^\prime = 14°$, $c^\prime = 0$). The reconstruction of the geometric disposition of the cited material inside the topsoil has not been possible, for the lack of previous in-situ surveys.

With the help of the back analysis method it has been possible to verify that the mentioned material can trigger a landslide only when placed in a rather continuous level, being therefore, able to influence in a remarkable way the slope stability.

24.6
Landslide Trigger

Studies on thickness and critical rainfall rates, with reference to return periods, using the "Pradel and Raad Model" have been achieved in order to complete the present survey.

This model finds application only when the rainfall is considered the unique responsible for the seepage triggering mechanism, with no consideration to the uphill recharge.

The following necessary conditions are needed for the application of the model:

- Constantly wet soil surface;
- Constant permeability in the saturated terrain;
- Linear saturation front;
- Constant negative pressure above saturation limits.

Parameters valid for both the events are the following ones:

- $D_{10} = 0.1$ mm;
- $P_{ris} = 0.18$;
- $N_{sat} = 0.2$;
- $K = 5.8\ 10^{-7}$ m/s.

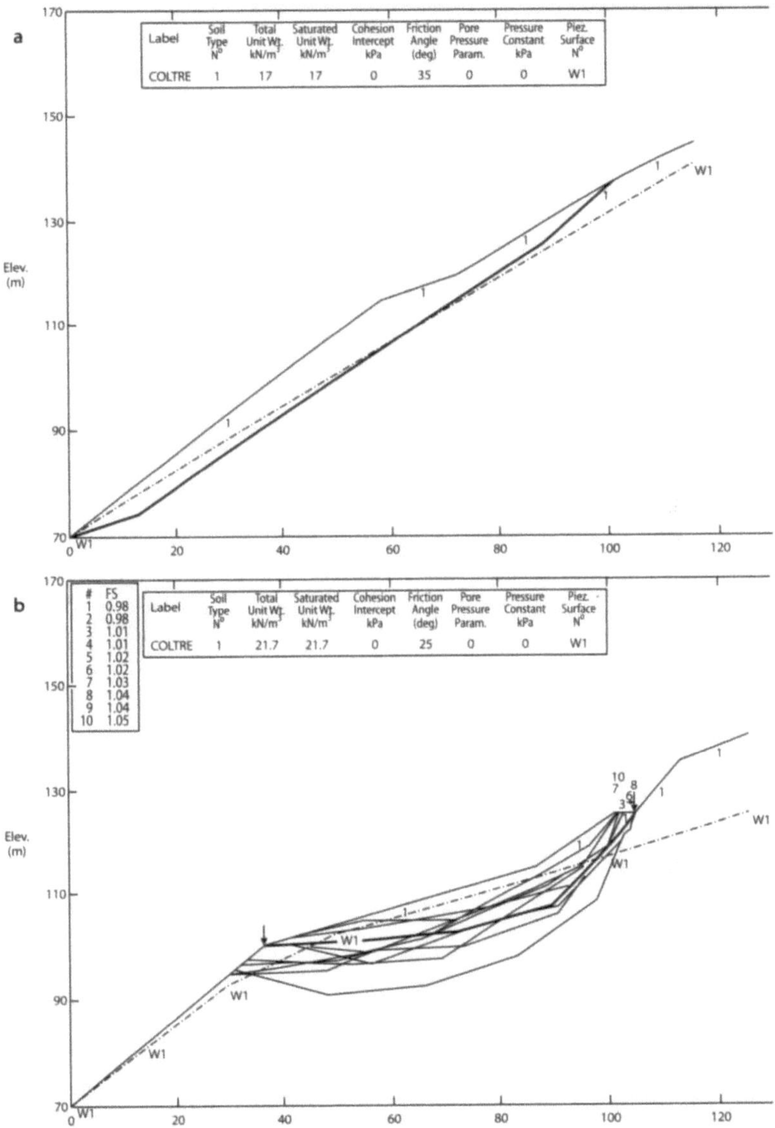

Fig. 24.6. Back analysis **a** Chiesino landslide: hypothesis of a slope with a water table; **b** Carpenara landslide: hypothesis of a slope with a water table

By varying the terrain critical thickness rates, and by keeping the permeability rate (k) constant, it has been possible to calculate the intensity and duration minimum values necessary to the saturation, with reference to the different thickness rates, and afterwards to deduce the saturation conditions line. The envelop lines of maximum rainfall intensities, referred to the various return periods, have been derived from the

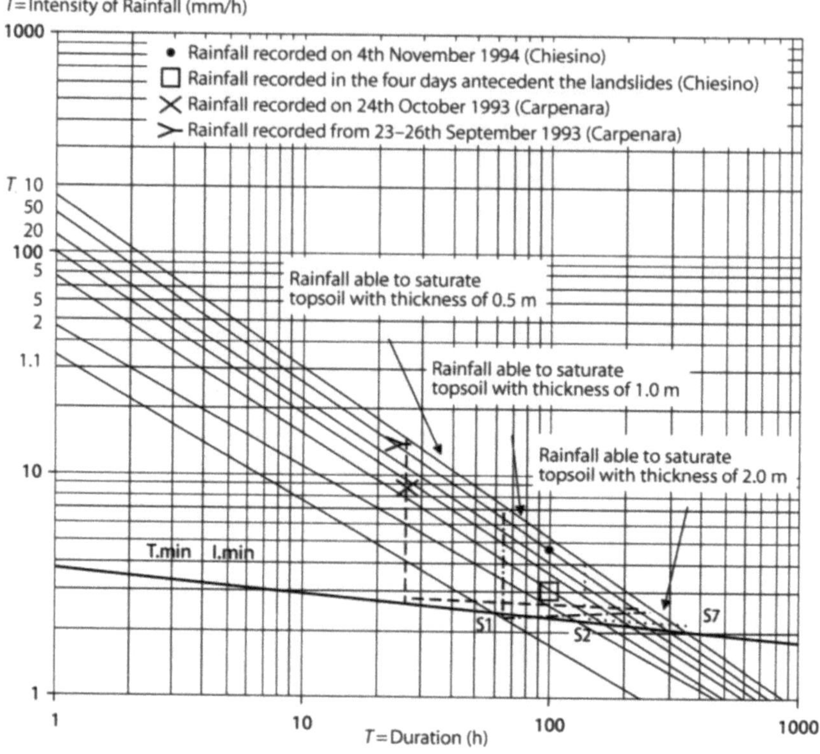

Point	B (m)	T_{lim} (h)	I_{lim} (mm/h)	T_r (a)
S1	0.9	58	2.50	1.1
S2	2.2	162	2.25	2
S3	2.6	201	2.23	5
S4	3.3	265	2.20	10
S5	3.7	301	2.18	20
S6	4.3	356	2.17	50
S7	4.3	400	2.16	100

Fig. 24.7. Diagram "Intensity/Duration of rainfall" with minimum intensity (I_{min}) and duration (T_{min}) values for different topsoil thickness and correlation with the several return (T_r). In this diagram are indicated the graphic areas representing the relative critical rainfall able to saturate topsoil of equivalent thickness (0.5; 1; 2)

"Frequence-Rainfall" graphic, (Gumbel Pluviogram). By transferring the obtained figures on the bilogarithmic scale diagram, it has been possible to build up the "Intensity-Duration" graphic (Fig. 24.7) from which a first indication on the geological hazard for different topsoil thickness rates with all those geotechnical characteristics is obtainable. The line showing duration and minimum intensity values necessary for the saturation (I, T, K) intercepts those related to several return periods, in points S1, S2, S3 and S4 (Time thresholds) and singling out thickness rates for which minimum saturation conditions and return periods are strictly related.

Figure 24.7 indicates, for different thickness rates, (0.5 m, 1 m, 2 m) the graphic areas representing the relative critical rainfall able to saturate topsoil of equivalent thickness.

It should be noted how highly intensive but brief rainfalls are capable of saturating only thin topsoils ($B = 0.5$ m) while lesser intense but long-lasting rains saturate thicker soils.

For instance, terrains with a thickness ranging from 2–4 m may be saturated only by rains with a return period varying from 10–20 years, while soils thicker than 5 m need rain with a return period longer than 100 years.

Graphic analysis has made possible the detection, for both the examined landslides, of critical thickness rates related to the precipitation of the days foregoing the events.

In fact, 208.6 mm in the 24 h (T_r 5–10 years), equivalent to the rain values recorded on the 4th November 1994, may bring to saturation a stratum with the thickness of 0.5 m but considering also the four days antecedent the landslide event ($P_{max} = 287.8$ mm; $T = 96$ h) the thickness rate increases up to 1.5 m

The rainfall data related to Carpenara landslide show a high intensity average figure rainfall event at the end of September ($P_{max} = 350$ mm in four days and a peak equal to 350.8 mm in the 24 h).

The graphic clearly shows how these figure fall in the area representing the critical precipitation for the return period within 20 and 30 years, able to saturate soil with a thickness between 1 and 1.5 m. The low intensity rains of the following days maintained the soil saturated by adding weight and enabling a deeper water infiltration. Unlike the Chiesino landslide for which even the immediately previous rainfall of high intensity is held responsible, the Carpenara landslide has been essentially triggered by long lasting previous precipitation.

24.7
Conclusions

The figures obtained with the use of Pradel and Raad Model show that only small portions of topsoil are liable to be saturated (ranging from 0.5–1.5 m). An empirical assessment is also possible, since the severe downpours cause damages only to the colluvial cover, which is quickly brought to saturation by reason of an insufficient drainage of the underlying strata (almost impermeable and prone to behave as preferential slide surfaces).

A comparison between the former values and the ones obtained through in-situ surveys shows that the two examined landslides displaced a larger amount of material (4–7 m in Chiesino, 6–7 m in Carpenara).

A possible explanation lies in the fact that high intensity rainfalls trigger first-time landslides, above all when in presence of two different strata with a sharp contrast in the permeability coefficient, so that a seepage water flow may run down parallel to the slope.

Therefore, other causes must be held responsible for deeper landslides triggering. The two events suggest the following observations:

- Chiesino landslide: there is no evidence of poor quality materials along the slide surface. The landslide mechanism is therefore, to be put down to the intense rainfall occurred the previous day, which brought about a double increase in the water table and in the Varenna Torrent level, the latter responsible for the slope toe undermining.
- Carpenara landslide may have been triggered by the action of two contributory causes, the former consisting in a long-lasting rainfall (during the whole month of October) and the latter being the presence of an altered and poor-quality material which may have acted as impermeable stratum and responsible for the saturation-front deepening.

In the examined basins, geologic and geomorphologic studies lead to assert that the most remarkable events always originate from tectonic lines in which serpentinites come in contact with other lithologies. A detailed geotechnical survey on the tectonised terraces areas, together with a hydrologic verification, allow the definition of both the propensity to landslide for all the areas with homogenous geologic and geomorphologic features, and the landslide trigger threshold values for each individual rainfall rate, to a detailed scale. This methodology is particularly useful when researches need to define the criteria aiming at the reduction of flood and landslide integrated hazard.

References

Benini G (1990) Sistemazioni idraulico-forestali. UTET, Torino

Brandolini P, Ramella A (1994) Eventi alluvionali e dissesti idrogeologici: il caso della Val Varenna (Liguria). Studi Geografici in onore di D. Ruocco, Loffredo Editore, Napoli

Brandolini P, Ramella A (1997) Processi erosivi e fenomeni di dissesto nei versanti terrazzati delle valli costiere genovesi, Atti Convegno Geografico Internazionale "I valori dell'agricoltura nel tempo e nello spazio", Rieti 1–4 nov. 1995, vol II, pp 339–355

Bromhead EN (1986) La stabilità dei pendii. Dario Flaccovio (ed), Blackie and Son LTA

Carrara A, D'elia B, Semenza E (1985) Classificazione e nomenclatura dei fenomeni franosi, Università di Bari, Ist. Geologia Applicata e Geotecnica – Geologia Applicata e Idrogeologia, vol XX., parte II

Castagny G (1985) Idrogeologia, principi e metodi., Libreria Dario Floccovio (ed)

Gostelow TP (1991) Rainfall and landslides, Preventio and control of landslides and other mass moviments, Almeida-Texeia ME et al. (ed)

Ministero dei Lavori Pubblici (1953–1985) Annali Idrografici. Servizio Idrografico, Sez. Autonoma di Genova

Pradel D, Raad G (1993) Effect of permeability on surficial stability of homogeneus slopes. Journal Geotech. Enging., ASCE

Raviolo PL (1993) Il laboratorio Geologico. Procedure di prova: Elaborazione, Acquisizione dati, Editrice Controls, Milano

Renard KG, Freimund JR (1994) Using monthly precipitation data to estimate the R-factor in revised USLE, Journal of Hydrology

Skempton AW, DE Lory FA (1957) Stability of natura slopes in London Clay. Proc 4th Int. Conf. on Soil Mech. and Found. Eng

Tropeano D et al. (1993) Gli eventi alluvionali del 22–27 Settembre 1992 in Liguria. Studio idrologico e geomorfologico. Supplemento a GEAM – Geoingegneria Ambientale e Mineraria, Anno XXX, N° 4

Index

Springer and the environment

At Springer we firmly believe that an international science publisher has a special obligation to the environment, and our corporate policies consistently reflect this conviction.

We also expect our business partners – paper mills, printers, packaging manufacturers, etc. – to commit themselves to using materials and production processes that do not harm the environment. The paper in this book is made from low- or no-chlorine pulp and is acid free, in conformance with international standards for paper permanency.

If you have any concerns about our products,
you can contact us on
ProductSafety@springernature.com

In case Publisher is established outside the EU,
the EU authorized representative is:
Springer Nature Customer Service Center GmbH
Europaplatz 3, 69115 Heidelberg, Germany

Printed by Libri Plureos GmbH
in Hamburg, Germany